Lecture Notes in Mathematics

Edited by A. Dold and B. Eckmann

AF003562

962

Category Theory

Applications to Algebra, Logic and Topology
Proceedings of the International Conference
Held at Gummersbach, July 6 – 10, 1981

Edited by K.H. Kamps, D. Pumplün, and W. Tholen

Springer-Verlag
Berlin Heidelberg New York 1982

Editors

Klaus Heiner Kamps
Dieter Pumplün
Walter Tholen
Fachbereich Mathematik und Informatik
Fernuniversität – Gesamthochschule –
Lützowstr. 125, 5800 Hagen
Federal Republic of Germany

AMS Subject Classifications (1980): 18-06, 03D, 05C, 06D, 08A, 13C, 13E, 16A, 18A, 18B, 18C, 18D, 18F, 18G, 20L, 26E, 46A, 46B, 46G, 46M, 54B, 54D, 54E, 55F, 55N, 55P, 57M

ISBN 3-540-11961-2 Springer-Verlag Berlin Heidelberg New York
ISBN 0-387-11961-2 Springer-Verlag New York Heidelberg Berlin

This work is subject to copyright. All rights are reserved, whether the whole or part of the material is concerned, specifically those of translation, reprinting, re-use of illustrations, broadcasting, reproduction by photocopying machine or similar means, and storage in data banks. Under § 54 of the German Copyright Law where copies are made for other than private use, a fee is payable to "Verwertungsgesellschaft Wort", Munich.

© by Springer-Verlag Berlin Heidelberg 1982
Printed in Germany

Printing and binding: Beltz Offsetdruck, Hemsbach/Bergstr.
2146/3140-543210

During the last stages of the preparation of this volume the editors learnt of the tragic death of our colleague Graciela Salicrup. Her personality and her work will always be remembered by all of us.

PREFACE

The International Conference on Category Theory - Applications to Algebra, Logic and Topology - was held in Gummersbach, July 6-10, 1981; it was attended by 93 mathematicians from 19 different countries.

Financial support for this conference was provided by a grant of the Deutsche Forschungsgemeinschaft (grant no. 4851/140/80) and by additional means of the Minister für Wissenschaft und Forschung des Landes Nordrhein-Westfalen. The organizers would like to express their sincere thanks for this financial assistance, without which this conference would not have been possible.

The conference had been divided into three sections: General category theory, category theory and logic, and applications of category theory to analysis, topology and computer science. It was very much appreciated by the organizers that John Gray agreed to be chairman of this conference and special thanks are due to him for his essential contribution to its success. The organizers are also very grateful to Horst Herrlich for his help as chairman for the section on applications of category theory to analysis, topology and computer science.

The organizers would like to express their thanks to the Rektor of the Fernuniversität, Prof. Dr. Dr. h.c. O. Peters, for his opening of the conference and for the welcome he extended to the participants on behalf of the Fernuniversität. During the conference and during its preparation essential and effective help was given by the administration of the Fernuniversität, and this help has been gratefully acknowledged by the organizers. Especially Mr. Blümel from the university administration should be mentioned for his engagement for this conference.

Thanks are due to the Fachbereich Mathematik und Informatik of the Fernuniversität for supporting this conference in every respect. Many colleagues advised and assisted us during the conference and its preparation. We would like especially to thank the secretaries Mrs. I. Müller and Mrs. K. Topp for their most efficient work.

Last, but by no means least, we would like to express our thanks to Dr. G. Greve, Dr. W. Sydow, Dr. D. Brümmer, Dr. B. Hoffmann and Dr. T. Müller, all members of the Fachbereich Mathematik und Informatik of the Fernuniversität for their engagement. It is due to their efforts that there were no organizational difficulties during the conference and they did their best to make the participants of the conference feel at ease.

This volume of Springer Lecture Notes constitutes the proceedings of this conference. We would like to thank the editors of the Springer Lecture Notes in Mathematics for accepting the proceedings for this series. All contributions to this volume have been refereed and our sincere thanks go to all the referees for their work.

Klaus Heiner Kamps Dieter Pumplün Walter Tholen

PARTICIPANTS

M. Adelman
C. Anghel
H. Bargenda
M. Barr
J.M. Beck
H.L. Bentley
G.J. Bird
R. Börger
D. Bourn
H. Brandenburg
R.D. Brandt
R. Brown
C. Cassidy
Y. Diers
G. Dubrule
A. Duma
J.W. Duskin
R. Dyckhoff
A. Frei
P. Freyd
A. Frölicher
J.W. Gray
C. Greither
G. Greve
R. Guitart
R. Harting
M. Hébert
H. Herrlich
P.J. Higgins
M. Höppner
B. Hoffmann
R.-E. Hoffmann
M. Hušek
J. Isbell
B. Jay
P.T. Johnstone
K.H. Kamps
G.M. Kelly
H. Kleisli
A. Kock
J. Lambek
H. Lindner
F.E.J. Linton
H. Lord
R.B. Lüschow
J. MacDonald
S. MacLane
L. Mårki
G. Maury
A. Möbus
T. Müller
C.J. Mulvey
A. Mysior
R. Nakagawa
G. Naudé
L.D. Nel
S.B. Niefield
A. Obtułowicz
B. Pareigis
J. Penon
M. Pfender
A.M. Pitts
H.-E. Porst
T. Porter
A. Pultr
D. Pumplün
R. Reiter
G. Richter
R. Rosebrugh
J. Rosický
G. Salicrup
B.M. Schein
D. Schumacher
F. Schwarz
Z. Semadeni
T. Spircu
G.E. Strecker
R. Street
T. Swirszcz
W. Sydow
M. Thiébaud
T. Thode
W. Tholen
V.V. Topentcharov
V. Trnková
K. Ulbrich
R.F.C. Walters
H. Weberpals
S. Weck
R. Wiegandt
A. Wiweger
R.J. Wood
O. Zurth

AUTHORS' ADDRESSES

H.L. Bentley	Department of Mathematics University of Toledo Toledo, Ohio 43605 U.S.A.
R. Betti	Istituto Matematico Università di Milano Via Saldini 50, Milano Italy
F. Borceux	Université Catholique de Louvain 1348-Louvain-La-Neuve Belgium
D. Bourn	Université de Picardie U.E.R. de Mathématiques 33, rue St Leu 80039 Amiens France
H. Brandenburg	Institut für Mathematik I Freie Universität Berlin Arnimallee 2-6 1000 Berlin 33 Fed. Rep. of Germany
R. Brown	School of Mathematics and Computer Science University College of North Wales Bangor, Gwynedd LL57 2UW U.K.
Y. Diers	Département de Mathématiques U.E.R. des Sciences Université de Valenciennes 59326 Valenciennes France
A. Frei	Mathematics Department University of British Columbia Vancouver, B.C. Canada V6T 1Y4
A. Frölicher	Section de Mathématiques Université de Genève 2-4, rue du Lièvre 1211 Genève 24 Switzerland

J.W. Gray

Department of Mathematics
University of Illinois
Urbana, Ill. 61801
U.S.A.

G. Greve

Fachbereich Mathematik und Informatik
Fernuniversität
5800 Hagen
Fed. Rep. of Germany

P.J. Higgins

Department of Mathematics
University of Durham
Science Laboratories
South Road
Durham DH1 3LE
U.K.

R.-E. Hoffmann

Fachbereich Mathematik
Universität Bremen
2800 Bremen 33
Fed. Rep. of Germany

M. Höppner

Fachbereich Mathematik-Informatik
Universität-Gesamthochschule-Paderborn
4790 Paderborn
Fed. Rep. of Germany

M. Hušek

Matematický Ustav
University Karlova
Sokolovská 83
18600 Praha
Czechoslovakia

S. Kaijser

Uppsala University
Uppsala
Sweden

J. Lambek

Department of Mathematics
McGill University
805 Sherbrooke St. West
Montreal, PQ
Canada H3A 2K6

J. MacDonald

Mathematics Department
University of British Columbia
Vancouver, B.C.
Canada V6T 1Y4

L. Márki

Mathematical Institute
Hungarian Academy of Sciences
Reáltanoda u. 13-15
1053 Budapest
Hungary

A. Melton
Department of Computer Science
Wichita State University
Wichita, Kansas 67208
U.S.A.

A. Mysior
Institute of Mathematics
University of Gdansk
80952 Gdansk
Poland

L.D. Nel
Department of Mathematics
Carleton University
Ottawa, Ontario
Canada K1S 5B6

S.B. Niefield
Union College
Schenectady, N.Y. 12308
U.S.A.

J.W. Pelletier
Faculty of Arts
York University
4700 Keele Street
Downsview, Ontario
Canada M3J 1P3

M. Pfender
MA 7-1
Technische Universität Berlin
Str. des 17. Juni 135
1000 Berlin
Fed. Rep. of Germany

H.-E. Porst
Fachbereich Mathemaik
Universität Bremen
2800 Bremen 33
Fed. Rep. of Germany

T. Porter
School of Mathematics and
Computer Science
University College of
North Wales
Bangor, Gwynedd LL57 2UW
U.K.

A. Pultr
Matematický Ustav
University Karlova
Sokolovská 83
18600 Praha
Czechoslovakia

R. Reiter
Fachbereich Mathematik
Technische Universität Berlin
Str. des 17. Juni 135
1000 Berlin
Fed. Rep. of Germany

G. Richter

Fakultät für Mathematik
Universität Bielefeld
Universitätsstr. 25
4800 Bielefeld 1
Fed. Rep. of Germany

M. Sartorius

Fachbereich Mathematik
Technische Universität Berlin
Str. des 17. Juni 135
1000 Berlin
Fed. Rep. of Germany

T. Spircu

National Institute
for Scientific and Technical
Creation
Department of Mathematics
Bdul Păcii 220
79622 Bucharest
Romania

A. Stone

Mathematics Department
UC Davis
Davis, California
U.S.A.

G.E. Strecker

Department of Mathematics
Kansas State University
Manhattan, Kansas 66506
U.S.A.

R. Street

School of Mathematics and
Physics
Macquarie University
North Ryde, N.S.W. 2113
Australia

W. Sydow

Fachbereich Mathematik und
Informatik
Fernuniversität
5800 Hagen
Fed. Rep. of Germany

J. Taylor

Department of Mathematics
University of Durham
Science Laboratories
South Road
Durham DH1 3LE
U.K.

W. Tholen

Fachbereich Mathematik und
Informatik
Fernuniversität
5800 Hagen
Fed. Rep. of Germany

V. Trnková

Matematický Ustav
University Karlova
Sokolovská 83
18600 Praha
Czechoslovakia

R.F.C. Walters

Department of Pure Mathematics
University of Sydney
N.S.W. 2006
Australia

R. Wiegandt

Mathematical Institute
Hungarian Academy of Sciences
Reáltanoda u. 13-15
1053 Budapest
Hungary

A. Wiweger

Institute of Mathematics
Polish Academy of Sciences
Śniadeckich 8
00-950 Warszawa
Poland

CONTENTS

H.L. Bentley
A note on the homology of regular nearness spaces 1

R. Betti and R.F.C. Walters
The symmetry of the Cauchy-completion of a category 8

F. Borceux
On algebraic localizations 13

D. Bourn
A canonical action on indexed limits. An application to coherent homotopy 23

H. Brandenburg and M. Hušek
A remark on cartesian closedness 33

R. Brown and P.J. Higgins
Crossed complexes and non-abelian extensions 39

Y. Diers
Un critère de représentabilité par sections continues de faisceaux 51

A. Frei
Kan extensions and systems of imprimitivity 62

A. Frölicher
Smooth structures 69

J.W. Gray
Enriched algebras, spectra and homotopy limits 82

G. Greve
General construction of monoidal closed structures in topological, uniform and nearness spaces 100

P.J. Higgins and J. Taylor
The fundamental groupoid and the homotopy crossed complex of an orbit space 115

R.-E. Hoffmann
Minimal topological completion of ${}_{\mathbb{K}}Ban_1 \longrightarrow {}_{\mathbb{K}}Vec$ 123

M. Höppner
On the freeness of Whitehead-diagrams 133

M. Hušek
Applications of category theory to uniform structures 138

S. Kaijser and W. Pelletier
A categorical framework for interpolation theory 145

J. Lambek
Toposes are monadic over categories 153

J. MacDonald and A. Stone
Essentially monadic adjunctions 167

J. MacDonald and W. Tholen
Decomposition of morphisms into infinitely many factors 175

L. Márki and R. Wiegandt
Remarks on radicals in categories 190

A. Melton and G.E. Strecker
On the structure of factorization structures 197

A. Mysior
A remark on scattered spaces 209

L.D. Nel
Bornological L_1-functors as Kan extensions and Riesz-like representations 213

S.B. Niefield
Exactness and projectivity 221

M. Pfender, R. Reiter, and M. Sartorius
Constructive arithmetics 228

H.-E. Porst
Adjoint diagonals for topological completions 237

T. Porter
Internal categories and crossed modules 249

A. Pultr
Subdirect irreducibility and congruences 256

G. Richter
Algebraic categories of topological spaces 263

T. Spircu
Extensions of a theorem of P. Gabriel 272

R. Street
Characterization of bicategories of stacks 282

W. Sydow
On hom-functors and tensor products of topological vector spaces 292

V. Trnková
Unnatural isomorphisms of products in a category 302

A. Wiweger
Categories of kits, coloured graphs, and games 312

A Note on the Homology of Regular Nearness Spaces

H. L. Bentley

Abstract: It is shown that the homology and cohomology groups of a regular nearness space can be defined by means of a variation on the Čech method, which uses nerves of uniform covers: the variation involves associating with each uniform cover, not the nerve, but a complex, called the vein, defined by means of nearness

In a recent paper, the author showed that the Čech homology and cohomology groups (= Vietoris homology and Alexander cohomology groups) of merotopic and nearness spaces satisfy, in a variant form, all the axioms of Eilenberg-Steenrod. For definitions of these groups and for historical information, the reader is referred to that paper [1]. We are interested here in regular nearness spaces (for the definition, see Herrlich [5]) and in the possibility of using what is, formally, a different definition of the homology and cohomology groups, but a definition which we prove gives rise to the usual Čech groups.

By a pair (X, Y) of nearness spaces we mean a nearness space X together with a nearness subspace Y of X. A uniform cover of (X, Y) is a pair $\mathcal{O} = (\mathcal{O}_1, \mathcal{O}_2)$ where $\mathcal{O}_1$ is a uniform cover of X, $\mathcal{O}_2 \subset \mathcal{O}_1$, and $\mathcal{O}_2 \cup \{X - Y\}$ is a uniform cover of Y.

The <u>nerve</u> $K(\mathcal{O})$ of a uniform cover $\mathcal{O} = (\mathcal{O}_1, \mathcal{O}_2)$ of (X, Y) is a pair of simplicial complexes $K(\mathcal{O}) = (K_1(\mathcal{O}), K_2(\mathcal{O}))$. The vertices of $K_1(\mathcal{O})$ are the elements of $\mathcal{O}_1$; a simplex of $K_1(\mathcal{O})$ is a finite subset $\mathcal{G}$ of $\mathcal{O}_1$ such that $\cap \mathcal{G} \neq \phi$. The vertices of $K_2(\mathcal{O})$ are the elements of $\mathcal{O}_2$; a simplex of $K_2(\mathcal{O})$ is a finite subset $\mathcal{G}$ of $\mathcal{O}_2$ such that $Y \cap \cap \mathcal{G} \neq \phi$.

Recall that a collection $\mathcal{G}$ of subsets of a nearness space X is said to be <u>near</u> in X if for each uniform cover $\mathcal{C}$ of X there exists $C \in \mathcal{C}$ such that for all $G \in \mathcal{G}$, $C \cap G \neq \phi$. Recall also that if Y is a nearness subspace of X then a collection $\mathcal{G}$ of subsets of Y is near in Y if and only if $\mathcal{G}$ is near in X.

Now we are ready to make our main definition; it is a variation on the definition of the nerve.

The vein $J(\mathfrak{A})$ of a uniform cover $\mathfrak{A} = (\mathfrak{A}_1, \mathfrak{A}_2)$ of (X, Y) is a pair of simplicial complexes $J(\mathfrak{A}) = (J_1(\mathfrak{A}), J_2(\mathfrak{A}))$. The vertices of $J_1(\mathfrak{A})$ are the elements of $\mathfrak{A}_1$; a simplex of $J_1(\mathfrak{A})$ is a finite subset $\mathfrak{g}$ of $\mathfrak{A}_1$ such that $\mathfrak{g}$ is near in X. The vertices of $J_2(\mathfrak{A})$ are the elements of $\mathfrak{A}_2$; a simplex of $J_2(\mathfrak{A})$ is a finite subset $\mathfrak{g}$ of $\mathfrak{A}_2$ such that $\mathfrak{g} \wedge \{Y\}$ is near in Y.

If $\mathfrak{A} = (\mathfrak{A}_1, \mathfrak{A}_2)$ and $\mathfrak{B} = (\mathfrak{B}_1, \mathfrak{B}_2)$ are uniform covers of the pair (X, Y) of nearness spaces then we say that $\mathfrak{A}$ is a refinement of $\mathfrak{B}$ if $\mathfrak{A}_1$ is a refinement of $\mathfrak{B}_1$ and $\mathfrak{A}_2$ is a refinement of $\mathfrak{B}_2$. Under this relation of refinement, the set of all uniform covers of a pair of nearness spaces becomes a directed set.

Thus, there is a spectrum of complexes

$$K(\mathfrak{A}) \longrightarrow K(\mathfrak{B})$$

and of complexes

$$J(\mathfrak{A}) \longrightarrow J(\mathfrak{B})$$

for $\mathfrak{A}$ a refinement of $\mathfrak{B}$. From these spectra there arise two spectra of homology groups and two of cohomology groups.

From now on, let G be a fixed abelian group. G will be the coefficient group of our homology and cohomology theories but explicit denotation of G will be suppressed.

The direct spectrum of cohomology groups

$$\alpha_{\mathfrak{B}}^{\mathfrak{A}} : H^n(K(\mathfrak{B})) \longrightarrow H^n(K(\mathfrak{A}))$$

has for its limit group the n-dimensional Čech cohomology group of (X, Y) which we will denote by $\check{H}^n(X, Y)$. The inverse spectrum of homology groups

$$\beta_{\mathfrak{B}}^{\mathfrak{A}} : H_n(K(\mathfrak{A})) \longrightarrow H_n(K(\mathfrak{B}))$$

has for its limit group the n-dimensional Čech homology group of (X, Y) which we will denote by $\check{H}_n(X, Y)$.

The direct spectrum of cohomology groups

$$\gamma_{\mathfrak{B}}^{\mathfrak{A}} : H^n(J(\mathfrak{B})) \longrightarrow H^n(J(\mathfrak{A}))$$

has for its limit group the n-dimensional vascular cohomology group of (X, Y) which we will denote by $\mathring{H}^n(X, Y)$. The inverse spectrum of homology groups

$$\tau^{\mathfrak{A}}_{\mathfrak{B}} : H_n(J(\mathfrak{A})) \longrightarrow H_n(J(\mathfrak{B}))$$

has for its limit group the <u>n-dimensional vascular homology group</u> of (X, Y) which we will denote by $\dot{H}_n(X, Y)$.

We are now ready for the statement of our main result.

<u>Theorem</u>. If (X, Y) is a pair of regular nearness spaces then the Čech and vascular homology, and cohomology, groups coincide, i.e.

$$\check{H}_n(X, Y) = \dot{H}_n(X, Y)$$

and

$$\check{H}^n(X, Y) = \dot{H}^n(X, Y)$$

for all n.

Proof: We give a proof only for the homology groups; the proof for cohomology is dual. With each collection $\mathcal{D}$ of subsets of X, we associate the collection

$$\mathcal{D}^* = \{E \subset X \mid \text{for some } D \in \mathcal{D}, \ E < D\}.$$

Of course, as usual we are using the notation $E < D$ to mean that $\{D, X - E\}$ is a uniform cover of X. For each uniform cover $\mathfrak{A} = (\mathfrak{A}_1, \mathfrak{A}_2)$ of (X, Y) we will write $\mathfrak{A}^* = (\mathfrak{A}_1^*, \mathfrak{A}_2^*)$; note that because X is regular then $\mathfrak{A}^*$ is again a uniform cover of (X, Y). (To show that $\mathfrak{A}_2^* \cup \{X - Y\}$ is a uniform cover of X, note that $(\mathfrak{A}_2 \cup \{X - Y\})^*$ refines $\mathfrak{A}_2^* \cup \{X - Y\}$.)

For each uniform cover $\mathfrak{A}$ of X, there exists a simplicial map

$$g_{\mathfrak{A}} : J(\mathfrak{A}^*) \longrightarrow K(\mathfrak{A})$$

which, on vertices $E \in \mathfrak{A}_1^*$, satisfies $g_{\mathfrak{A}}(E) \in \mathfrak{A}_1$ and $E < g_{\mathfrak{A}}(E)$. Of course, $g_{\mathfrak{A}}$ is not determined by this condition but any two such simplicial maps have to be contiguous and so, at the homology level, a unique homomorphism

$$f_{\mathfrak{A}} = (g_{\mathfrak{A}})_* : H_n(J(\mathfrak{A}^*)) \longrightarrow H_n(K(\mathfrak{A}))$$

is determined, which depends only on $\mathfrak{A}$ and not on the particular choice of $g_{\mathfrak{A}}$. Before going on, it should be noted that the fact that $g_{\mathfrak{A}}$ is a simplicial map arises from the fact that whenever $\mathcal{E}$ is a finite subset of $\mathfrak{A}_1^*$, $\mathcal{E}$ is near if and only if $\bigcap\{g_{\mathfrak{A}}(E) \mid E \in \mathcal{E}\} \neq \emptyset$. Also, since the set of all covers of the form $\mathfrak{A}^*$ is a cofinal subset of the set of all uniform covers of (X, Y), it follows that the $f_{\mathfrak{A}}$ form a homomorphism of the inverse spectrum.

For each uniform cover $\mathcal{A}$ of (X, Y), K($\mathcal{A}$) is a subcomplex of J($\mathcal{A}$) so we have the homomorphism

$$k_{\mathcal{A}} : H_n(K(\mathcal{A})) \longrightarrow H_n(J(\mathcal{A}))$$

induced by the inclusion map.

Turning our attention now to the limit groups, we have the projection homomorphisms

$$u_{\mathcal{A}} : \dot{H}_n(X, Y) \longrightarrow H_n(J(\mathcal{A}))$$

and

$$v_{\mathcal{A}} : \check{H}_n(X, Y) \longrightarrow H_n(K(\mathcal{A})),$$

as well as the limit homomorphisms

$$f_\infty : \dot{H}_n(X, Y) \longrightarrow \check{H}_n(X, Y)$$

and

$$k_\infty : \check{H}_n(X, Y) \longrightarrow \dot{H}_n(X, Y).$$

Consider the following diagram:

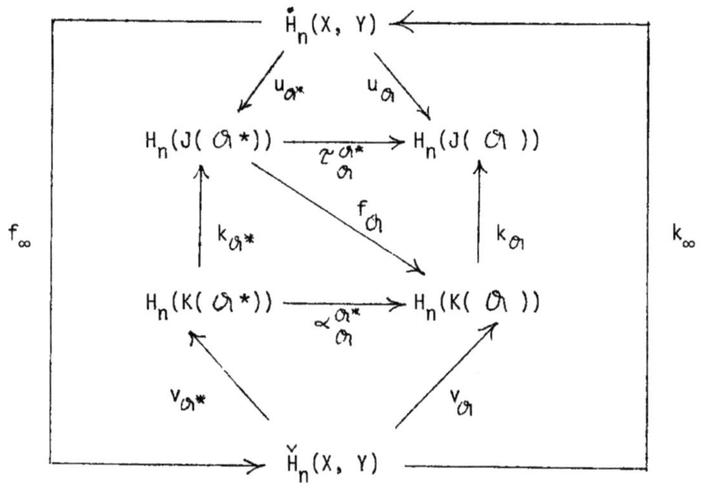

It is clear that each of the inner triangles is commutative, because each homomorphism is induced either by a projection of refinements or by an inclusion map. To show that $f_\infty \circ k_\infty = 1$, let $x \in \check{H}_n(X, Y)$ and compute as follows:

$$\begin{aligned} v_{\mathfrak{A}}\, f\;\; k\;\; x &= f_{\mathfrak{A}}\, u_{\mathfrak{A}^*}\, k_\infty\, x \\ &= f_{\mathfrak{A}}\, k_{\mathfrak{A}^*}\, v_{\mathfrak{A}^*}\, x \\ &= \alpha_{\mathfrak{A}}^{\mathfrak{A}^*}\, v_{\mathfrak{A}^*}\, x \\ &= v_{\mathfrak{A}}\, x. \end{aligned}$$

Consequently, $f_\infty k_\infty x = x$.

An equally pleasant computation shows that $k_\infty o f_\infty = 1$ and the proof of the theorem is complete.

For regular nearness spaces, the above theorem provides an alternative method of computing the Čech groups: one can compute by means of the vascular theory. If X is a regular nearness space and Y is a dense nearness subspace of X and if $\mathfrak{A}$ is a collection of subsets of Y which satisfies $\cap \{cl_X A \mid A \in \mathfrak{A}\} \neq \emptyset$ then $\mathfrak{A}$ is near in Y. This observation, together with the knowledge that, in the above situation, the homology groups of X are the same as those of Y, indicates that, instead of passing to the extension X and using the Čech theory, one could stay inside Y and use the vascular theory.

Of course, not every nearness space Y is a subspace of a topological nearness space X so, even if X is the completion of Y, there may exist collections $\mathfrak{A}$ of subsets of Y such that $\cap \{cl_X A \mid A \in \mathfrak{A}\} = \emptyset$. In such a case, it might also be advantageous to use the vascular theory.

We will now present an example using thevascular homology groups $\dot{H}_n(X, Y)$. Consider the Euclidean plane as a nearness subspace (= uniform subspace) of its Alexandrov one-point compactificaton. Let X be the nearness subspace (= uniform subspace) induced on the subset

$$\{(1, y) \mid -1 \leq y \leq 1\} \cup \{(x, \tfrac{1}{x}) \mid 1 \leq x\} \cup \{(x, -\tfrac{1}{x}) \mid 1 \leq x\} .$$

The completion of X is a circle S^1 on a 2-sphere S^2. Thus, by the fact proved in [2], the homology of X is the same as that of S^1.

The point here though is that the homology of X can be computed without going outside X. The details are as follows.

The set of all finite uniform covers of X is a cofinal subset of the set of all uniform covers of X. So, consider an arbitrary finite uniform cover $\mathcal{A}$ of X. Let

$$\mathcal{L} = \{A \in \mathcal{A} \mid A \text{ is unbounded}\}$$

and let $\varepsilon > 0$ be such that

$$\{G \subset X \mid \operatorname{diam} G \leq \varepsilon\} \quad \text{refines } \mathcal{A}.$$

Let x^+ be the supremum of the set

$$\{1\} \cup \{x \in R \mid \text{for some } y > 0 \text{ and for some } A \in \mathcal{A}-\mathcal{L},\ (x, y) \in A\}$$

and let x^- be the supremum of the set

$$\{1\} \cup \{x \in R \mid \text{for some } y < 0 \text{ and for some } A \in \mathcal{A}-\mathcal{L},\ (x, y) \in A\}.$$

Let $\mathcal{G}$ be a set of intervals on X of diameter at most ε such that

$$\{(1, y) \mid -1 \leq y \leq 1\} \cup \{(x, \tfrac{1}{x}) \mid 1 \leq x \leq x^+\} \cup \{(x, -\tfrac{1}{x}) \mid 1 \leq x \leq x^-\}$$

is covered by the union of the intervals in $\mathcal{G}$ and such that if $\mathcal{H} \subset \mathcal{G}$ with $\cap\mathcal{H} \neq \emptyset$ then $\mathcal{H}$ has cardinal at most 2. Obviously, such a collection $\mathcal{G}$ of intervals exists.

Now consider the uniform cover $\mathcal{G} \cup \mathcal{L}$ of X. Clearly it refines $\mathcal{A}$. And it is easy to see what the vein $J(\mathcal{G} \cup \mathcal{L})$ looks like. From $(x^-, -1/x^-)$ to $(x^+, 1/x^+)$ it is just a finite sequence of straight line segments which touch at consecutive endpoints and nowhere else. Furthermore, $J(\mathcal{G} \cup \mathcal{L})$ contains $\mathcal{L}$ itself as a simplex because $\mathcal{L}$ is near in X. Thus, pictorially $J(\mathcal{G} \cup \mathcal{L})$ looks as follows:

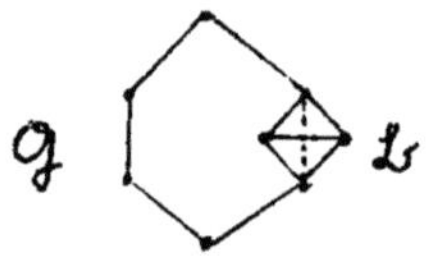

So we see that $J(\mathcal{G} \cup \mathcal{L})$ is uniformly (= topologically) homotopic to a circle S^1. Passing to the limit, we have

$$\dot{H}_n(X) = H_n(S^1).$$

In conclusion, we pose the following question: For non-regular nearness spaces, do we still have the isomorphisms between the Čech and the vascular groups? If so, a new proof will need to be found since it is clear that the proof presented above works only for regular nearness spaces. (For non-regular spaces, there exist uniform covers $\mathcal{A}$ for which the collection $\mathcal{A}^*$ fails to be a uniform cover.)

Bibliography

1. H. L. Bentley, Homology and cohomology for merotopic and nearness spaces, preprint.

2. D. Czarcinski, The Čech Homology Theory for Nearness Spaces, Dissertation, University of Toledo (1975).

3. S. Eilenberg and N. Steenrod, Foundations of Algebraic Topology, Princeton (1952).

4. H. Herrlich, A concept of nearness, General Topology and Appl. 4 (1974) 191-212.

5. H. Herrlich, Topological structures, In: Topological Structures, Math. Centre Tracts 52, Amsterdam (1974) 59-122.

THE SYMMETRY OF THE CAUCHY-COMPLETION OF A CATEGORY

R. Betti and R.F.C. Walters

Cauchy-completion for enriched categories was introduced by F.W. Lawvere [3], generalizing the notion for metric spaces. In this paper, we are concerned with the case of categories based on a bicategory which is locally partially-ordered ([5],[6],[1]). A natural question that arises is whether the Cauchy-completion of a symmetric category is again symmetric. This is true for metric spaces (although the contrary is claimed in [4]), but false in general. We prove that if the base bicategory satisfies the "modular law" (as defined by P. Freyd) then symmetry is preserved by Cauchy-completion. An application is the description of sheafification in terms of Cauchy-completion.

We refer to [5],[6] for definitions not given here.

1. The modular law

Let B be a bicategory. Objects of B will be denoted u,v,w,..., and arrows $\rho,\sigma,\tau,\ldots$. Throughout this paper we will assume that B is locally a complete poset, and that suprema in each B(u,v) are preserved by intersection in B(u,v) and by composition with arrows (or both sides). We will also suppose given an involution $(\ )^o\colon B^{op} \to B$ (reversing arrows, but not order) which is the identity on objects.

Examples. The main examples we have in mind are (i) Lawvere's monoidal category R regarded as a bicategory with one object (see [3]), (ii) the bicategory, Rel, of sets and relations, and, more generally (iii) the bicategories Rel(C,J) of relations arising from a category C with a topology J, as defined in [1] or [6].

Definition. (P. Freyd) B satisfies the *modular law* if, for arrows $\rho\colon u \to v$, $\sigma\colon v \to w$ and $\tau\colon u \to w$, we have

$$\tau \wedge (\sigma\rho) \le \sigma(\sigma^o\tau \wedge \rho).$$

Remark. The bicategories in examples (ii) and (iii) satisfy the modular law, though not example (i).

We need the following technical

Lemma. If B satisfies the modular law and $(a_i)_{i\in I}$ is a family of arrows $a_i\colon u \to u$ satisfying $1_u \le \bigvee_{i\in I} a_i$ then $1_u \le \bigvee_{i\in I} (a_i \wedge a_i^o)$.

Proof. First we prove that if $b \le 1_u\colon u \to u$ then $b = b^o$. To see this note that

$$b = 1_u \wedge (b1_u) \le b(b^o 1_u \wedge 1_u) \qquad \text{(modular law)}$$
$$\le 1_u(b^o \wedge 1_u) \qquad (b \le 1_u)$$
$$\le b^o.$$

But $b \le 1_u$ implies $b^o \le 1_u^o = 1_u$ and hence, as above, $b^o \le b$.

Now applying this result to $b = 1 \wedge a_i$ we get that $1_u \wedge a_i = 1_u \wedge a_i^o \le a_i^o$ and so $1_u \wedge a_i \le a_i \wedge a_i^o$.

Finally notice that $1_u \le \bigvee_{i\in I} a_i$ implies that

$$1_u = 1_u \wedge 1_u \le 1_u \wedge \left(\bigvee_{i\in I} a_i\right) = \bigvee_{i\in I}(1_u \wedge a_i) \le \bigvee_{i\in I} (a_i \wedge a_i^o) \ . \qquad \text{Q.E.D.}$$

2. Adjoint bimodules

Let X be a B-category. The *Cauchy-completion* PX of X (described in [1]) is defined by

(i) elements of PX over u are adjoint pairs of bimodules $\phi: \hat{u} \longrightarrow X$, $\psi: X \longrightarrow \hat{u}$, $\phi \dashv \psi$;

(ii) $d((\phi_1,\psi_1),(\phi_2,\psi_2)) = \psi_2 \cdot \phi_1$.

Now the main result of the paper depends on the following representation theorem for adjoint pairs of bimodules:

Theorem. If B satisfies the modular law and $\phi \dashv \psi: u \rightleftarrows X$ is an adjoint pair of bimodules, then

(i) $\phi(x) = \bigvee_{y\in X} d(y,x) \cdot (\psi^o(y) \wedge \phi(y))$,

and (ii) $\psi(x) = \bigvee_{y\in X} (\psi(y) \wedge \phi^o(y)) \cdot d(x,y)$.

Proof of (i).

Firstly,
$$\phi(x) = \bigvee_y d(y,x) \cdot \phi(y) \qquad \text{(Yoneda)}$$
$$\ge \bigvee_y d(y,x)(\psi^o(y) \wedge \phi(y)).$$

Now, from the adjunction we have that $1_u \le \bigvee_y \psi(y)\phi(y)$, and hence by our lemma that $1_u \le \bigvee_y (\psi(y)\phi(y) \wedge \phi^o(y)\psi^o(y))$.

Hence
$$\phi(x) = \phi(x) \cdot 1_u$$
$$\le \phi(x) \cdot \bigvee_y (\psi(y)\phi(y) \wedge \phi^o(y)\psi^o(y))$$
$$\le \phi(x) \cdot \bigvee_y [\psi(y)(\psi^o(y)\phi^o(y)\psi^o(y) \wedge \phi(y))] \ . \qquad \text{(modularity)}$$

But

$$\psi^{o}(y)\phi^{o}(y)\psi^{o}(y) = (\psi(y)\phi(y)\psi(y))^{o}$$
$$\leq (\psi(y)d(y,y))^{o} \qquad \text{(adjunction)}$$
$$\leq \psi^{o}(y).$$

Hence

$$\phi(x) \leq \bigvee_{y} \phi(x)\psi(y)(\psi^{o}(y) \wedge \phi(y))$$
$$\leq \bigvee_{y} d(y,x)(\psi^{o}(y) \wedge \phi(y)). \qquad \text{(adjunction)}$$

The proof of (ii) is similar. Q.E.D.

Remark. Although the bicategory R of example (i) does not satisfy the modular law, a very similar (though simpler) calculation to that above shows that adjoint bimodules can be represented in the same way in that case. We do not have a natural proof that includes both calculations.

3. Symmetry

For X to be *symmetric* we require that $d(y,x) = d(x,y)^{o}$ for all $x,y \in X$.

Theorem. If B satisfies the modular law and X is symmetric, then so is PX, the Cauchy-completion of X.

Proof. It is sufficient to prove that if $\phi \dashv \psi : \hat{u} \rightleftarrows X$ is an adjoint pair of bimodules then $\psi = \phi^{o}$, because then

$$d((\phi_1,\psi_1),(\phi_2,\psi_2)) = \psi_2 \cdot \phi_1 = \phi_2^{o} \cdot \phi_1$$
$$= (\phi_1^{o} \cdot \phi_2)^{o} = (\psi_1 \cdot \phi_2)^{o}$$
$$= d((\phi_2,\psi_2),(\phi_1,\psi_1))^{o}.$$

Now from the theorem in §2, assuming the symmetry of X, we have

$$\phi^{o}(x) = [\bigvee_{y} d(y,x)\cdot(\psi^{o}(y) \wedge \phi(y))]^{o}$$
$$= \bigvee_{y} (\psi(y) \wedge \phi^{o}(y))\cdot d(y,x)^{o}$$
$$= \bigvee_{y} (\phi^{o}(y) \wedge \psi(y))\cdot d(x,y) \qquad \text{(symmetry of X)}$$
$$= \psi(x).$$

Q.E.D.

That symmetry is not always preserved is shown by the following very simple counterexample suggested by S. Kasangian.

Counterexample. Let $G = \{1,a,b\}$ be the group with three elements. Let B be the bicategory with one object whose arrows are subsets of G. Let $(\ )^{o}$ be the identity. Let I be the one element B-category. Then adjoint pairs of bimodules $I \rightleftarrows I$ correspond to elements of G, and under this correspondence $d(g,h) = h^{-1}g$. Hence in PI, $d(1,a) = a^{-1} \neq a = d(a,1)$, so PI is not symmetric.

4. Sheafification

In [1] and [5], sheaves on a site (C,J) are shown to be symmetric Cauchy-complete Rel(C,J)-categories. In particular, presheaves on C are sheaves for a topology J_0 on C and there is an obvious functor

$$\mathrm{Rel}(C,J_0) \to \mathrm{Rel}(C,J)$$

which induces a functor

$$\mathrm{Rel}(C,J_0)\text{-cat} \to \mathrm{Rel}(C,J)\text{-cat} \qquad \text{(change of base)}$$

This preserves symmetry.

Now from the description of sheafification in [1], using the fact that symmetry is preserved by Cauchy-completion, we get that sheafification is the composite

$$\mathrm{Preshv}(C) \subseteq \mathrm{sym\ Rel}(C,J_0)\text{-cat} \to \mathrm{sym\ Rel}(C,J)\text{-cat} \xrightarrow[\text{Cauchy-completion}]{} \mathrm{Shv}(C,J).$$

So we have a description of sheafification in terms of standard constructions of enriched category theory.

REFERENCES

1. R. Betti and A. Carboni, Cauchy-completion and the associated sheaf, to appear in *Cahiers top. et géom. diff.*

2. Denis Higgs, A category approach to boolean valued set theory, *Lecture Notes, University of Waterloo,* 1973.

3. F.W. Lawvere, Metric spaces, generalized logic, and closed categories, *Rendiconti del Seminario Matematico e Fisico di Milano* 43 (1973) 135-166.

4. H. Lindner, Morita equivalences of enriched categories, *Cahiers top. et géom. diff.* 15 (1974), 377-397.

5. R.F.C. Walters, Sheaves and Cauchy-complete categories, *Cahiers top. et géom. diff.* 22 (1981), 283-286.

6. R.F.C. Walters, Sheaves on a site as Cauchy-complete categories, to appear in *Journal of Pure and Applied Algebra.*

ON ALGEBRAIC LOCALIZATIONS

by

Francis BORCEUX
Université Catholique de Louvain
1348-Louvain-La-Neuve - Belgium

Throughout this paper, $\mathbb{E}$ denotes a topos of sheaves on a locale $\mathbb{H}$ and T denotes a finitary algebraic theory internally defined with respect to this topos $\mathbb{E}$. (i.e. a sheaf of finitary algebraic theories on $\mathbb{H}$). We are interested in studying the localizations of the category $\mathbb{E}^T$ of T-algebras in $\mathbb{E}$ (i.e. the full exact reflective subcategories of $\mathbb{E}^T$).

For any integer n and any element α in $\mathbb{H}$, let us denote by $\phi_n(\alpha)$ the set of n-ary operations of the theory $T(\alpha)$. ϕ_n turns out to be a sheaf on H, i.e. an object in $\mathbb{E}$. As the composite of two 1-ary operations is again a 1-ary operation, the object ϕ_1 is in fact a monoid in $\mathbb{E}$.

We denote by $\mathbb{E}_T$ the topos of objects in $\mathbb{E}$ on which the monoid ϕ_1 acts. There is a trivial forgetful functor $\mathbb{E}^T \to \mathbb{E}_T$ which allows us to consider the object in $\mathbb{E}_T$ which underlies a T-algebra. We keep the same notation for a T-algebra and its underlying ϕ_1-object.

The topos $\mathbb{E}^T$ has a subobject classifier Ω_T. In this paper we define in $\mathbb{E}_T$ an inf-semi-lattice ω_T with 0 and 1 which plays, with respect to $\mathbb{E}^T$, the same role as Ω_T with respect to the $\mathbb{E}_T$. ω_T classifies the subobjects in $\mathbb{E}^T$ and the topologies on ω_T classify the localizations of $\mathbb{E}^T$.

If A is some T-algebra, we define the notion of a characteristic map on A. This is a map $A \to \omega_T$ in the topos $\mathbb{E}_T$ which satisfies some compatibility condition with respect to the theory T. We prove that for a commutative theory T, the characteristic maps on A in $\mathbb{E}_T$ are in one to one correspondence with the subobjects of A in the algebraic category $\mathbb{E}^T$.

ω_T is in $\mathbb{E}_T$ an inf-semi-lattice with greatest element. So it makes sense to speak of a map $j : \omega_T \to \omega_T$ which satisfies the three Lawvere-Tierney conditions for a topology. This is what we call a T-topology. We prove that for a commutative theory T, the T-topologies on ω_T in $\mathbb{E}_T$ classify exactly the localizations of the algebraic category $\mathbb{E}^T$.

It is also possible to generalize in this context the notion of Gabriel-Grothendieck topology. The notion of crible becomes that of a subobject of the free algebra generated by a representable sheaf. The three conditions for a Gabriel-Grothendieck topology translate in this context but a fourth condition appears which takes care of the fact that we start with a topos $\mathbb{E}$ of sheaves and not of presheaves. These generalized Gabriel-Grothendieck topologies classify again the localizations of the algebraic category $\mathbb{E}^T$.

We treat explicitly an example : the topos of sets and the theory of abelian groups. In this case the classifying object ω_T is exactly the set of natural numbers. We describe the characteristic map of a subgroup and we give some examples of Ab-topologies on $\mathbb{N}$.

Finally we give a counterexample to our results when the commutativity of the theory T is lacking. This is the case of sets and the theory of groups.

In this paper, the proofs are only sketched. The details of them can be found in [2].

§1 - Some lemmas

Some more notation. $U : \mathbb{E}^T \to \mathbb{E}$ is the forgetful functor and $F : \mathbb{E} \to \mathbb{E}^T$ is its left adjoint which, of course, preserves monomorphisms. If α is some element in H, $h_\alpha : H^{op} \to \text{Sets}$ is the corresponding representable sheaf and $\alpha\!\downarrow$ is the sublocale of H consisting of all those β such that $\beta \leqslant \alpha$. $\alpha\!\downarrow$ is again a locale and $\mathbb{E}^T(\alpha)$ is the corresponding algebraic category. The object of this paragraph is to describe the relations between $\mathbb{E}^T$ and the various $\mathbb{E}^T(\alpha)$ for $\alpha \in H$.

LEMMA 1. *If* A *is any* T*-algebra and* $\alpha \in H$,

$$A(\alpha) \cong \mathbb{E}^T(Fh_\alpha, A) \qquad \square$$

PROPOSITION 2. $\mathbb{E}^T$ *is a complete, cocomplete and regular category in which finite limits commute with filtered colimits. The objects* Fh_α*, for* $\alpha \in H$*, form a regular set of generators of* $\mathbb{E}^T$. $\square$

PROPOSITION 3. *For any* $\alpha \in H$, *the restriction functor*

$$\alpha^* : \mathbb{E}^T \to \mathbb{E}^T(\alpha)$$

has a right adjoint α_* *and a left adjoint* $\alpha!$. α^* *and* $\alpha!$ *are full and faithful and* $\alpha!$ *preserves and creates monomorphisms. Moreover, if* $A \in \mathbb{E}^T$ *and* $\phi_o \to A$ *is monic, the canonical morphism* $\alpha!\,\alpha^*\,A \to A$ *is also monic.*

α_* is simply defined by

$$(\alpha_* A)(\beta) = A(\alpha \wedge \beta)$$

for any A in $\mathbb{E}^T(\alpha)$ and β in H. Now consider

$$(\alpha' A)(\beta) = \begin{cases} A(\beta) & \text{if } \beta \leqslant \alpha \\ \phi_o(\beta) & \text{if not} \end{cases}$$

$\alpha!(A)$ is the sheaf universally associated to the presheaf $\alpha'(A)$. α_* is full and faithful and thus the same holds for $\alpha!$ ([12] - 16 - 8 - 9). Now point out that ϕ_o (= the free algebra on 0) is the initial object in $\mathbb{E}^T$. The facts concerning monomorphisms follow easily from this remark and the component-by-component description of α^* and $\alpha!$. □

LEMMA 4. *For any* $\alpha \in H$, $\phi_o \to Fh_\alpha$ *is monic.* □

COROLLARY 5. *If* $\alpha \leqslant \beta$ *in* H, $\beta!\beta_*(Fh_\alpha) = Fh_\alpha$. □

LEMMA 6. *If* $\alpha = \bigvee_{i\in I} \alpha_i$ *in* H, $Fh_\alpha = \bigcup_{i\in I} Fh_{\alpha i}$ □

LEMMA 7. *For any* α *in* H, $Fh_\alpha \cong \alpha!\alpha^*\phi_1$ □

LEMMA 8. *If* $\phi_o \to A$ *is monic in* $\mathbb{E}^T$, $S \rightarrowtail A$ *is a subobject in* $\mathbb{E}^T$ *and* $\alpha \in H$

$$\alpha!\alpha^*\,S = S \cap \alpha!\alpha^* A$$ □

LEMMA 9. *If* $\phi_o \to A$ *is monic in* $\mathbb{E}^T$, $S \rightarrowtail A$ *and* $S_i \rightarrowtail A$ *are subobjects and* α, α_i *are elements in* H,

$$\alpha!\alpha^* A \cap (\bigcup_{i\in I} S_i) \cong \bigcup_{i\in I} \alpha!\alpha^*\,S_i$$

$$S \cap (\bigcup_{i\in I} \alpha_i!\,\alpha_i^* A) \cong \bigcup_{i\in I} \alpha_i!\,\alpha_i^*\,S$$ □

LEMMA 10. *Let* $f : A \to B$ *be a morphism and* $\phi_o \to B$ *a monomorphism in* $\mathbb{E}^T$. *Let* $1 = \bigvee_{i\in I} \alpha_i$ *in* H.

$$A = \bigcup_{i\in I} f^{-1}(\alpha_i!\,\alpha_i^*\,B)$$ □

COROLLARY 11. *If* $1 = \bigvee_{i\in I} \alpha_i$ *in* H *and* $\phi_o \to A$ *is monic in* $\mathbb{E}^T$,

$$A = \bigcup_{i\in I} \alpha_i ! \, \alpha_i^* \, A.$$ □

LEMMA 12. *If* T *is commutative and* A *is any object in* $\mathbb{E}^T$, *the canonical morphism* $\phi_o \to A$ *is monic.* □

§2 - The classifying pair $(\mathbb{E}_T, \omega_T)$.

At each level $\alpha \in H$, the composite of two 1-ary operations is a 1-ary operation. This provides ϕ_1 with the structure of a monoid in $\mathbb{E}$. We denote by $\mathbb{E}_T$ the topos of ϕ_1-objects in $\mathbb{E}$.

PROPOSITION 13. *For any* α *in* H

$$\phi_1(\alpha) \cong \mathbb{E}^T \, (Fh_\alpha, Fh_\alpha).$$

If $\beta \leqslant \alpha$, *the restriction* map $\phi_1(\alpha) \to \phi_1(\beta)$ *is given by the action of* $\beta ! \beta^*$. □

PROPOSITION 14. *Any algebra* $A \in \mathbb{E}^T$ *is canonically provided with the structure of a* ϕ_1*-object. Any morphism in* $\mathbb{E}^T$ *becomes in this way a morphism in* $\mathbb{E}^T$.

For any α in H, define the action by composition

$$\phi_1(\alpha) \times A(\alpha) \cong \mathbb{E}^T \, (Fh_\alpha, Fh_\alpha) \times \mathbb{E}^T \, (Fh_\alpha, A) \to \mathbb{E}^T \, (Fh_\alpha, A) \cong A(\alpha).$$ □

We turn now into the definition of the classifying object ω_T in $\mathbb{E}_T$ (which is not, in general, the Ω-object of this topos $\mathbb{E}_T$). For any α in H we define $\omega_T(\alpha)$ = {subobjects of Fh_α in $\mathbb{E}^T$}. Now if $\beta \leqslant \alpha$, we define a restriction map

$$\omega_T(\alpha) \to \omega_T(\beta)$$

by pulling back along the canonical inclusion

$$Fh_\beta \to Fh_\alpha.$$

We also define an action of ϕ_1 on ω_T

$$\phi_1(\alpha) \times \omega_T(\alpha) \cong \mathbb{E}^T(Fh_\alpha, Fh_\alpha) \times \omega_T(\alpha) \to \omega_T(\alpha)$$

again by pulling back along the morphisms $Fh_\alpha \to Fh_\alpha$. □

PROPOSITION 15. ω_T *is an inf-semi-lattice in* $\mathbb{E}_T$.

ω_T is a separated presheaf (corollary 11 applied to $\mathbb{E}^T(\alpha)$ if $\alpha = \bigvee_{i\in I} \alpha_i$ in H).

Now let $\alpha = \bigvee_{i\in I} \alpha_i$ in H and $R_i \in \omega_T(\alpha_i)$ a compatible family. Each R_i is in particular a subobject of Fh_α; we define R as being the union of the R_i's, as subobjects of Fh_α. For any j in I one has

$$\begin{aligned} Fh_{\alpha_j} \cap R &= Fh_{\alpha_j} \cap \Big(\bigcup_{i\in I} R_i\Big) \\ &= \bigcup_{i\in I} (Fh_{\alpha_j} \cap R_i) \\ &= \bigcup_{i\in I} (Fh_{\alpha_i} \cap R_j) \\ &= \bigcup_{i\in I} \alpha_i!\, \alpha_i^* (R_j) \\ &= R_j \end{aligned}$$

by lemmas 8, 9 and the compatibility of the R_i's. So ω_T is a sheaf and thus an object in $\mathbb{E}_T$.

ω_T is provided with mappings in $\mathbb{E}_T$

$$\wedge_T : \omega_T \times \omega_T \to \omega_T$$

$$t_T : 1 \to \omega_T$$

where $\wedge_T$ acts by intersection of subobjects and t_T chooses the maximal subobject Fh_α in each component. So ω_T is an inf-semi-lattice with 1. □

We are now able to introduce the notion of a characteristic map on an object of $\mathbb{E}^T$; this is a map in $\mathbb{E}_T$ with values in ω_T. This technical notion will be useful in classifying the localizations of $\mathbb{E}^T$.

DEFINITION 16. *Let* A *be any object in* $\mathbb{E}^T$. *A characteristic map on* A *is a morphism* $\varphi : A \to \omega_T$ *in* $\mathbb{E}_T$ *such that for any* α *in* H, n *in* $\mathbb{N}$, θ *in* $\phi_n(\alpha)$ *and* $x_1,\dots,x_n$ *in* $A(\alpha)$

$$\bigwedge_{i=1}^{n} \varphi_\alpha(x_i) \leq \varphi_\alpha(\theta(x_1,\dots,x_n)).$$

THEOREM 17. *If* T *is a commutative theory and* A *an object in* $\mathbb{E}_T$, *there is a bijection between*

(1) the subobjects of A *in* $\mathbb{E}^T$

(2) the characteristic maps on A *in* $\mathbb{E}_T$.

Moreover, if some subobject $S \rightarrowtail A$ *has a characteristic map* $\varphi : A \to \omega_T$, *the following square is a pullback in* $\mathbb{E}_T$

$$\begin{array}{ccc} S & \longrightarrow & 1 \\ \downarrow & & \downarrow t_T \\ A & \xrightarrow{\ \varphi\ } & \omega_T \end{array}$$

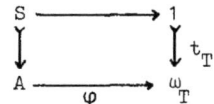

If $S \rightarrowtail A$ in $\mathbb{E}^T$, we define $\varphi : A \to \omega_T$ at the level $\alpha \in H$ by pulling S back along the morphisms $Fh_\alpha \to A$:

$$A(\alpha) \cong \mathbb{E}^T(Fh_\alpha, A) \to \omega_T(\alpha).$$

If $\theta \in \phi_n(\alpha)$ and $f_1, \ldots f_n : Fh_\alpha \to A$ are morphisms in $\mathbb{E}^T$, the commutativity of the theory allows us to define $f = \theta(f_1, \ldots f_n) : Fh_\alpha \to A$ and for any subobject $R \rightarrowtail A$ one has

$$\bigcap_{i=1}^{n} f_i^{-1}(R) \subseteq f^{-1}(R).$$

This implies that φ is a characteristic map.

Now if the characteristic map $\varphi : A \to \omega_T$ is given, define $S \rightarrowtail A$ in $\mathbb{E}^T$ by pulling back $t_T : 1 \to \omega_T$ along φ in $\mathbb{E}_T$. The conditions on φ imply that S is in fact a subobject of A in $\mathbb{E}^T$.

To check the one-to-one correspondance, it useful to point out that for $\alpha \leqslant \beta$ in H, any morphism $Fh_\alpha \to Fh_\beta$ in $\mathbb{E}^T$ factors through the canonical inclusion $Fh_\alpha \rightarrowtail Fh_\beta$ and a morphism $Fh_\alpha \to Fh_\alpha$, i.e. an element of $\phi_1(\alpha)$. □

EXAMPLE 18. Let $\mathbb{E}$ be the topos of sets and T the theory of abelian groups. $\mathbb{E}^T$ is the category of abelian groups and $\mathbb{E}_T$ is the topos of $(\mathbb{Z},x)$-sets. ω_T is the set $\mathbb{N}$ provided with the action

$$\mathbb{Z} \times \mathbb{N} \to \mathbb{N} \; ; \; z * n = \frac{n}{n \wedge |z|}$$

where $\wedge$ denotes the greatest common divisor. If $S \rightarrowtail A$ is an inclusion of abelian groups, the corresponding characteristic map $\varphi : A \to \mathbb{N}$ is defined by

$$\varphi(a) = \inf \{n \geqslant 1 | n\, a \in S\}$$

and $\varphi(a)$ is zero if this set is empty. Thus if $\varphi(a) \neq 0,1$, $\varphi(a)$ is exactly the order of a in A/S.

§3 - Classifying the algebraic localizations.

We generalize now the notions of "Lawvere-Tierney topology" and "Gabriel-Grothendieck topology" into our algebraic context. We prove that for a commutative theory T, the localizations of $\mathbb{E}^T$ can be exactly classified by these topologies and also by the universal closure operations on $\mathbb{E}^T$. (cfr. [8]-3 - 13). But let us start with an arbitrary T.

DEFINITION 19. *A Lawvere-Tierney* T*-topology on* ω_T *is a morphism* $j : \omega_T \to \omega_T$ *in* $\mathbb{E}_T$ *such that*

(1) $j \cdot t_T = t_T$

(2) $j \cdot j = j$

(3) $j \cdot \wedge_T = \wedge_T \cdot (j \times j)$.

DEFINITION 20. *A Gabriel-Grothendieck* T*-topology on* $\mathbb{E}^T$ *consists in given, for any* $\alpha \in H$, *a family* $J(\alpha)$ *of subobjects of* Fh_α *in* $\mathbb{E}^T$ *such that*

(1) $Fh_\alpha \in J(\alpha)$

(2) $R \in J(\alpha)$ and $f : Fh_\beta \to Fh_\alpha \Rightarrow f^{-1}(R) \in J(\beta)$

(3) $R \in J(\alpha)$ and $S \rightarrowtail Fh_\alpha$ and $\forall f : Fh_\beta \to R\ f^{-1}(R \cap S) \in J(\beta) \Rightarrow S \in J(\alpha)$

(4) $R \rightarrowtail Fh_\alpha$ and $\alpha = \bigvee_{i \in I} \alpha_i$ and $R \cap Fh_{\alpha i} \in J(\alpha_i) \Rightarrow R \in J(\alpha)$.

PROPOSITION 21. *Let* $\mathbb{C} \overset{l}{\hookrightarrow} \mathbb{E}^T$ *be any localization of* $\mathbb{E}^T$. *Define*

$$J(\alpha) = \{R \overset{r}{\rightarrowtail} Fh_\alpha \mid lr \text{ iso}\}.$$

J is a Gabriel-Grothendieck T*-topology on* $\mathbb{E}^T$ *and this process describes an injection of the set of localizations into the set of Gabriel-Grothendieck* T*-topologies.*

Proof analogous to that given in the sets-case. □

COUNTEREXAMPLE 22. This counterexample shows that, for an arbitrary T, there is no bijection between the localizations of $\mathbb{E}^T$ and the Gabriel-Grothendieck T-topologies.

Consider the topos $\mathbb{E}$ of sets and the theory T of (abelian) groups. $H = \{0,1\}$, $Fh_0 = (0)$ and $Fh_1 = \mathbb{Z}$. It follows that in this case the Gabriel-Grothendieck T-topologies coincide for both theories of groups and abelian groups. Moreover axiom 4 of definition 20 vanishes in this context and thus the T-topologies coincide with the usual Gabriel topologies on $\mathbb{Z}$. We know that there are infinitely many such topologies.

On the other hand, the category of groups admits only the two obvious localizations : (0) and itself. Indeed, if $\mathbb{C} \overset{l}{\hookrightarrow} Gr$ is a localization and if $n\mathbb{Z} \in J(1)$ $(n \neq 0)$, consider the canonical inclusion $n\mathbb{Z} \coprod n\mathbb{Z} \hookrightarrow \mathbb{Z} \coprod \mathbb{Z}$ which is taken by l into an isomorphism. If x and y are the two generators of $\mathbb{Z} \coprod \mathbb{Z}$, pull $n\mathbb{Z} \coprod n\mathbb{Z}$ back along the morphism $\mathbb{Z} \to \mathbb{Z} \coprod \mathbb{Z}$ which sends 1 on the word xy. You get an element of J(1) which is nothing but (0); thus J(1) is the set of all subgroups of $\mathbb{Z}$ and the localization is obvious.

If we turn again to a commutative T, we get the main result of this paper which generalizes to an algebraic context various results on toposes.

THEOREM 23. *If* T *is a commutative theory, there is a bijection between*

(1) the localizations of $\mathbb{E}^T$

(2) the universal closure operations on $\mathbb{E}^T$

(3) the Lawvere-Tierney T*-topologies on* ω_T

(4) the Gabriel-Grothendieck T*-topologies on* $\mathbb{E}^T$.

If $\mathbb{C} \underset{i}{\overset{l}{\leftrightarrows}} \mathbb{E}^T$ is a localization and $S \rightarrowtail A$ a subobject in $\mathbb{E}^T$, define $\bar{S}$ by the pullback

$$\begin{array}{ccc} \bar{S} & \longrightarrow & i\,lS \\ \downarrow & & \downarrow \\ A & \longrightarrow & i\,lA \end{array}$$

where $A \to i\,lA$ is the canonical morphism arising from the adjunction $l \dashv i$.

From a closure operation, define $j : \omega_T \to \omega_T$ by $j_\alpha(S) = \bar{S}$ for any α in H.

From a Lawvere-Tierney T-topology j define

$$J(\alpha) = \{S \to Fh_\alpha \mid j_\alpha(S) \cong Fh_\alpha\}$$

for any α in H.

If a Gabriel-Grothendieck topology J is given, define a localization $\mathbb{C} \underset{i}{\overset{l}{\leftrightarrows}} \mathbb{E}^T$ as follows : $A \in |\mathbb{C}|$ iff for any $\alpha \in H$ and $S \in J(\alpha)$, the canonical morphism $\mathbb{E}^T(Fh_\alpha, A) \to \mathbb{E}^T(S,A)$ is a bijection. Now put $l'A(\alpha) = \varinjlim_{S \in J(\alpha)} \mathbb{E}^T(S,A)$ and $lA = l'l'A$. This makes sense because T is commutative and thus $\mathbb{E}^T(S,A)$ is naturally provided with the structure of a $T(\alpha)$-algebra.

The proof is rather long; its sketch is very close to that of the usual proof in the case of the localizations of a topos of presheaves. In order to write it down, it is useful to point out the following facts.

(1) condition (4) of definition 20 takes care of the fact that we are working with sheaves and not with presheaves.

(2) if $\varphi : A \to \omega_T$ is a characteristic map and $j : \omega_T \to \omega_T$ is a Lawvere-Tierney T-topology, $j \circ \varphi$ is again a characteristic map and thus defines a subobject of A in $\mathbb{E}^T$.

(3) if a Lawvere-Tierney T-topology j and a Gabriel-Grothendieck T-topology J correspond to each other by the constructions above, the mapping "$J : H^{op} \to \text{Sets};\ \alpha \leadsto J(\alpha)$" is in fact a sheaf provided with an action of ϕ_1 given by pulling back. Thus J is a subobject of ω_T in $\mathbb{E}_T$ and the following diagram is a pullback in $\mathbb{E}_T$.

$$\begin{array}{ccc} J & \longrightarrow & 1 \\ \downarrow & & \downarrow{\scriptstyle t_T} \\ \omega_T & \xrightarrow{\;j\;} & \omega_T \end{array}$$

(4) let $\alpha, \beta \in H$ with $\alpha \nleq \beta$. Let $f : Fh_\alpha \to Fh_\beta$ be any morphism in $\mathbb{E}^T$ and $S \rightarrowtail Fh_\alpha$ any subobject in $\mathbb{E}^T$. Necessarily, $f^{-1}(S) = Fh_\alpha$. To see this, denote by $F'h_\alpha$ the free presheaf of T-algebras on h_α. If f is of the form ag (a = associated sheaf functor) with $g : F'h_\alpha \to F'h_\beta$, then $F'h_\beta(\beta) = \phi_o(\alpha)$ and therefore g, and thus f, factors through ϕ_o. So $f^{-1}(S) = Fh_\alpha$. Now if $f : Fh_\alpha \to Fh_\beta$ is any morphism, it can be described locally by morphisms of the form ag and the result follows from the glueing construction for sheaves. □

EXAMPLE 24. Let us go back to the case of the topos of sets and the theory T of abelian groups (example 18).

The localization of the category of abelian groups given by the usual localization process at some prime ideal $p\mathbb{Z}$ of $\mathbb{Z}$ is classified by the Lawvere-Tierney T-topology $j : \mathbb{N} \to \mathbb{N}$ given by :

$$j(n) = \text{the greatest power of p dividing n.}$$

The localization of the category of abelian groups corresponding to the thick subcategory of abelian groups all of whose element have torsion is just the category of rational vector spaces. The corresponding Lawvere-Tierney T-topology is given by $j : \mathbb{N} \to \mathbb{N}$:

$$j(n) = \begin{cases} 1 \text{ if } n \neq 0 \\ 0 \text{ if } n = 0. \end{cases}$$

BIBLIOGRAPHY.

[1] BORCEUX F., Sheaves of algebras for a commutative theory. Ann. Soc. Scient. Bruxelles; 95, I, pp. 3-19 (1981).

[2] BORCEUX F., Sur les localisations algébriques. Preprint; Rapp. Sém. Math. Pures, Université de Louvain, Belgium (1981).

[3] GABRIEL, Des catégories abéliennes. Bull. Soc. Math. France, 90, pp. 323-448, (1962).

[4] GABRIEL-ULMER, Lokal präsentierbare Kategorien. Lect. Notes in Math., 221, Springer (1971).

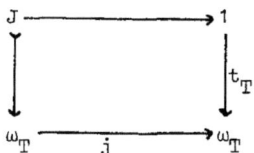

[5] GRILLET, Regular categories. Lect. Notes in Math., 226, Sringer (1972).

[6] GROTHENDIECK, Théorie des topos et cohomologie étale des shémas. Lect. Notes in Math., 269, Springer (1972).

[7] HELLER-ROWE, On the category of sheaves. Am. Jour. of Math., 84, pp. 205-216, (1962).

[8] JOHNSTONE, Topos theory. Academic Press (1977).

[9] LAWVERE, Functorial semantics of the algebraic theories. Proc. Nat. Acad. Sc., 50, pp. 869-872, (1963).

[10] LINTON, Autonomous categories and duality of functors. Journ. alg., 2, pp. 315-349, (1965).

[11] POPESCU, Abelian categories with applications to rings and modules. Academic Press (1973).

[12] SCHUBERT, Categories. Springer (1972).

[13] TIERNEY, Axiomatic sheaf theory, cime, varenna (1971). Ed. Cremoneze, pp. 249-326, (1973).

[14] ULMER, On the existence and exactness of the associated sheaf functor. J. Pure Appl. Algebra, 3, pp. 295-306, (1973).

A CANONICAL ACTION ON INDEXED LIMITS

AN APPLICATION TO COHERENT HOMOTOPY

by Dominique BOURN

Université de Picardie
U.E.R. de Mathématiques
33, rue St Leu
80039 - Amiens - France

It is well known that the lax limit of a monad (seen as a 2-functor from the simplicial 2-category $\mathbb{D}$ to Cat) is its category of algebras, and that there is a comonad on this category (i.e. an action of the 2-category $\mathbb{D}^{co}$). There are other examples of such a situation. For instance, if $\mathbb{H}$ is the sub 2-category of $\mathbb{D}$, generated by the 2-cell which gives the multiplication of the monads, then it is easy to see that there is an action of a certain 2-category on the lax limit of every 2-functor from $\mathbb{H}$ to Cat, namely the sub-2-category $\mathbb{J}$ of $\mathbb{D}^{co}$ generated by the 2-cell determining the co-unit of the comonads. The purpose of this work is to show that this fact is very general : given a $\mathbb{V}$-category $\mathbb{A}$ and an indexation $\phi : \mathbb{A} \dashrightarrow \mathbb{1}$ there is a $\mathbb{V}$-monoïd which acts on each ϕ-indexed limit. The proof is of the same kind as the proof that Kan extensions are shape invariant.

On the other hand, Bousfield-Kan homotopy limits are another example of indexed limits. Recent strong shape theory tries to preserve traces of higher homotopy coherences. For instance Dwyer-Kan [10] give a standard resolution in order to do simplicial localizations. To each category C , they associate a simplicial category (i.e. enriched in the category of simplicial sets) $F_* C$ in order to take higher simplicial coherences into account. A category being a particular simplicial set, via its nerve, a 2-category is a special simplicial category. Actually their resolution applied to the idempotent-category (with a single object and a single non trivial arrow t such that $t^2 = t$) is just the 2-category $\mathbb{H}$.

I call a simplicial functor F from $\mathbb{H}$ to Top (with its simplicial category structure) a coherent homotopy idempotent. Choosing the indexation $\mathbb{H} \dashrightarrow \mathbb{1}$ which gives rise to what Gray [14] considers as a generalization of the homotopy limits, we describe the canonical action of the canonical simplicial monoïd on this homotopy limit of F . In particular we show that the associated homotopy idempotent (idempotent in the homotopy category Ho-Top) splits, although homotopy idempotents do not split in general (Freyd-Heller [12] , Dydak-Minc [11]).

I choose to use $\mathbb{V}$-profunctors, introduced by Benabou [2] and Lawvere [15] in this work. Though it is necessary to recall some results at the beginning (Part I : composition, right Kan extension, representability) the proofs of this paper are so much easier in this context and the methods so intuitive and powerful that it seems justified to use them. They illuminate the notion of indexed limits, introduced by Auderset [1] and Borceux-Kelly [3], to which Part II is devoted with particular studies of the Bousfield-Kan homotopy limits [7], the lax limits of Gray [13] [4] (we give a new profunctorial description of the associated indexation of Street [17]) and specially the lax limit of a semiad (a 2-functor from $\mathbb{H}$ to Cat), the main tool for the application. Part III contains the main result about the canonical action and Part IV the application to coherent homotopy.

I. The $\mathbb{V}$-profunctors.

Let $\mathbb{V}$ be a symmetric monoïdal closed category, complete and cocomplete. Let us recall from [2] and [14] that if $\phi : \mathbb{A} \dashrightarrow \mathbb{B}$ and $\Psi : \mathbb{B} \dashrightarrow \mathbb{C}$ are two $\mathbb{V}$-profunctors between the $\mathbb{V}$-category $\mathbb{A}$, $\mathbb{B}$, $\mathbb{C}$ (i.e. : $\mathbb{V}$-functors $\phi : \mathbb{B}^{op} \otimes \mathbb{A} \to \mathbb{V}$ and $\Psi : \mathbb{C}^{op} \otimes \mathbb{B} \to \mathbb{V}$), there is a

Composition of profunctors.

The composite $\Psi \otimes \phi$ is defined by the coends : $\Psi \otimes \phi(C, A) = \int^{B} \Psi(C,B) \otimes \phi(B,A)$ that are the cokernels of the maps d_0, d_1 :

$$\coprod_{B,B'} \Psi(C, B) \otimes \mathbb{B}(B, B') \otimes \phi(B', A) \overset{d_0}{\underset{d_1}{\rightrightarrows}} \coprod_{B} \Psi(C, B) \otimes \phi(B, A)$$

where d_0 and d_1 are respectively induced by : $\Psi_{BB'} \otimes \phi(B',A)$ and $\Psi(C,B) \otimes \phi_{BB'}$

$$\Psi(C,B) \otimes \mathbb{B}(B,B') \otimes \phi(B',A) \longrightarrow \Psi(C,B') \otimes \phi(B',A)$$

$$\Psi(C,B) \otimes \mathbb{B}(B,B') \otimes \phi(B',A) \longrightarrow \Psi(C,B) \otimes \phi(B',A)$$

the maps $\phi_{BB'}$ and $\Psi_{BB'}$ being respectively the actions of $\mathbb{B}(B, B')$ on ϕ and Ψ.

This composition can be clearly extended to natural transformations of profunctors, and is associative and unitary up to isomorphism, if the unit associated to $\mathbb{A}$ is the $\mathbb{V}$-functor $\mathbb{A}(-, -) : \mathbb{A}^{op} \otimes \mathbb{A} \to \mathbb{V}$.

Right Kan extensions of profunctors.

The main property of the (bi-)category of profunctors, is that there always exist right Kan extensions. If $\phi : \mathbb{A} \dashrightarrow \mathbb{B}$ and $\chi : \mathbb{A} \dashrightarrow \mathbb{D}$ are two profunctors, then the right Kan extension $[\![\phi, \chi]\!]$ of χ along ϕ is given by the ends :

$$[\![\phi, \chi]\!] \ (D, B) = \int_{A} [\phi(B, A), \chi(D, A)]$$

where $[X, Y]$ is the value at Y of the right adjoint of the functor $- \otimes X : \mathbb{V} \to \mathbb{V}$, that are the kernels of the maps t_0, t_1 :

$$\prod_{A} [\phi(B, A), \chi(D, A)] \overset{t_0}{\underset{t_1}{\rightrightarrows}} \prod_{AA'} [\phi(B, A) \otimes \mathbb{A}(A,A'), \chi(D,A')]$$

where t_0 and t_1 are respectively induced (via the above adjunction) by :

$$[\phi(B,A), \chi(D,A)] \otimes \phi(B,A) \otimes \mathbb{A}(A,A') \xrightarrow{ev \otimes 1} \chi(D,A) \otimes \mathbb{A}(A,A') \xrightarrow{\chi_{AA'}} \chi(D,A')$$

$$[\phi(B,A'), \chi(D,A')] \otimes \phi(B,A) \otimes \mathbb{A}(A,A') \xrightarrow{1 \otimes \phi_{AA'}} [\phi(B,A'), \chi(D,A')] \otimes \phi(B,A') \xrightarrow{ev} \chi(D,A')$$

Thus $[\![\phi, \chi]\!]$ (D, B) is nothing but $\mathbb{V}$-Nat$(\phi(B, -), \chi(D, -))$.

Representability of profunctors.

There is an embedding from $\mathbb{V}(A, B)$, the $\mathbb{V}$-category of $\mathbb{V}$-functors between $\mathbb{A}$ and $\mathbb{B}$, to $\mathbb{V}$- prof $(\mathbb{A}, \mathbb{B})$ the $\mathbb{V}$-category of $\mathbb{V}$-profunctors between $\mathbb{A}$ and $\mathbb{B}$ which associates to each functor $F : \mathbb{A} \to \mathbb{B}$, the profunctor $\phi F : \mathbb{A} \dashrightarrow \mathbb{B}$, where

$$\phi F(B, A) = \mathbb{B}(B, F A).$$

Definition. A profunctor $\phi : \mathbb{A} \dashrightarrow \mathbb{B}$ is called representable if there is a functor $F : \mathbb{A} \to \mathbb{B}$ such that ϕF is naturally isomorphic to ϕ.
Hence we have the classical result :

Proposition. Let $F : \mathbb{A} \to \mathbb{B}$ and $G : \mathbb{A} \to \mathbb{D}$ be two $\mathbb{V}$-functors, then G admits a point wise (i.e. preserved by all representable functors $\mathbb{D}(D, -) : \mathbb{D} \to \mathbb{V}$) right Kan extension, if and only if the right Kan extension of profunctors $[\![\phi F, \phi G]\!]$ is representable.

II. The ϕ-indexed limits.

Let $\mathbb{1}$ be the $\mathbb{V}$-category with only one object $\star$, and such that $\mathbb{1}(\star, \star)$ is the unit object of $\mathbb{V}$. If $\mathbb{A}$ is a $\mathbb{V}$-category, there is no canonical $\mathbb{V}$-functor $\mathbb{A} \to \mathbb{1}$ in general, so no canonical notion of limit. Let $\phi : \mathbb{A} \dashrightarrow \mathbb{1}$ be a $\mathbb{V}$-profunctor (i.e. a $\mathbb{V}$-functor $\mathbb{A} \to \mathbb{V}$) and $F : \mathbb{A} \to \mathbb{B}$ a $\mathbb{V}$-functor. Then ([1], [3]), we have the following.

Definition. A projective ϕ-indexed cone from an object B of $\mathbb{B}$ to F is an element of $[\![\phi, \phi F]\!]$ $(B, \star)$. The $\mathbb{V}$-functor F admits a projective (or inverse) ϕ-indexed limit, if the right Kan extension $[\![\phi, \phi F]\!]$ is representable by an object of $\mathbb{B}$ (denoted by ϕ-lim F) ; that is, if and only if $[\![\phi, \phi F]\!]$ $(B, \star) \simeq \mathbb{B}(B, \phi\text{-lim } F)$. Likewise a $\mathbb{V}$-functor $L : \mathbb{A}^{op} \to \mathbb{B}$ has an (inductive or direct) ϕ-indexed colimit (denoted by ϕ-colim L), if $L^{op} : \mathbb{A} \to \mathbb{B}^{op}$ has a ϕ-indexed limit.

We shall call ϕ the indexation of the limit.

Examples. If F is a $\mathbb{V}$-functor $: \mathbb{A} \to \mathbb{V}$, then it is clear that $\phi\text{-lim}F \simeq \mathbb{V}\text{-Nat}(\phi,F)$ since $[X, \mathbb{V}\text{-Nat}(\phi, F)] \simeq \mathbb{V}\text{-Nat}(X \otimes \phi, F)$ for each object X of $\mathbb{V}$.
If L is a $\mathbb{V}$-functor $: \mathbb{A}^{op} \to \mathbb{V}$, then ϕ-colim $L = \phi \otimes L$ where L is seen as a profunctor $\mathbb{1} \dashrightarrow \mathbb{A}$, since $[\phi \otimes L, X] = \text{Nat}(\phi, [L(-),X])$ for each object X of $\mathbb{V}$.
On the other hand, if Y is the Yoneda embedding $: \mathbb{A}^{op} \to \mathbb{V}^{\mathbb{A}}$ then ϕ-colim $Y = \phi$

since $V^A(\phi, L) = Nat(\phi, L)$ and $Nat(\phi, V^A(Y(\ ), L)) = Nat(\phi, Nat(Y(\ ), L))$. Now, by Yoneda ! , L is isomorphic to $Nat(Y(\ ), L)$.

We shall study more precisely two particular important examples for our purpose.

The Bousfield-Kan homotopy limits.

Let Δ be the well known category of ordered sets $[n] = \{0, 1, 2, \dots, n\}$, $n \in \mathbb{N}$ and non decreasing maps. Let $S = set^{\Delta^{op}}$ be the category of simplicial sets, which is cartesian closed as is every presheaf category. Let I be an ordinary small category, X and Y two functors : $I \to S$, then Bousfield - Kan [7] define Hom(X, Y) as the kernel of :

$$\prod_{i\in I}[X_i,Y_i] \rightrightarrows \prod_{i\xrightarrow{j} i'\in I}[X_i,Y_i] \quad . \text{ But } \quad \prod_{i\xrightarrow{j} i\in I}[X_i,Y_{i'}] \approx \prod_{i,i'\in I}[X_i \times I(i,i'), Y_{i'}]$$

where the set $I(i, i')$ is considered as a trivial (constant) simplicial set. Therefore this Hom(X, Y) is nothing but Nat(X, Y), where I, X and Y are trivially enriched in S .

There is a canonical functor $I/- : I \to Cat$ which associates the category I/i over i to each object i of I . A category C being a particular simplicial set (often noted Ner C , but here we shall forget, as in [7] , the notation Ner), the functor $I/-$ can be regarded as $I \to S$.

Definition (Bousfield-Kan) : Let $X : I \to S$ be a functor, then

$$\operatorname{holim} X = \operatorname{Hom}(I/-, X) .$$

So, with the previous remark, holim X is the $I/-$ indexed limit of X .

The lax limits of Gray.

There are several ways to describe lax limits either from lax transformations [13] , [4] which are generalized natural transformations, or as indexed limits [17] . We shall give a new description of the indexation, which links the two points of view and is more appropriate to our aim.

The Total category.

Let Δ_2 be the full subcategory of Δ , formed by all the objects $[i]$ such that $i \leqslant 2$. Each functor $\Delta_2 \to Cat$ determines a diagram :

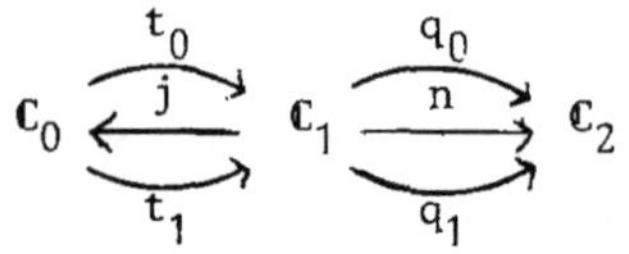

with : $j.t_0 = j.t_1 = 1$, $\quad q_1.t_0 = q_0.t_1$, $\quad n.t_0 = q_0.t_0$, $\quad n.t_1 = q_1.t_1$.

Let us define the total category Tot of this diagram as the category whose objects are the pairs $(C,\ f : t_0(C) \to t_1(C))$ consisting of an object C of $\mathbb{C}_0$ and a morphism f of $\mathbb{C}_1$ such that

$$j(f) = 1_C \quad \text{and} \quad q_1(f) \,.\, q_0(f) = n(f)$$

and whose morphisms $(C, f) \to (C', f')$ are the morphisms $C \xrightarrow{g} C'$ such that $f' \,.\, t_0(g) = t_1(g) \,.\, f$.

The total category has a clear universal property. Precisely let $Y_2 : \Delta_2 \to \mathrm{Cat}$ the inclusion ([n] can be seen as a category). Then if $F: \Delta_2 \to \mathrm{Cat}$, the category Tot F is the Y_2-indexed limit of F, since clearly :

$$\mathrm{Tot}\ F \simeq \mathrm{Nat}(Y_2, F).$$

<u>The laxcones.</u>

Let $\mathbb{A}$ and $\mathbb{B}$ be 2-categories (i.e. categories enriched in Cat), and F a 2-functor $\mathbb{A} \to \mathbb{B}$. A lax cone from an object X to F is the following data :

for each $A \in \mathbb{A}$: a morphism $X \xrightarrow{\tau_A} F(A)$

for each $f : A \to A'$: a 2-morphism $\tau_f : \tau_A \Rightarrow F(f)\,.\,\tau_{A'}$, i.e. a 2-cell $\Downarrow \tau_f$ between $X \to F(A) \xrightarrow{F(f)} F(A')$ and $X \to F(A')$

satisfying obvious coherences, for compositions and 2-morphisms. These coherences are such that a lax cone is exactly an object of the total category of the following diagram :

$$\prod_A \mathbb{B}(X,FA) \underset{t_1}{\overset{t_0}{\rightrightarrows}} \prod_{A,A'} [\mathbb{A}(A,A'),\mathbb{B}(X,FA')] \overset{q_0,\ n,\ q_1}{\rightrightarrows} \prod_{A,A',A''} [\mathbb{A}(A,A') \times \mathbb{A}(A',A''),\mathbb{B}(X,FA'')]$$

(with also an arrow j going back from the second to the first term)

with $t_0((\tau_A)_{A\in A}) = F(f) \,.\, \tau_A$, $\quad [q_0(\Theta_{A,A'})]_{A,A',A''}(f,g) = F(g).\Theta_{A,A'}(f)$

$t_1((\tau_A)_{A\in A}) = \tau_{A'}$, $\quad [(n(\Theta_{A,A'})]_{A,A',A''}(f,g) = \Theta_{A,A''}(g.f)$

$[j(\Theta_{A,A'})]_A = \Theta_{A,A}(1_A)$, $\quad [(q_1(\Theta_{A,A'})]_{A,A',A''}(f,g) = \Theta_{A',A''}(g)$

Actually, this diagramm is determined by the right Kan extension of ϕ F along $\mathbb{A}/- : \mathbb{A} \dashrightarrow \Delta_2$, where $\mathbb{A}/\alpha$ is the internal category in Cat :

$$\coprod_A \mathbb{A}(A,\alpha) \underset{d_1}{\overset{d_0}{\leftleftarrows}} \coprod_{A,A'} \mathbb{A}(A,A') \times \mathbb{A}(A',\alpha) \overset{p_0,\ m,\ p_1}{\leftleftarrows} \coprod_{A,A',A''} \mathbb{A}(A,A') \times \mathbb{A}(A',A'') \times \mathbb{A}(A'',\alpha)$$

(with also an arrow i going from the first to the second term)

where $d_0(A \xrightarrow{f} A' \xrightarrow{g} \alpha) = A' \xrightarrow{g} \alpha$

$d_1(A \xrightarrow{f} A' \xrightarrow{g} \alpha) = A \xrightarrow{g.f} \alpha$

$$i(A \overset{f}{\to} \alpha) = (A = A \overset{f}{\to} \alpha)$$

$$p_0(A \overset{f}{\to} A' \overset{g}{\to} A'' \overset{h}{\to} \alpha) = (A \overset{f}{\to} A' \overset{h.g}{\to} \alpha)$$

$$p_1(A \overset{f}{\to} A' \overset{g}{\to} A'' \overset{h}{\to} \alpha) = (A' \overset{g}{\to} A'' \overset{h}{\to} \alpha)$$

$$m(A \overset{f}{\to} A' \overset{g}{\to} A'' \overset{h}{\to} \alpha) = A \overset{g.f}{\to} A'' \overset{h}{\to} \alpha) \ .$$

Whence the lax limits are the $Y_2 \otimes \mathbb{A} / -$ limits.
We shall denote $Y_2 \otimes A / -$ by $L(\mathbb{A})$.

Remark. This new description of lax limit leads to a generalization, studied in [6] , of the Bousfield-Kan homotopy limits : if $\mathbb{A}$ is a simplicial category, then $\mathbb{A}/\alpha$ is an internal category in S and $\mathrm{Ner}(\mathbb{A}/\alpha)$ is a cosimplicial space. The simplicial indexation for these homotopy limits is the profunctor $H(\mathbb{A}) = \mathbb{A} \dashrightarrow \mathbb{1}$, defined by $H(\mathbb{A})\ (\alpha) = \mathrm{Diag}(\mathrm{Ner}(\mathbb{A}/\alpha))$. This indexation allows us to generalize the replacement scheme of [7] to simplicial categories.

The monads and the semiads.

Let $\mathbb{D}$ be the 2-category with a single object $*$, generated by a 1-cell, and two 2-cells $\lambda : * \to t$ and $\mu : t^2 \to t$ satisfying the well-known relations :

$$\mu \cdot \lambda\, t = \mu \cdot t\, \lambda = t, \qquad \mu \cdot \mu\, t = \mu \cdot t\, \mu$$

Then it is clear that a monad on a category C is a 2-functor from $\mathbb{D}$ to Cat such that its value at $*$ is C . It is well known too [13] , [4] that the lax limit of this 2-functor is the category of algebras of the monad and that there is a cotriple on this category of algebras, that is an action of $\mathbb{D}^{co}$ (the dual of $\mathbb{D}$ for the 2-cells).
Let $\mathbb{H}$ be the sub 2-category of $\mathbb{D}$ generated by the 2-cell μ . Let us call a semiad a 2-functor from $\mathbb{H}$ to Cat, that is a category C with an endo-functor T and a natural transformation $\mu : T^2 \to T$ such that $\mu \cdot \mu\, T = \mu \cdot t\, \mu$.
Then the lax limit of a semiad is the category of algebras C^T whose objects are the pairs $(c,\ b : c \leftarrow Tc)$ such that $b \cdot \mu(c) = b \cdot T\, b$ and whose morphisms $(c, b) \to (c', b')$ are the morphisms $f: c \to c'$ such that $f \cdot b = b' \cdot T\, f$.

The universal lax cone is given by the forgetful functor $U : C^T \to C$ $U(c, b) = c$, and the 2-cell $\beta : T \cdot U \to U$ given by $\beta(c, b) = b : T\, c \to c$.
There is no longer an adjunction between C and C^T, but it is clear too, that we have a functor $F : C \to C^T$ with $F(c) = (T(c), \mu(c))$, such that $U \cdot F = T$. Further more there is a natural transformation $\eta : F \cdot U \to 1\, C^T$.

Indeed $\eta(c, b) = b : (T(c), \mu(c)) \to (c, b)$ defines a natural transformation. Let $\mathbb{J}$ be the 2-category with only one object $*$, generated by a 1-morphism t and a 2-cell $\eta : t \to *$. We can sum up this result in the following :

Proposition. There is a canonical action of $\mathbb{J}$ on the category of algebras of a semiad.

III. The canonical action on ϕ-indexed limits.

These two last examples raise the question: is this fact general, is there always an action on the category of algebras? Let $\mathbb{A}$ be a $\mathbb{V}$-category, $\mathbb{A} \overset{\phi}{\dashrightarrow} \mathbb{1}$ an indexation. The profunctor ϕ can be viewed as a functor $\mathbb{A} \to \mathbb{V}$.

Proposition. The ϕ-indexed limit of ϕ is a $\mathbb{V}$-monoïd and this monoïd has a canonical action on each ϕ-indexed limit.

Proof. $\phi\text{-lim}\ \phi = [\![\phi, \phi]\!] = \mathrm{Nat}(\phi, \phi)$ has an obvious structure of $\mathbb{V}$-monoïd. Let $F : \mathbb{A} \to \mathbb{V}$ be a $\mathbb{V}$-functor. So $\phi\text{-lim}\ F = [\![\phi, F]\!] = \mathrm{Nat}(\phi, F)$ and it is clear that $\phi\text{-lim}\ \phi$ acts on the ϕ-indexed limit of F . More generally, let $F : \mathbb{A} \to \mathbb{B}$ be a $\mathbb{V}$-functor and L the ϕ-indexed limit of F . So we have $\mathbb{B}(B,L) \simeq \mathrm{Nat}(\phi, \mathbb{B}(B,F\ -))$ and the canonical action $\phi\text{-lim}\ \phi \times \mathbb{B}(-, L) \to \mathbb{B}(-, L)$ which is equivalent by the Yoneda lemma to a morphism : $\phi\text{-lim}\ \phi \to \mathbb{B}(L, L)$.
It is easy to see that it is a morphism of $\mathbb{V}$-monoïds.

Examples. The monad case. We have seen that the indexation $\mathbb{D} \overset{L(\mathbb{D})}{\dashrightarrow} \mathbb{1}$ is the lax limit of $Y : \mathbb{D}^{op} \to \mathrm{Cat}^{\mathbb{D}}$ so that we can exhibit it as a monad on the Kleisli category of the monad $\mathbb{D}(-, *)$ on $\mathbb{D}(*, *)$. A simple but tedious computation of its category of algebras shows us that this category is exactly $\mathbb{D}^{co}(*, *)$ and the monoïd structure is that of $\mathbb{D}^{co}$. The canonical action on the category of algebras of a monad is the usual comonad.

In the same way, we can study the semiad case. We must calculate the category of algebras of a semiad on the lax colimit of the semiad $\mathbb{H}(-, *)$ on $\mathbb{H}(*, *)$, which is just $\mathbb{J}(*, *)$ and the monoid structure is that of $\mathbb{J}$. The canonical action is the one described by the former proposition. More details will be given in the proof of the next proposition.

Remark. I choose this proof for sake of simplicity and quickness. But it is a very general result that (as in the case of $\mathbb{V}$-functors [9]) right Kan extensions of profunctors are equipped with an action of the codensity monad (which always exists since we deal with profunctors) so that they can be factorized through the Kleisli category [16] of that codensity monad of profunctors, which in our case has only one object and so is a monoïd. This general result is used in [5] to show in a very simple and categorical way that Kan extensions are shape invariant, so we could say that $\phi\text{-lim}\ \phi$ is the "shape" monoïd of ϕ .

IV. An application to coherent homotopy.

Following Dwyer-Kan [10] , the standard resolution $F_* C$ of the idempotent category C (a single object $*$ with a single non trivial morphism t such that $t^2 = t$) is a simplicial category.

But a category is a particular simplicial set (via its nerve), so a 2-category is a particular simplicial category. Now for this category C , no composite of non identity maps is an identity, so $F_{*}C$ is actually a 2-category and it is not hard to see that this 2-category is just $\mathbb{H}$.

We are now going to study the consequences of the higher homotopy coherences involved in the data of a simplicial functor from $\mathbb{H}$ to a simplicial category $\mathbb{B}$ which I keep on calling a semiad. In the special case $\mathbb{B}$ = Top , I shall speak of a coherent homotopy idempotent.

The 2-pro-functor $L(\mathbb{H}) : \mathbb{H} \dashrightarrow \mathbb{1}$ indexing lax limits can be considered as a simplicial profunctor. The simplicial embedding $\mathrm{Cat} \hookrightarrow S$ has a simplicial adjoint K preserving products, preserves exponentiations, ends and $L(\mathbb{H})$ - indexed limits. So the right Kan extension of a 2-profunctor χ along $L(\mathbb{H})$ is also the right Kan extension of χ considered as a simplicial profunctor along $L(\mathbb{H})$. Thus the lax limit of a 2-functor $F : \mathbb{H} \to \mathbb{B}$ in a 2-category $\mathbb{B}$ is the simplicial $L(\mathbb{H})$-indexed limit of F considered as a simplicial functor. Therefore the simplicial monoïd acting on the simplicial $L(\mathbb{H})$-indexed limits (homotopy limits in the sense of Gray [14]), being $L(\mathbb{H})\text{-lim}\ L(\mathbb{H})$, is still the 2-monoïd $\mathbb{J}$. In the case of a 2-functor F we studied previously the action of $\mathbb{J}$. We have the following result about this action in the general situation of $\mathbb{B}$ a simplicial category and $F : \mathbb{H} \to \mathbb{B}$ a simplicial functor.

Proposition. If the simplicial semiad F has an $L(\mathbb{H})$-indexed limit L, then there exists a map $\nu : F(*) \to L$ such that, if $\sigma(*)$ is the canonical projection $L \to F(*)$, we have $\sigma(*) . \nu = F(t)$, and a 2-cell between $\nu . \sigma(*)$ and 1_L .

Proof. The proof will be given by a careful study of the action of $\mathbb{J}$ on the limit L . Firstly let us consider the 2-enriched situation. We saw that $L(\mathbb{H})$ is the lax colimit of $Y : \mathbb{H}^{op} \to \mathrm{Cat}^{\mathbb{H}}$. On the other hand, we have a lax cone Θ between Y and the constant 2-functor on $\mathbb{H}(*, -)$ given by $\Theta(*) = \mathbb{H}(t,-) : \mathbb{H}(*,-) \to \mathbb{H}(*,-)$

$$\Theta(t) = \mathbb{H}(\mu, -) : \quad \begin{array}{ccc} \mathbb{H}(*, -) & \xrightarrow{\Theta(*)} & \\ \mathbb{H}(t,-)\big\downarrow & \mathbb{H}(\mu, -)\Uparrow & \mathbb{H}(*, -) \\ \mathbb{H}(*, -) & \xrightarrow[\Theta(*)]{} & \end{array}$$

since we can verify that :

$\Theta(t) . \Theta(*)\,\mathbb{H}(\mu, -) = \mathbb{H}(\mu, -) . \mathbb{H}(t, -)\,\mathbb{H}(\mu, -) = \mathbb{H}(\mu . t\,\mu, -)$

and $\Theta(t^2) = \Theta(t) . \Theta(t)\,\mathbb{H}(t, -) = \mathbb{H}(\mu, -) . \mathbb{H}(\mu\, t, -) = \mathbb{H}(\mu . \mu\, t, -)$

Whence a natural transformation

$$\ell : L(\mathbb{H}) \to \mathbb{H}(*, -) \qquad \text{such that} \qquad \begin{array}{l} \ell\,\tau(*) = \mathbb{H}(t, -) \\ \ell\,\tau(t) = \mathbb{H}(\mu, -) \end{array}$$

if τ is the universal lax cone associated to the lax colimit of Y. Furthermore, there is a 2-lax cone δ between $\tau(*)\theta$ and τ, given by

$$\delta(*)=\tau(t) : \tau(*)\theta(*) = \tau(*)\, \mathbb{H}(t,-) \Longrightarrow \tau(*)$$

since the second members of the following equalities

$$\delta(*).\tau(*)\theta(t) = \tau(t).\tau(*)\mathbb{H}(\mu,-) , \qquad \tau(t).\delta(*)\mathbb{H}(t,-) = \tau(t).\tau(t)\, \mathbb{H}(t,-)$$

are equal, because of the coherence of the lax cone τ with respect to the 2-cell μ. Whence a 2-natural transformation $d\colon \tau(*)\ell \Longrightarrow 1_{L(\mathbb{H})}$.

Now let us begin with a semiad $F\colon \mathbb{H} \longrightarrow \mathcal{S}$. Let L be the $L(\mathbb{H})$-indexed limit of F, that is $L = \mathrm{Nat}(L(\mathbb{H}),F)$. The universal $L(\mathbb{H})$-indexed cone σ associated to L is given by

$$\sigma(*) = \mathrm{Nat}(\tau(*),F) : \mathrm{Nat}(L(\mathbb{H}),F) \longrightarrow \mathrm{Nat}(\mathbb{H}(*,-),F) = F(*) , \quad \sigma(t) = \mathrm{Nat}(\tau(t),F)$$

and so on. We have a map of simplicial sets $\nu\colon F(*) \to L$, that is:

$$\mathrm{Nat}(\ell,F)\colon \; F(*) = \mathrm{Nat}(\mathbb{H}(*,-),F) \longrightarrow \mathrm{Nat}(L(\mathbb{H}),F) = L$$

and we verify $\sigma(*).\nu = F(t)$, since

$$\sigma(*).\nu = \mathrm{Nat}(\tau(*),F).\mathrm{Nat}(\ell,F) = \mathrm{Nat}(\tau(*)\ell,F) = \mathrm{Nat}(\mathbb{H}(t,-),F) = F(t).$$

Furthermore we have a 2-cell between $\nu.\sigma(*)$ and 1_L given by $\mathrm{Nat}(d,F)$ and so F "splits" at L.

More generally let $F\colon\mathbb{H} \to \mathbb{A}$ be a semiad in a simplicial category $\mathbb{A}$. Let L be the $L(\mathbb{H})$-indexed limit of F and σ the universal $L(\mathbb{H})$-indexed cone. Thus we have the following commutative diagram with natural isomorphisms:

$$\begin{array}{rcl}
\mathrm{Nat}(\tau(*),\mathbb{A}(X,F-))\colon \mathrm{Nat}(L(\mathbb{H}),\mathbb{A}(X,F-)) & \longrightarrow & \mathrm{Nat}(\mathbb{H}(*,-),\mathbb{A}(X,F-)) \\
\wr\!\mid & & \wr\!\mid \\
\mathbb{A}(X,\sigma *) : \mathbb{A}(X,L) & \longrightarrow & \mathbb{A}(X,F(*))
\end{array}$$

Furthermore we get a natural (in X) transformation:

$$\mathbb{A}(X,F(*)) \simeq \mathrm{Nat}(\mathbb{H}(*,-)),\mathbb{A}(X,F-)) \xrightarrow{\mathrm{Nat}(\ell,\mathbb{A}(X,F-))} \mathrm{Nat}(L(\mathbb{H}),\mathbb{A}(X,F-)) \simeq \mathbb{A}(X,L)$$

and so by the Yoneda lemma a morphism $v\colon F(*) \longrightarrow L$ such that this natural transorma-tion is just $\mathbb{A}(-,v)$. Then it is clear that $\sigma(*).v$ is $F(t)$, since $\ell.\tau(*) = \mathbb{H}(t,-)$. And now the 2-(natural) cell

$$\begin{array}{c}
\mathbb{A}(X,L) \simeq \mathrm{Nat}(L(\mathbb{H}),\mathbb{A}(X,F-)) \\
\mathrm{Id}\downarrow \quad \xLeftarrow{\mathrm{Nat}(d,\mathbb{A}(X,F-))} \quad \curvearrowright\ \mathrm{Nat}(\tau(*)\ell,\mathbb{A}(X,F-)) \\
\mathbb{A}(X,L) \simeq \mathrm{Nat}(L(\mathbb{H}),\mathbb{A}(X,F-))
\end{array}$$

determines, by Yoneda ! , a 2-cell in $\mathbb{A}(L, L)$ between $\nu \cdot \sigma(*)$ and 1_L .

Corollary. The homotopy idempotent (i.e. idempotent in the homotopy category H_0-Top) associated to a coherent homotopy idempotent, splits.

References.

1. C. Auderset, Adjonctions et monades au niveau des 2-catégories, Cahiers de Top. et Géom. Diff., XV (1974), 3-20.

2 J. Bènabou, les distributeurs, Inst. Math. Pures et Appl. Univ. Louvain la Neuve Rapport n° 33 (1973).

3 F. Borceux and G.M.Kelly, A notion of limit for enriched categories, Bull. of the Australian Math. Soc., 12 (1975) 45-72.

4 D. Bourn, Natural anadeses and catadeses, Cahiers de Top. et Géom. Diff. XIV (1974) 371-480.

5 D. Bourn and J.M. Cordier, Distributeurs et théorie de la forme, Cahiers de Top. et Géom. Diff., XXI (1980) 161-189.

6 D. Bourn and J.M. Cordier, Une formulation générale des limites homotopiques, Notes, Univ. Amiens (1980).

7 A.K. Bousfield and D.M. Kan, Homotopylimits, completions and localizations, Springer Lecture Notes in Math., 304 (1972).

8 J.M. Cordier, Sur la notion de diagramme homotopiquement coherent , Proceedings 3ème colloque sur les catégories Amiens 1980 (à paraître).

9 E.J. Dubuc, Kan extensions in enriched category, Springer Lecture Notes in Math., 106 (1969).

10 W.G. Dwyer and D.M. Kan, Simplicial localizations of categories, Journal of P.A. Algebra, 17 (1980) 267-284 .

11 J. Dydak, A simple proof that pointed FANR-spaces are regular fundamental retracts of ANR's, Bull. Acad. Polon. Sci. Math., 25(1977) 55-62.

12 P. Freyd and A. Heller (in preparation).

13 J.W. Gray, Formal category theory, Springer Lecture Notes in Math., 391 (1974).

14 J.W. Gray, Closed categories, lax limits, homotopylimits, Journal of P.A. Algebra, 19 (1980) 127-158.

15 F.W. Lawvere, Teoria delle categorie sopra un topos di base, Mimeographed notes, Perugia (1973).

16 G.B. Segal, Categories and cohomology theories, Topology,13,(1974), 293-312.

17 R. Street, Limits indexed by category valued 2-functors, Journal of P.A. Algebra, 8 (1976), 149-181.

18 M. Thiebaud, Self-dual structure semantics and algebraic categories, Dalhousie Univ., Halifax, N.S., (1971).

A REMARK ON CARTESIAN CLOSEDNESS

H. Brandenburg and M. Hušek

1. A category $\underline{A}$ with finite products is called *cartesian closed* if for every $\underline{A}$-object X the functor X × - has a right adjoint [3]. For background concerning cartesian closedness and its importance for certain aspects of algebraic topology, analysis, and topological algebra we refer to [2], [5], [14], [15], and the literature cited there. Although the categories $\underline{\text{TOP}}$ of topological spaces and $\underline{\text{UNIF}}$ of uniform spaces are not cartesian closed, it is known that they contain some nice non-trivial cartesian closed subcategories (for $\underline{\text{UNIF}}$ see [12]). All these categories are coreflective in $\underline{\text{TOP}}$ or $\underline{\text{UNIF}}$, and they have the disadvantage that their products are different from the usual products. Motivated by this fact, F. Schwarz has recently asked (at the 1980 Ottawa Conference on Categorical Aspects of Topology and Analysis) whether $\underline{\text{TOP}}$ or $\underline{\text{UNIF}}$ have non-trivial cartesian closed epireflective subcategories, i.e. cartesian closed subcategories containing a non-indiscrete space which are closed with respect to the formation of usual subspaces and products in $\underline{\text{TOP}}$ or $\underline{\text{UNIF}}$. Since his question can be answered negatively (see remark (c) below), we are interested here in the more general problem whether there exists a non-trivial reflective subcategory of $\underline{\text{TOP}}$ or $\underline{\text{UNIF}}$ which is cartesian closed. Note that this problem is included in H. Herrlich's recent survey article on Categorical Topology (see [7], Problem 11). For the case of $\underline{\text{TOP}}$ we obtain the following theorem, which will be proved in section 2:

THEOREM 1. *If a reflective subcategory of* $\underline{\text{TOP}}$ *contains the two-point discrete space, then it is not cartesian closed.*

As a consequence of Theorem 1 every cartesian closed reflective subcategory of $\underline{\text{TOP}}$ must consist of connected spaces. If X is a strongly rigid Hausdorff space of cardinality ≥2 (i.e. the identity is the only non-constant continuous mapping from X into X), then all powers of X form an example of a reflective subcategory $\underline{A}_X$ of $\underline{\text{TOP}}$ which contains only connected spaces [4]. However, one can easily show that all categories $\underline{A}_X$ obtained in this way are not cartesian closed. We conjecture that there exists no non-trivial cartesian closed reflective subcategory of TOP. Our second theorem shows the validity of the corresponding statement for uniform spaces.

THEOREM 2. *If a reflective subcategory of* UNIF *contains a non-indiscrete space, then it is not cartesian closed.*

2. Throughout this note all subcategories are assumed to be full and isomorphism-closed. A subcategory A of a category C is *reflective* in C if the embedding functor has a left adjoint. For additional information about reflective subcategories we refer to [8]. In the proof of Theorem 1 we will use the following proposition which is interesting for itself.

PROPOSITION. *Let* A *be a subcategory of* TOP *which is closed with respect to countable products in* TOP *and contains the countable infinite discrete space* ω. *Then* A *is not cartesian closed.*

Proof: Since A contains a singleton, it is easy to verify that if A is cartesian closed, then for each pair Y,Z of A-objects there exists a topology τ on the set C(Y,Z) of continuous mappings from Y into Z with the following properties:

(i) τ is *admissible*, i.e. the evaluation map $e:Y \times (C(Y,Z),\tau) \to Z$ is continuous, where $e(y,g)=g(y)$.

(ii) τ is A-proper, i.e. for every A-object X and for every continuous mapping $f:X \times Y \to Z$ the mapping $\varphi(f):X \to (C(Y,Z),\tau)$ is continuous, where $\varphi(f)(x)(y)=f(x,y)$.

Hence it suffices to show that there exist two spaces Y,Z in A such that every admissible topology on C(Y,Z) is not A-proper. To this end let Y be the space of irrationals with the usual topology and let Z be the product space $Y \times Y_o$, where Y_o is the subspace $Y \cup \{o\}$ of the reals with the usual metric. Being homeomorphic to ω^ω, both Y and Z belong to A. Let τ be an arbitrary admissible topology on C(Y,Z). In order to prove that there exists a space X in A and a continuous mapping $f:X \times Y \to Z$ such that $\varphi(f):X \to (C(Y,Z),\tau)$ is not continuous we start with an arbitrary continuous mapping $g:Y \to Z$ satisfying $|p_1(g(y))| < 1$ and $p_2(g(y)) = 0$ for each $y \in Y$, where $p_1:Z \to Y$ and $p_2:Z \to Y_o$ are the projections. Consider an arbitrary point y_o in Y which has no compact neighborhood. By the continuity of the evaluation map e we can find neighborhoods V and V' of y_o and a $U \in \tau$ such that $\text{cl}\, V \subset V'$ and $(y_o,g) \in V' \times U \subset e^{-1}[W]$, where $W = \{z \in Z \mid |p_i(z)| < 1 \text{ for } i=1,2\}$. Since cl V is not compact, there exists a sequence $(y_n)_{n\in\mathbb{N}}$ of distinct points in cl V such that $\{y_n \mid n \in \mathbb{N}\}$ is closed in Y. Now let X be the subspace of Y_o consisting of all non-negative numbers. Being homeomorphic to Y, the space X belongs to $\underline{A}$. As a closed subspace of $X \times Y$ the space $A = (\{0\} \times Y) \cup (X \times \{y_n \mid n \in \mathbb{N}\})$ is a retract of $X \times Y$ (see [11], §26 II, Corollary 2). Consequently the continuous mapping from A into Z defined by

$$(x,y) \to \begin{cases} g(y) & \text{if } x = 0 \\ (p_1(g(y_n)),nx) & \text{if } x > 0 \text{ and } y=y_n \end{cases}$$

has a continuous extension $f:X \times Y \to Z$.

Suppose that $\varphi(f):X \to (C(Y,Z),\tau)$ is continuous. Then there exists an open neighborhood B of 0 in X such that $\varphi(f)[B] \subset U$. Moreover there exists an $x \in B$ and an $m \in \mathbb{N}$ satisfying $mx > 1$. On the other hand we conclude from $e(y_m,\varphi(f)(x)) = \varphi(f)(x)(y_m) = f(x,y_m) = (p_1(g(y_m)),mx)$ and $e(y_m,\varphi(f)(x)) \in W$ that $mx < 1$. This contradiction shows that $\varphi(f)$ is not continuous, which completes the proof.$\square$

Proof of Theorem 1: Let $\underline{A}$ be a reflective subcategory of $\underline{TOP}$ containing the two-point discrete space $\{0,1\}$. If $\underline{A}$ contains the countable discrete space ω, then $\underline{A}$ is not cartesian closed by virtue of the proposition. Hence we assume that ω does not belong to $\underline{A}$.

Suppose that $\underline{A}$ is cartesian closed. Then the $\underline{A}$-reflection of $\omega \times \omega$ is given by $r_\omega \times r_\omega:\omega\times\omega \to r\omega \times r\omega$, where $r_\omega:\omega \to r\omega$ denotes the $\underline{A}$-reflection of ω. To verify this fact one only has to note that the $\underline{A}$-reflection of $\omega\times\omega$ always has the form $r(\omega\times\omega) = \coprod_{n\in\omega} r\omega \times \{n\}$, and that $\coprod_{n\in\omega} r\omega \times \{n\} = r\omega \times \coprod_{n\in\omega}\{n\} = r\omega \times r\omega$ by the cartesian closedness of $\underline{A}$, where $\coprod$ denotes the coproduct in $\underline{A}$.

Now consider the continuous mapping $f:\omega\times\omega \to \{0,1\}$ defined by

$$f(n,m) = \begin{cases} 1 & \text{if} \quad n = m \quad \text{and} \quad m \neq 1 \\ 1 & \text{if} \quad n \neq m \quad \text{and} \quad m = 1 \\ 0 & \text{otherwise} \end{cases}$$

According to the preceding observation there exists a unique continuous mapping $g:r\omega\times r\omega \to \{0,1\}$ such that $g\circ r_\omega\times r_\omega = f$. We claim that $r\omega\setminus r_\omega[\omega] \neq \emptyset$ and that $g(x,x) = 1$ for every $x \in r\omega\setminus r_\omega[\omega]$. In fact, $r\omega = r_\omega[\omega]$ would imply that $r_\omega:\omega \to r\omega$ is a homeomorphism (since $\{0,1\} \in \underline{A}$), i.e. that $\omega \in \underline{A}$, contrary to our assumption. To prove the second assertion assume that there exists an $x_o \in r\omega\setminus r_\omega[\omega]$ such that $g(x_o,x_o) = 0$. Then the continuous mapping $h:r\omega \to r\omega$ defined by

$$h(x) = \begin{cases} x & \text{if} \quad g(x,x) = 1 \\ r_\omega(1) & \text{if} \quad g(x,x) = 0 \end{cases}$$

satisfies $h\circ r_\omega = r_\omega = id_{r\omega}\circ r_\omega$ and $h \neq id_{r\omega}$, which is impossible.

Applying essentially the same argument to the subspaces $r\omega\times\{r_\omega(n)\}$ of $r\omega\times r\omega$ we can show that

$$g(x,r_\omega(n)) = \begin{cases} 0 & \text{if} \quad n \neq 1 \\ 1 & \text{if} \quad n = 1 \end{cases}$$

for each $x \in r\omega\setminus r_\omega[\omega]$. Repeating the argument again for the subspaces $\{x\}\times r\omega$ of $r\omega\times r\omega$ we obtain that $g(x,y) = 0$ for each pair $x,y \in r\omega\setminus r_\omega[\omega]$, in particular that $g(x,x) = 0$ for each $x \in r\omega\setminus r_\omega[\omega]$ - a contradiction! Consequently $\underline{A}$ cannot be cartesian closed. □

We conclude this section with some remarks.

(a) Theorem 1 implies that every reflective subcategory of $\underline{\text{TOP}}$ containing a finite space X which is not indiscrete cannot be cartesian closed. In fact, it has to contain the reflective hull of X, which - by [9], Theorem 1.1 - always contains the two-point discrete space.

(b) If a reflective subcategory $\underline{A}$ of $\underline{TOP}$ contains a non-indiscrete space X which is not T_1, then it is not cartesian closed. In case that X is not T_o this follows from the fact that $\underline{A}$ must contain the bireflective hull of X (e.g. see [10]) and hence a non-indiscrete finite space. If X is a T_o-space, then the category of sober spaces is contained in $\underline{A}$ ([10], Theorem 1.3). In particular, the two-point discrete space belongs to $\underline{A}$.

(c) No non-trivial epireflective subcategory of $\underline{TOP}$ can be cartesian closed, since it has to contain the two-point discrete space. This answers Schwarz's question mentioned in the introduction. However, a simpler proof of this fact results from the observation that there exist zero-dimensional T_1-spaces X,Y,Z and a coequalizer $q:Y \to Z$ such that $id_X \times q: X\times Y \to X\times Z$ is not a quotient mapping in $\underline{TOP}$ (e.g. see [1], Example 4.3.4). Essentially the same argument shows that there is no non-trivial epireflective subcategory of the category of pseudotopological spaces [13].

(d) Every epireflective subcategory of an epireflective subcategory of $\underline{TOP}$ is reflective in $\underline{TOP}$. Hence Theorem 1 applies, for example, to epireflective subcategories of the categories of Hausdorff spaces or completely regular spaces.

3. In order to prove Theorem 2 let $\underline{A}$ be a reflective subcategory of $\underline{UNIF}$ containing a non-indiscrete space X. If $\underline{A}$ is cartesian closed, then $X^\omega \times \coprod_{n\in\omega}\{n\} = \coprod_{n\in\omega} X^\omega \times \{n\}$, where $\coprod$ denotes the coproduct in $\underline{A}$. For each $n \in \omega$ let $i_n: X^\omega\times\{n\} \to X^\omega \times \coprod_{n\in\omega}\{n\}$ be the injection, and let $p_n: X^\omega\times\{n\} \to X$ be defined by $((x_i),n) \mapsto x_n$. Since p_n is uniformly continuous, there exists a uniformly continuous mapping $f: X^\omega \times \coprod_{n\in\omega}\{n\} \to X$ satisfying $f\circ i_n = p_n$ for each $n \in \omega$. Moreover there are two points $x,y \in X$ and a uniform cover $\mathcal{U}$ of X such that no element of $\mathcal{U}$ contains both x and y. By the uniform continuity of f there exist uniform covers $\mathcal{V}$ of X and $\mathcal{W}$ of $\coprod_{n\in\omega}\{n\}$ such that $\mathcal{V}\times\mathcal{W}$ refines $f^{-1}(\mathcal{U})$. It follows that $\mathcal{V}$ refines every $p_n^{-1}(\mathcal{U})$, contradicting the fact that the subspace $\{x,y\}^\omega$ of X^ω is not topologically discrete. Hence $\underline{A}$ cannot be cartesian closed which completes the proof of Theorem 2.

It is worth mentioning that our proof of Theorem 2 makes no use of star-refinements of uniform covers, hence Theorem 2 is valid also for the category NEAR of nearness spaces [6]. Moreover it follows that no non-trivial epireflective subcategory of an epireflective subcategory of NEAR is cartesian closed. This observation applies, for example, to the category of proximity spaces or to the category of contiguity spaces [6].

REFERENCES

[1] R. Brown, *Elements of Modern Topology*, McGraw-Hill, New York (1968).

[2] E.J. Dubuc and H. Porta, *Convenient categories of topological algebras and their duality theory*, J. pure appl. Algebra 1 (1971) 281-316.

[3] S. Eilenberg and G.M. Kelly, *Closed categories*, in: Proc. of the Conference on Categorical Algebra, La Jolla 1965, ed. by S. Eilenberg et.al., Springer-Verlag, Berlin-New York (1966).

[4] H. Herrlich, *On the concept of reflections in general topology*, in: Contributions to Extension Theory of Topological Structures , (Proc. Sympos., Berlin, 1967), 105-114, Deutscher Verlag d. Wissensch., Berlin (1969).

[5] H. Herrlich, *Cartesian closed topological categories*, Math. Colloquium Univ. Cape Town 9 (1974) 1-16.

[6] H. Herrlich, *A concept of nearness*, General Topol. Appl. 4 (1974) 191-212.

[7] H. Herrlich, *Categorical Topology 1971-1981*, in: General Topology and its Relations to Modern Analysis and Algebra V (Proc. of the Fifth Prague Topological Symposium, Prague 1981), Heldermann Verlag, Berlin, (to appear).

[8] H. Herrlich and G. Strecker, *Category Theory*, sec. ed., Heldermann Verlag, Berlin, (1979).

[9] R.-E. Hoffmann, *Reflective hulls of finite topological spaces*, Arch. der Math. 33 (1979) 258-262.

[10] R.-E. Hoffmann, *Charakterisierung nüchterner Räume*, Manuscripta Math. 15 (1975) 185-191.

[11] K. Kuratowski, *Topology*, Vol. I, Academic Press, New York, (1966).

[12] M.D. Rice and G.J. Tashian, *Cartesian closed coreflective subcategories of uniform spaces*, (preprint).

[13] F. Schwarz, *Cartesian closedness, exponentiality, and final hulls in pseudotopological spaces*, (preprint).

[14] U. Seip, *Kompakt erzeugte Vektorräume und Analysis*, Springer Lecture Notes in Math. 273 (1972).

[15] N.E. Steenrod, *A convenient category of topological spaces*, Michigan Math. J. 14 (1967) 133-152.

CROSSED COMPLEXES AND NON-ABELIAN EXTENSIONS

Ronald Brown and Philip J. Higgins

School of Mathematics and Computer Science, University College of North Wales. Bangor, Gwynedd, U.K.

Department of Mathematics, Science Laboratories, Durham University. Durham, U.K.

Introduction

Crossed complexes may be thought of as chain complexes with operators from a group (or groupoid) but with non-abelian features in dimensions one and two. We start by surveying briefly their use.

The definition of crossed complex is motivated by the standard example, the *homotopy crossed complex* $\pi\underline{X}$ of a filtered space $\underline{X} : X_0 \subset X_1 \subset \ldots \subset X_n \subset X_{n+1} \subset \ldots \subset X$. Here $\pi_1\underline{X}$ is the fundamental groupoid $\pi_1(X_1, X_0)$ of homotopy classes rel $\dot{I}$ of maps $(I, \dot{I}) \to (X_1, X_0)$, with the usual groupoid structure induced by composition of paths. For $n \geq 2$, $\pi_n\underline{X}$ is the family of relative homotopy groups $\pi_n(X_n, X_{n-1}, p)$ for all $p \in X_0$. For $n \geq 2$, there is an action of $\pi_1\underline{X}$ on $\pi_n\underline{X}$, and there is a boundary map $\delta : \pi_n\underline{X} \to \pi_{n-1}\underline{X}$; there are also the initial and final maps $\delta^0, \delta^1 : \pi_1\underline{X} \to X_0$. The rules which are satisfied by all such $\pi\underline{X}$ are taken as the defining rules for a crossed complex (§1). In particular, the rule $\delta\delta = 0$ shows the analogy with chain complexes. Of course the individual rules are commonly used in homotopy theory, without necessarily considering the total structure.

By a *reduced* crossed complex C we mean one in which C_0 is a point. These have been considered for some 35 years. They were called "group systems" by Blakers [2]. He writes that he follows a suggestion of Eilenberg in using these group systems to apply the homotopy addition lemma in his investigation of the relationship between the homology and homotopy groups of pairs. His proofs involve a functor from reduced crossed complexes to simplicial sets; the values of this functor have been shown recently by Ashley [1] to be *simplicial* T-*complexes*, and Ashley has proved the hard theorem that this functor gives an equivalence N between crossed complexes and simplicial T-complexes. This equivalence generalises the well known equivalence of chain complexes and simplicial abelian groups, due to Dold and Kan [27; Theorem 22.4], and the functor N generalises also the nerve of a groupoid, which we use in §3.

Reduced crossed complexes satisfying in each dimension a freeness condition were called "homotopy systems" by Whitehead [31, 32], and his main example was $\pi\underline{K}$ where $\underline{K}$ is the filtration of a CW-complex K by its skeletons. The paper [31] gives interesting relations between homotopy systems and chain complexes with operators: we shall generalise these results to crossed complexes in [10]. An overall consideration in the papers [30, 31, 32] is *realisability*. In §17 of [32] Whitehead sketches a proof of a theorem announced in §7 of [31], that if $\phi : C \to C'$ is a homotopy equivalence of finite dimensional homotopy systems, and C is realisable as $\pi\underline{K}$ for some

CW-complex K , then C' is also realisable as $\pi \underline{K}'$ and ϕ is realisable by a map $K \to K'$. The approach to simple homotopy theory in this section of [32] seems to have been ignored and indeed its predecessor [31] is not widely read.

Huebschmann, Holt and others (cf. [20, 17] and the historical note [26]) have shown how crossed complexes may be used to give an interpretation of all the cohomology groups $H^n(G; A)$ of a group G with coefficients in a G-module A . Lue has explained in [24] how related ideas had been developed earlier for varieties of algebras, rather than just for groups. However, the tie-up with classical cohomology was not made explicit (cf. p.172 of [24]).

We have given in [6, 7] a colimit theorem for the homotopy crossed complex of a union of filtered spaces. This theorem includes the usual Seifert-van Kampen theorem on the fundamental groupoid of a union of spaces; it also includes the Brouwer degree theorem ($\pi_n S^n = \mathbb{Z}$), the relative Hurewicz theorem, and a subtle theorem of J.H.C. Whitehead on free crossed modules [31; §16]. The proof of the colimit theorem in [7] involves in an essential way two other categories equivalent to crossed complexes, namely ω-groupoids and cubical T-complexes [6, 8]. With simplicial T-complexes [1], ∞-groupoids [9] and poly-T-complexes [22], there are now five categories known to be equivalent to crossed complexes, the proofs in each case being highly non-trivial.

The papers [16, 18] give other work on crossed complexes.

One of our aims here is to show how the *homotopy* addition lemma (which plays a key rôle in the work of Blakers [2] and of the authors [6, 7]) is also important in the *cohomology* of a group G . We do this by showing that the *standard crossed resolution* CG , which is constructed algebraically in [20] and applied further in [21], in fact arises as πBG , the homotopy crossed complex of the classifying space of G . The boundary maps in CG are determined by the homotopy addition lemma.

Our further aim is an exposition of the Schreier theory of non-abelian extensions.

Much has been written on non-abelian extensions and cohomology, (cf. [5, 12, 13 23] and the further references there), but it is notable that, while there are accounts in several books on group theory, texts on homological algebra remain largely silent on the subject, presumably because there is no known exposition using chain complexes, on which expositions of the abelian case are rightly based. Here we show that the non-abelian features of crossed complexes allow an exposition closer to the abelian case, involving morphisms and homotopies. We strengthen the theory, by presenting an equivalence of groupoids which on components induces the usual one-one correspondence of sets. We also generalise the theory, to extensions of groupoids rather than just groups, and to "free" equivalences of extensions.

1. Crossed Complexes

We recall from [6] the definition of the category (here denoted XC) of crossed complexes.

A crossed complex C (over a groupoid) is a sequence

$$\cdots \longrightarrow C_n \xrightarrow{\delta} C_{n-1} \longrightarrow \cdots \longrightarrow C_2 \xrightarrow{\delta} C_1 \underset{\delta^1}{\overset{\delta^0}{\rightrightarrows}} C_0$$

satisfying the following axioms:

(1.1) C_1 is a groupoid with C_0 as its set of vertices and δ^0, δ^1 as its initial and final maps.

We write $C_1(p,q)$ for the set of arrows from p to q $(p,q \in C_0)$ and $C_1(p)$ for the group $C_1(p,p)$.

(1.2) For $n \geq 2$, C_n is a family of groups $\{C_n(p)\}_{p \in C_0}$ (equivalently, C_n is a totally disconnected groupoid over C_0) and for $n \geq 3$ the groups $C_n(p)$ are abelian.

(1.3) The groupoid C_1 operates on the right of each $C_n (n \geq 2)$ by an action denoted $(x,a) \mapsto x^a$. Here if $x \in C_n(p)$ and $a \in C_1(p,q)$, then $x^a \in C_n(q)$. (Thus $C_n(p) \cong C_n(q)$ if p and q lie in the same component of the groupoid C_1.)

We use additive notation for all groups $C_n(p) (n \geq 2)$ and for the groupoid C_1, and we use the same symbol 0 for all their identity elements.

(1.4) For $n \geq 2$, $\delta : C_n \to C_{n-1}$ is a morphism of groupoids over C_0 and preserves the action of C_1, where C_1 acts on the groups $C_1(p)$ by conjugation: $x^a = -a + x + a$.

(1.5) $\delta\delta = 0 : C_n \to C_{n-2}$ for $n \geq 3$ (and $\delta^0\delta = \delta^1\delta : C_2 \to C_0$, as follows from (1.4)).

(1.6) If $c \in C_2$, then δc operates trivially on C_n for $n \geq 3$ and operates on C_2 as conjugation by c, that is

$$x^{\delta c} = -c + x + c \quad (x,\ c \in C_2(p)) .$$

In the case when C_0 is a single point, we call C a *reduced* crossed complex.

We observe that the above laws make each $C_2(p)$ a crossed module over $C_1(p)$; we take the laws up to dimension two as defining C_2 as a *crossed module over the groupoid* (C_1, C_0), or, simply, as a *crossed* C_1*-module*. Let $n \geq 3$. Then $C_n(p)$ is a module over $C_1(p)$, and we take the laws (1.1) - (1.3) as defining C_n as *module over the groupoid* (C_1, C_0), or, simply, as a C_1*-module*.

A *morphism* $f : C \to D$ of crossed complexes is a family of morphisms of groupoids $f_n : C_n \to D_n$, compatible with the boundary maps $C_n \to C_{n-1}$, $D_n \to D_{n-1}$ and the actions of C_1, D_1 on C_n, D_n. We denote by XC the resulting category of crossed complexes.

By restriction of structure, we have categories of modules, and of crossed modules (over groupoids). Let $f : C \to D$ be a morphism of crossed modules. If f_0 is the identity (as happens throughout §5) we write f as a pair (f_1, f_2). If f_0 and f_1 are the identity, we call f a *morphism of crossed* C_1*-modules*.

Suppose given a crossed module C, a set X and function λ from X to the union of the $C_1(p)$, $p \in C_0$. Then we say C is the *free crossed* C_1*-module on generators* $[x] \in C_2$ with $\delta[x] = \lambda x$ for all $x \in X$, if such elements $[x]$ are given, and for any other crossed C_1-module C' and elements $[x]' \in C_2'$ with $\delta[x]' = \lambda x$ for all $x \in X$, there is a unique morphism $f : C \to C'$ of crossed C_1-modules such that $f[x] = [x]'$ for all $x \in X$. This definition becomes the usual one in the reduced case [31]. Analogously to the group case, (an exposition of which is given in [11]), C_2 is constructed, given C_1, X and λ, as the groupoid with generators $x^a \in C_2(q)$ for all $x \in X$, $a \in C_1(\delta^0\lambda x, q)$, and $q \in C_0$, with the usual relations $-x^a + y^b + x^a = y^{b-a+\lambda x+a}$ where these make sense, and with $[x] = x^0$, $\delta(x^a) = -a + \lambda x + a$.

A module over C_1 can be regarded as a crossed C_1-module C with $\delta : C_2 \to C_1$ trivial. Such a C_1-module is *free on generators* $[x] \in C_2(p_x)$, $x \in X$, if it is a free crossed C_1-module on these generators with $\delta[x]$ equal to the zero at p_x, $x \in X$.

Let C be a crossed complex. Its *fundamental groupoid* $\pi_1 C$ is the quotient [15] of the groupoid C_1 by the normal, totally disconnected subgroupoid δC_2. The rules for a crossed complex give C_n, for $n \geq 3$, the induced structure of $\pi_1 C$-module.

A crossed complex C is *free* if C_1 is a free groupoid (on some graph X_1), C_2 is a free crossed C_1-module (for some $\lambda : X_2 \to C_1$), and for $n \geq 3$, C_n is a free $\pi_1 C$-module (on some X_n).

A crossed complex C is *exact* if for $n \geq 2$

$$\mathrm{Ker}\,(\delta : C_n \longrightarrow C_{n-1}) = \mathrm{Im}\,(\delta : C_{n+1} \longrightarrow C_n) .$$

If C is exact and G is a groupoid, then C together with an isomorphism $\pi_1 C \to G$ (or, equivalently, C with a quotient morphism $C_1 \to G$ whose kernel is δC_2) is called a *crossed resolution* of G. It is a *free* crossed resolution if also C is free.

Let G be a groupoid. A free crossed resolution of G may be constructed as follows. Let X be any subgraph of G generating G and let (C_1, C_0) be the free groupoid on X, with quotient morphism $\phi : C_1 \to G$. Let R be any set and let $w : R \to C_1$ be a function to the union of the $C_1(p)$, $p \in C_0$, such that the normal closure of the image of w is $\mathrm{Ker}\,\phi$. (The triple $(X; R, w)$ is a *presentation* of G.) Let C_2 be the free crossed C_1-module determined by w. Then $\kappa = \mathrm{Ker}(\delta : C_2 \to C_1)$ is the G-*module of identities* for the presentation (cf. [11]). Choose any free G-resolution $\to C_n \to \dots \to C_3 \to \kappa$ of κ by G-modules; this may be spliced into $\delta : C_2 \to C_1$ to give a free crossed resolution of G. (Such a construction for groups is used in [20, 21].)

As explained in the introduction, a key example of a crossed complex is the homotopy crossed complex $\pi\underset{\sim}{X}$ of a filtered space $\underset{\sim}{X}$. Let $\underset{\sim}{X}$ be the filtered space defined by the skeletons of a CW-complex X; then $\pi\underset{\sim}{X}$ is a free crossed complex. (This is due to Whitehead [31; §16] in the reduced case, from which the more general case follows. A simple proof of freeness in dimension two is given in [11].) Further $\pi_1\underset{\sim}{X}$ is $\pi_1(X, X_0)$, and the homology of $\pi\underset{\sim}{X}$ (i.e. $\mathrm{Ker}\,\delta/\mathrm{Im}\,\delta$) is for $n \geq 2$ isomorphic to the family of groups $H_n(\widetilde{X}_p)$, $p \in X_0$, where $\widetilde{X}_p$ is the universal cover of X based at p (cf. [32; Footnote 41]). In particular, if X is aspherical (i.e. $\pi_n X = 0$ for $n \geq 2$), then $\pi\underset{\sim}{X}$ is exact, and so it is a free crossed resolution of $\pi_1(X, X_0)$.

2. The homotopy addition lemma

This is a basic, but not so easy to prove, lemma in homotopy theory. Intuitively, it expresses the idea that "the boundary of a simplex is the composite of its faces". Its formulation involves all the structural elements of the homotopy crossed complex, and so for completeness we state it here.

Let Δ^n be the standard n-simplex with ordered set of vertices $\{v_0,v_1,\ldots,v_n\}$, and let Δ^n have its filtration by skeletons Δ^n_r. Then $\pi_n(\Delta^n, \Delta^n_{n-1}, v_n)$ is for $n > 1$ an infinite cyclic group with generator σ, say. The unique arrow of $\pi_1(\Delta^1, \Delta^1_0)$ from v_0 to v_1 is also written σ. The face maps $\partial_i : \Delta^{n-1} \to \Delta^n$ then determine elements $\partial_i\sigma \in \pi_{n-1}\underset{\sim}{\Delta}^n$, and the map $u : \Delta^1 \to \Delta^n$, which sends v_0, v_1 to v_{n-1}, v_n respectively, determines $u\sigma \in \pi_1\underset{\sim}{\Delta}^n$.

Proposition 1. *(The homotopy addition lemma) The elements* σ *may be chosen so that the boundary*

$$\delta : \pi_n(\Delta^n, \Delta^n_{n-1}, v_n) \longrightarrow \pi_{n-1}(\Delta^n_{n-1}, \Delta^n_{n-2}, v_n)$$

is given by

$$\delta\sigma = \begin{cases} -\partial_1\sigma + \partial_0\sigma + \partial_2\sigma & \text{if } n = 2, \\ \partial_0\sigma - (\partial_3\sigma)^{u\sigma} - \partial_1\sigma + \partial_2\sigma & \text{if } n = 3, \\ \sum_{i=0}^{n-1} (-1)^i\partial_i\sigma + (-1)^n(\partial_n\sigma)^{u\sigma} & \text{if } n \geq 4. \end{cases} \quad \Box$$

For a proof of this result, see for example [29]. A corresponding cubical form of the homotopy addition lemma is given for ω-groupoids as Lemma 7.1 of [6].

3. The standard crossed resolution

Let G be a groupoid. There is a well-known simplicial set NG [28], the *nerve* of G, in which N_nG is the set of composable elements $(u_1,\ldots,u_n)$ of G^n, i.e. of n-tuples of elements u_i of G such that $u_i + u_{i+1}$ is defined for $1 \leq i < n$. The geometric realisation $X = |NG|$ is known as the classifying space BG of G. The simplicial structure on NG induces a structure of CW-complex on X, and so the homotopy crossed complex $\pi\underset{\sim}{X}$ (for the skeletal filtration on X) is defined. We write CG for this crossed complex and call it the *standard crossed resolution* of G.

Proposition 2. *Let* G *be a groupoid. Then* CG *is a free crossed resolution of* G *and has the following structure.*

(i) $C_0G = G_0$; C_1G *is the free groupoid on the sub-graph* G^* *consisting of all the vertices and all the non-identity arrows of* G.
The basis element of C_1G *corresponding to* $u \in G^*$ *is written* $[u]$, *and this notation is extended to* G *by setting* $[0_p] = 0_p$.

(ii) C_2G *is the free crossed* C_1G*-module on generators* $[u,v] \in C_2G(\delta^1 v)$ *with*

$$\delta[u,v] = -[u + v] + [u] + [v]$$

for all $(u,v) \in N_2G^*$ *(the composable pairs of* G^**)*.

(iii) *For* $n \geq 3$, C_nG *is the free* G*-module on generators* $[u_1,\ldots,u_n] \in C_nG(\delta^1 u_n)$ *for all* $(u_1,\ldots,u_n) \in N_nG^*$.
We also let $[u_1,\ldots,u_n] \in C_nG$ *be the identity at* $\delta^1 u_n$ *if* $(u_1,\ldots,u_n) \in N_nG$ *and some* $u_i = 0$.
(iv) $\delta : C_3G \to C_2G$ *is given by*

$$\delta[u,v,w] = [v,w] - [u,v]^{[w]} - [u + v,w] + [u,v + w] ,$$

for all $(u,v,w) \in N_3G$.
(v) *For* $n \geq 4$, $\delta : C_nG \to C_{n-1}G$ *is given by*

$$\delta[u_1,\ldots,u_n] = [u_2,\ldots,u_n] + \sum_{i=1}^{n-1}(-1)^i[u_1,\ldots,u_i + u_{i+1},\ldots,u_n] + (-1)^n[u_1,\ldots,u_{n-1}]^{[u_n]} . \quad \square$$

This proposition follows from the homotopy addition lemma, the standard description of the face operators in NG , and the fact that BG is aspherical [28]. Note that if G is a group, then Proposition 2 shows CG to be the same as the standard (inhomogeneous) crossed resolution of G as defined in §9 of [20]. We have now shown how CG arises geometrically.

The curious formula for $\delta : C_3G \to C_2G$ should be noted; the values of this δ are in a family of (generally) non-abelian groups. There is a functor assigning to a crossed complex C a chain complex ΔC with operators from $\pi_1 C$ [10]; for this functor $\Delta(CG)$ is the bar resolution of G (cf. [25]), for the group case). However, Δ abelianises C_2 and so loses information.

The 3-simplices of NG may be pictured as

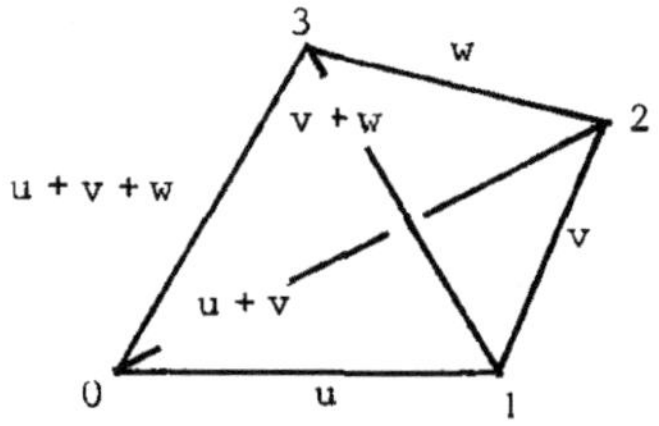

(cf. p.12 of [25]). Now NG is a T-complex in which every n-simplex is thin for $n > 1$. Every T-complex has a groupoid structure in dimension 1, and the above picture illustrates the 3-simplex used to prove associativity [1] of this groupoid. This suggests the link between $\delta : C_3G \to C_2G$ and associativity in extension theory.

4. Homotopies

The notion of homotopy has a similar importance for crossed complexes to that for chain complexes. However, because of the more complicated structure of crossed complexes, there are several possible conventions for the definition of homotopy, and there are also two levels of generality (corresponding to free and based homotopy in the topological case). Our definition follows from the cubical homotopy addition lemma in the algebra of ω-groupoids [6], applied to a natural notion of homotopy for ω-groupoids, a topic which we hope to develop elsewhere.

Let $f, g : C \to D$ be morphisms of crossed complexes. A *homotopy* $\theta : f \simeq g$ is a family of functions $\theta_n : C_n \to D_{n+1} (n \geq 0)$ with the following properties.

(4.1) If $p \in C_0$, then $\theta_0 p \in D_1(fp,gp)$. If $x \in C_1(p,q)$, then $\theta_1 x \in D_2(gq)$. If $n \geq 2$ and $x \in C_n(q)$, then $\theta_n x \in D_{n+1}(gq)$.

(4.2) $\theta_1 : C_1 \to D_2$ is a derivation over g_1, that is if $x + y$ is defined in C_1 then

$$\theta_1(x + y) = (\theta_1 x)^{gy} + \theta_1 y \ , \quad \text{where} \quad gy = g_1 y \ .$$

(4.3) For $n \geq 2$, $\theta_n : C_n \to D_{n+1}$ is an operator morphism over g_1, that is, if $a \in C_1(p,q)$, $x \in C_n(p)$, $y \in C_n(q)$, then

$$\theta_n(x^a + y) = (\theta_n x)^{ga} + \theta_n y \quad \text{where} \quad ga = g_1 a \ .$$

(4.4) If $x \in C_1(p,q)$ then

$$gx = -\theta_0 p + fx + \theta_0 q - (\delta\theta_1 x) \ .$$

(4.5) If $n \geq 2$, and $x \in C_n(q)$ then

$$gx = (fx)^{\theta q} - \theta_{n-1}\delta x - \delta\theta_n x \ , \quad \text{where} \quad \theta q = \theta_0 q \ .$$

(A similar definition, but with different conventions, is given in the reduced case by Whitehead [23]. For further comments, see Remark 4 at the end of the paper.)

A homotopy $\theta : f \simeq g$ is said to be *rel* C_0 if $\theta_0 p$ is an identity for all $p \in C_0$ (so that in consequence $f_0 = g_0$). (It is these homotopies, with different conventions, which are used by Huebschmann [20].) For emphasis, the more general kinds of homotopy are sometimes called *free* homotopies.

If $\theta : f \simeq g$, $\theta' : g \simeq h$ are (free) homotopies, their composite $\phi = \theta + \theta'$ is defined by $\phi_0 p = \theta_0 p + \theta'_0 p$, $p \in C_0$, and if $n \geq 1$ and $x \in C_1(p,q)$ or $x \in C_n(q)$, then $\phi_n x = \theta'_n x + (\theta_n x)^{\theta' q}$. It is easily checked that ϕ is a homotopy $f \simeq h$.

In the next section we will be considering only crossed complex morphisms $C \to D$ which are the identity on $C_0 = D_0$. Therefore we write

$$(C, D)_f \quad \text{and} \quad (C, D)$$

for the groupoids which have such morphisms as objects, and whose arrows are respectively the free, and the rel C_0, homotopies. The sets of components of these groupoids are thus the respective sets of homotopy classes of morphisms over $C_0 = D_0$, and they are written respectively

$$[C, D]_f \quad \text{and} \quad [C, D] \ .$$

5. Non-abelian extensions

Throughout this section, G and A will be groupoids such that $G_0 = A_0$ and A is totally disconnected (i.e. A is a family $A(p)$, $p \in A_0$, of groups). An *extension* of A by G is a pair

$$A \xrightarrow{i} E \xrightarrow{p} G$$

of morphisms of groupoids satisfying the following properties.

(5.1) $E_0 = G_0$ and i and p are the identity on objects.

(5.2) p is a quotient morphism of groupoids.

(5.3) i maps A isomorphically onto $\operatorname{Ker} p$.

(That p is a quotient morphism means that p induces an isomorphism $E/\operatorname{Ker} p \to G$; for more details see [15].) For such an extension, conjugation in E induces an action of E on A making A a crossed E-module. This can be extended trivially to a crossed complex $\underline{E} : \ldots \to 0 \to 0 \to A \to E$ which is (with the quotient morphism p) a crossed resolution of G .

A *free equivalence* of such extensions of A by G is a commutative diagram

$$\begin{array}{ccccc} A & \xrightarrow{i} & E & \xrightarrow{p} & G \\ \zeta \downarrow & & \downarrow \eta & & \downarrow = \\ A & \xrightarrow[i']{} & E' & \xrightarrow[p']{} & G \end{array} \qquad (*)$$

such that ζ is an isomorphism; this implies that η also is an isomorphism. Such a free equivalence is an *equivalence* if ζ is the identity. We can thus form two (large) groupoids

$$\underline{\mathrm{Ext}}_f(G, A) \quad \text{and} \quad \underline{\mathrm{Ext}}(G, A) ,$$

both having objects the extensions of A by G , but having arrows respectively the free equivalences and the equivalences of extensions.

For any groupoid A there is a groupoid $\operatorname{Act} A$ of actions on the vertex groups of A . Here $\operatorname{Act} A$ has the same objects as A , and an arrow in $\operatorname{Act} A$ from p to q is an isomorphism $A(p) \to A(q)$ of groups. There is a conjugation map $\partial : A \to \operatorname{Act} A$. Under our assumption that A is totally disconnected, this map and the action of $\operatorname{Act} A$ on A determine a crossed complex

$$\longrightarrow 0 \longrightarrow \ldots \longrightarrow 0 \longrightarrow A \xrightarrow{\partial} \operatorname{Act} A$$

which we write χA . If $A \xrightarrow{i} E \xrightarrow{p} G$ is an extension with associated crossed resolution $\underline{E}$ of G , then the action of E on A by conjugation induces a morphism $(\sigma, 1) : \underline{E} \to \chi A$ (where $\sigma : E \to \operatorname{Act} A$) . A free equivalence as in (*) induces an isomorphism $(\xi, \zeta) : \chi A \to \chi A$ where $\xi : \operatorname{Act} A \to \operatorname{Act} A$ is given by $a^{\xi\beta} = \zeta((\zeta^{-1}a)^{\beta})$; further $\sigma'\eta = \xi\sigma : E \to \operatorname{Act} A$.

<u>Theorem 3</u>. *There are canonical equivalences of groupoids*

$$e_f : (CG, \chi A)_f \longrightarrow \underline{\mathrm{Ext}}_f(G, A) ,$$

$$e : (CG, \chi A) \longrightarrow \underline{\mathrm{Ext}}(G, A) .$$

Proof. The morphism e is the restriction of e_f . We give the proof only for e_f . Also, since this result is a reformulation of standard theory, we do not give full details. [Some of the calculations are given in [14], §15.1 for the group case, and for equivalence rather than free equivalence, but with differences in notation as follows : for Hall's H, N, G read our G, A, E ; his factor set $(u,v) \in N$ becomes our morphism $k : C_2G \to A$; his automorphism $a \mapsto a^u$ of N for $u \in G$ becomes our morphism $h : C_1G \to \mathrm{Act}\ A$; his choice $u \mapsto \bar{u}$ of coset representatives becomes our morphism $\ell : C_1G \to E$; his function $\alpha : H \to N$ becomes our derivation $\alpha : C_1G \to A$.]

A morphism $CG \to \chi A$ over $G_0 = A_0$ is determined by a pair of morphisms $h : C_1G \to \mathrm{Act}\ A$, $k : C_2G \to A$ such that k is an operator morphism over h , and such that the equations

$$h\delta = \partial k \ , \ k\delta = 0$$

hold. (These equations are equivalent to the first two equations in Theorem 15.1.1 of [14], and indeed $k\delta = 0$ is, by Proposition 2, equivalent to the "factor set" condition

$$k[u + v,w] + k[u,v]^{h[w]} = k[u,v + w] + k[v,w] \ ,$$

for all $(u,v,w) \in N_3G^*$.) Given such a morphism $CG \to \chi A$, an extension E of A by G is defined by setting $E_0 = G_0$ and for $p, q \in G_0$, letting $E(p,q)$ be the set of pairs (u,a) such that $u \in G(p,q)$, $a \in A(q)$, with addition

$$(u,a) + (v,b) = (u + v,\ k[u,v] + a^{h[v]} + b) \ ,$$

for $v \in G(q,r)$, $b \in A(r)$. The verification that E is a groupoid is left to the reader (cf. p.220 of [14]). We write $E = e(h,k)$.

Suppose now given two morphisms $CG \to \chi A$ over G_0 , which we write as pairs (h,k) , (h',k') as above. Let $\theta : (h,k) \simeq (h',k')$ be a (free) homotopy, and write $\beta = \theta_0$, $\alpha = \theta_1$. Then α is a derivation over h' and if $u \in G(p,q)$, $v \in G(q,r)$, we have

$$h'[v] = -\beta q + h[v] + \beta r - \partial\alpha[v] \ ,$$
$$k'[u,v] = k[u,v]^{\beta r} - \alpha\delta[u,v] \ .$$

A straightforward calculation shows that

$$k'[u,v] + \alpha\delta[u,v] = -\alpha[u + v] + k'[u,v] + (\alpha[u])^{h'[v]} + \alpha[v]$$

(and this verifies that our definition of equivalence agrees with that on p.221 of [14]). Define

$$e_f(\theta) : e(h,k) \longrightarrow e(h',k')$$
$$(u,a) \longmapsto (u,\ \alpha[u] + a^{\beta q}) \ , \ u \in G(p,q) \ , \ a \in A(q) \ .$$

Then $e_f(\theta)$ is an isomorphism of groupoids which, with the automorphism $A \to A$ given by $a \mapsto a^{\beta q}$, $a \in A(q)$, defines a free equivalence of extensions. Conversely, any free equivalence

$$\begin{array}{ccccc} A & \longrightarrow & e(h,k) & \longrightarrow & G \\ \downarrow{\scriptstyle \zeta} & & \downarrow{\scriptstyle \eta} & & \downarrow{\scriptstyle =} \\ A & \longrightarrow & e(h',k') & \longrightarrow & G \end{array}$$

arises in the above way if $\beta : G_0 \to \mathrm{Act}\ A$ is defined by $\beta(q) = \zeta|A(q)$, and $\alpha : C_1G \to A$ is defined by extending to a derivation over h' the function $\alpha' : G \to A$ defined by $\eta(u,0) = (u,\alpha'u)$.

Finally, we show that any extension $A \xrightarrow{i} E \xrightarrow{p} G$ of A by G is equivalent to some $e(h,k)$. Let $\phi : C_1G \to G$ be the quotient morphism and consider the crossed complex $\underset{\approx}{E}$ obtained by trivial extension of the crossed E-module A . Consider the diagram

$$\begin{array}{ccccccc}
C_3G & \xrightarrow{\delta} & C_2G & \xrightarrow{\delta} & C_1G & \xrightarrow{\phi} & G \\
\downarrow & & \downarrow k & & \downarrow \ell & & \downarrow = \\
0 & \longrightarrow & A & \longrightarrow & E & \longrightarrow & G \\
 & & \downarrow = & & \downarrow \sigma & & \\
 & & A & \xrightarrow[\partial]{} & \mathrm{Act}\ A & & .
\end{array}$$

The crossed complex CG is free, while $\underset{\sim}{E}$ is exact, and both have G as fundamental groupoid. So the identity on G has a lift $(\ell,k) : CG \to E$. Let $h = \sigma\ell$. Then (h,k) is a morphism $CG \to \chi A$ over G_0 and an equivalence of extensions $e(h,k) \to E$ is defined by $(u,a) \mapsto \ell[u] + ia$. □

Thus the crossed complex approach is successful because some of the difficulties in non-abelian extension theory are so-to-speak compressed into the standard crossed resolution (a kind of universal example) and in particular into the formula for $\delta : C_3G \to C_2G$.

By standard homotopy arguments, we obtain from Theorem 3;

Corollary 4. *Let* C *be any free crossed resolution of the groupoid* G . *Then there are equivalences of groupoids*

$$e'_f : (C, \chi A)_f \longrightarrow \underline{\mathrm{Ext}}_f(G, A),$$

$$e' : (C, \chi A) \longrightarrow \underline{\mathrm{Ext}}(G, A) . \quad \square$$

Corollary 5. *Let* $N \xrightarrow{i} F \xrightarrow{p} G$ *be an extension of groupoids such that* F *is free. Let* $\underset{\sim}{P}$ *denote the crossed resolution of* G *obtained by trivial extension of the crossed* F*-module* N . *Then there are equivalences of groupoids*

$$e''_f : (\underset{\sim}{P}, \chi A)_f \longrightarrow \underline{\mathrm{Ext}}_f(G, A) ,$$

$$e'' : (\underset{\sim}{P}, \chi A) \longrightarrow \underline{\mathrm{Ext}}(G, A) .$$

Proof. Let C be a free crossed resolution of G such that $C_1 = F$ and $N = \delta(C_2)$. Then the projection $C \to \underset{\sim}{P}$ induces isomorphisms

$$(\underset{\sim}{P}, \chi A)_f \to (C, \chi A)_f \ , \ (\underset{\sim}{P}, \chi A) \to (C, \chi A) . \quad \square$$

An interesting special case of Corollary 5 is when A is centreless, i.e. when $\partial : A \to \mathrm{Act}\ A$ is injective. Then a morphism $\underset{\sim}{P} \to \chi A$ is determined by a morphism $h : F \to \mathrm{Act}\ A$ of groupoids such that $h(r)$ is a conjugation of A for each r in a set normally generating N .

The above methods also enable one to give a crossed complex version of a generalisation of Dedecker's work on non-abelian cohomology and extensions [12]. Let G , A be as above and suppose given a crossed Π-module A (where Π is a groupoid with $\Pi_0 = G_0$) . A Π*-extension* of A by G is an extension $A \xrightarrow{i} E \xrightarrow{p} G$ as above together with a morphism of crossed modules

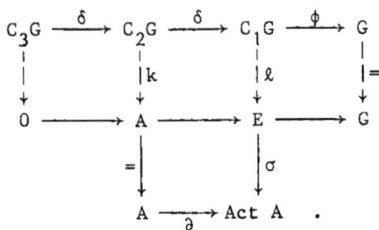

$$\begin{array}{ccc} A & \xrightarrow{\;i\;} & E \\ \Big\| & & \Big\downarrow \sigma \\ A & \xrightarrow[\partial]{} & \Pi \end{array} \quad .$$

In fact if, by extending trivially, we regard these crossed modules as crossed complexes $\underline{E}$ and $\chi_\Pi A$ respectively, then the above diagram is a morphism $(\sigma,1) : \underline{E} \to \chi_\Pi A$ of crossed complexes.

Define a *conjugation* $\chi_\Pi A \to \chi_\Pi A$ to be an isomorphism (ξ,ζ) for which there is a function β from Π_0 to the union of the $\Pi(q)$, $q \in \Pi_0$, such that $\xi(x) = -\beta p + x + \beta q$, $x \in \Pi(p,q)$ and $\zeta(a) = a^{\beta q}$, $a \in A(q)$. Define a *free equivalence* of Π-extensions to be an isomorphism $(\eta,\zeta) : \underline{E} \to \underline{E}'$ over the identity on G , and a conjugation $(\xi,\zeta) : \chi_\Pi A \to \chi_\Pi A$ such that $\sigma'\eta = \xi\sigma$. Such free equivalences form under composition a groupoid $\underline{\mathrm{Ext}}_f^\Pi(G, A)$. The *(strict) equivalences* are those in which (ξ,ζ) is the identity.

We have the following generalisation of Theorem 3.

Theorem 6. *There are equivalences of groupoids*

$$e_f : (CG, \chi_\Pi A)_f \longrightarrow \underline{\mathrm{Ext}}_f^\Pi(G, A) ,$$
$$e : (CG, \chi_\Pi A) \longrightarrow \underline{\mathrm{Ext}}^\Pi(G, A) . \quad \square$$

The proof is similar to that of Theorem 3. (Dedecker's result is the bijection induced by e on components when G , A , Π are groups.)

Remark 1. Huebschmann has proved related results for the group case. On p.309 of [20] he shows (for groups) that given a morphism $\underline{P} \to \chi_\Pi A$ (where $\underline{P}$ is as in Corollary 5), one can define an extension $A \to E \to G$ by taking E to be the coequaliser of two maps $A \rightrightarrows F \ltimes A$ (the semi direct-product). In a letter to Dedecker [19] he relates such morphisms $\underline{P} \to \chi_\Pi A$ to Dedecker's 2-cocycles.

Remark 2. A theory of extensions and cohomology of T-algebras is given in [23], using internal category objects for T-algebras. This generalises Dedecker's theory, and also includes the above equivalence e of groupoids. The results do not include extensions of groupoids, nor free equivalences, and crossed complexes are not used.

Remark 3. If X is a CW-complex, and C is a crossed complex, it seems reasonable to define the *cohomology of* X *with coefficients in* C simply as $[\pi\underline{X}, C]$. (A similar idea for chain complexes is developed in [2] and applied to Postnikov invariants of function spaces in [3].) It would be interesting to have applications in homotopy theory of such a non-abelian cohomology.

Remark 4. A *homotopy* f_t of filtered maps $f_0, f_1 : \underline{X} \to \underline{Z}$ is a homotopy $f_t : X \to Z$ such that $f_t(X_n) \subset Z_{n+1}$ for $n \geq 0$. We will prove elsewhere that if X_0 and Z_0 are discrete, then such a homotopy induces a homotopy $\pi f_0 \simeq \pi f_1$ of morphisms of crossed complexes. Consequently, the non-abelian cohomology suggested in Remark 3 above for CW-complexes, is a homotopy invariant.

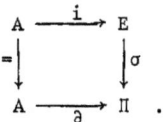

REFERENCES

0. H. ANDO*, A note on the Eilenberg-MacLane invariant, Tohoku Math. J. 9 (1957), 96-104.

1. N.K. ASHLEY, *Crossed complexes and T-complexes*, Ph.D. Thesis, University of Wales, (1978).

2. A.L. BLAKERS, Some relations between homology and homotopy groups, Ann. of Math., (49) 2 (1948), 428-461.

3. R. BROWN, Cohomology with chains as coefficients, Proc. Lond. Math. Soc., (3) 14 (1964), 545-565.

4. R. BROWN, On Künneth suspensions, Proc. Camb. Phil. Soc., (1964), 60, 713-720.

5. R. BROWN, Groupoids as coefficients, Proc. Lond. Math. Soc., (3) 25 (1072), 413-426.

6. R. BROWN and P.J. HIGGINS, The algebra of cubes, J. Pure Appl. Alg. 21 (1981), 233-260.

7. R. BROWN and P.J. HIGGINS, Colimit theorems for relative homotopy groups, J. Pure Appl. Alg. 22 (1981), 11-41.

8. R. BROWN and P.J. HIGGINS, The equivalence of ω-groupoids and cubical T-complexes, Cah. Top. Géom. Diff., (3e Coll. sur les catégories, dédié a Charles Ehresmann), 22 (1981), 349-370.

9. R. BROWN and P.J. HIGGINS, The equivalence of crossed complexes and ∞-groupoids, Cah. Top. Géom. Diff., (3e Coll. sur les catégories, dédié a Charles Ehresmann), 22 (1981), 371-386.

10. R. BROWN and P.J. HIGGINS, On the relation between crossed complexes and chain complexes with operators, (in preparation).

11. R. BROWN and J. HUEBSCHMANN, Identities among relations, in *Low-Dimensional Topology*, Ed. R. Brown and T.L. Thickstun, Lond. Math. Soc. Lecture Note Series 48 (1982).

12. P. DEDECKER, Les foncteurs Ext_{Π}, $\underline{H}^2_{\Pi}$ et H^2_{Π} non abéliens, C.R. Acad. Sci. Paris, 258 (1964), 4891-4894.

13. P. DEDECKER and A. FREI, Généralisation de la suite exacte de cohomologie non abélienne, C.R. Acad. Sci. Paris, 263 (1966), 203-206.

14. M. HALL, JR., *The theory of groups*, MacMillan (1959).

15. P.J. HIGGINS, *Categories and groupoids*, van Nostrand Math. Studies, 32 (1971).

16. P.J. HIGGINS and J. TAYLOR, The fundamental groupoid and homotopy crossed complex of an orbit space, (these proceedings).

17. D.F. HOLT, An interpretation of the cohomology groups $H^n(G, M)$, J. Alg., 60 (1979), 307-318.

18. J. HOWIE, Pullback functors and crossed complexes, Cah. Top. Géom. Diff., 20 (1979), 281-295.

19. J. HUEBSCHMANN, Letter to P. Dedecker, (4th June, 1977).

20. J. HUEBSCHMANN, Crossed N-fold extensions of groups and cohomology, Comm. Math. Helv., 55 (1980), 302-314.

21. J. HUEBSCHMANN, Automorphisms of group extensions and differentials in the Lyndon-Hochschild-Serre spectral sequence, J. Algebra, 72 (1981), 296-334.

22. D.W. JONES, *Poly-T-complexes*, Ph.D. Thesis, University of Wales, (in preparation).

23. R. LAVENDHOMME and J.R. ROISIN, Cohomologie non-abélienne de structures algébriques, J. Algebra, 67 (1980), 385-414.

24. A.S-T. LUE, Cohomology of groups relative to a variety, J. Algebra, 69 (1981), 155-174.

25. S. MACLANE, Topology and logic as a source of algebra, Bull. Amer. Math. Soc., 82 (1976), 1-40.

26. S. MACLANE, Historical note, J. Algebra, 60 (1979), 319-320.

27. J.P. MAY, *Simplicial objects in algebraic topology*, van Nostrand Math. Studies 11 (1967).

28. G. SEGAL, Classifying spaces and spectral sequences, Publ. Math. I.H.E.S., 34 (1968), 105-112.

29. G.W. WHITEHEAD, *Elements of homotopy theory*, Graduate texts in Maths. No. 61, Springer, Berlin-Heidelberg-New York, (1978).

30. J.H.C. WHITEHEAD, Combinatorial homotopy I, Bull. Amer. Math. Soc., (55) 3 (1949), 213-245.

31. J.H.C. WHITEHEAD, Combinatorial homotopy II, Bull. Amer. Math. Soc., 55 (1949), 453-496.

32. J.H.C. WHITEHEAD, Simple homotopy type, Amer. J. Math., 72 (1950), 1-57.

* *Note.* Reference [0] continues work of [2].

UN CRITERE DE REPRESENTABILITE PAR SECTIONS CONTINUES DE FAISCEAUX

Yves DIERS

Département de Mathématiques, U.E.R. des Sciences
Université de Valenciennes, 59326 VALENCIENNES

0. Introduction. Etant donné un foncteur $U : \mathbb{A} \to \mathbb{B}$, on détermine dans quelles conditions chaque objet B de $\mathbb{B}$ est isomorphe à l'objet des sections globales continues d'un faisceau à valeurs dans $\mathbb{B}$ et fibres dans $\mathbb{A}$, universel pour le foncteur sections globales défini sur la catégorie $\mathbb{F}\mathrm{ais}\mathbb{B}\mathbb{A}$ des faisceaux à valeurs dans $\mathbb{B}$ et fibres dans $\mathbb{A}$. La catégorie $\mathbb{B}$ peut alors être plongée d'une façon pleinement fidèle coréflexive dans la catégorie $\mathbb{F}\mathrm{ais}\mathbb{B}\mathbb{A}$ si bien que chaque objet de $\mathbb{B}$ peut s'identifier à son faisceau représentant.

On utilise la construction universelle des spectres, topologies sepctrales et faisceaux structuraux donnée dans [6] et on est ramené à déterminer dans quelle condition le morphisme canonique de chaque objet de $\mathbb{B}$ vers l'objet des sections globales continues de son faisceau structural, est un isomorphisme. On montre qu'une condition nécessaire et suffisante est que le foncteur U soit cogénérateur finiment régulier. Cette notion, plus forte que celle de foncteur cogénérateur propre [7] et plus faible que celle de foncteur codense [12], est obtenue à partir des notions de famille monomorphique stricte ou effective [8] ou régulière [5] de morphismes, de famille d'objets cogénératrice par monomorphismes stricts [8] et de morphismes de présentation finie relative [7], et est décrite de plusieurs façons différentes.

Dans certaines conditions, un foncteur est cogénérateur finiment régulier si et seulement si il est cogénérateur. Ainsi si $\mathbb{B}$ est une catégorie arithmétique [5] et [15], et si $\mathbb{A}$ est une sous-catégorie de $\mathbb{B}$ fermée pour les ultraproduits et dont les morphismes sont exactement les monomorphismes de $\mathbb{B}$ dont le but est dans $\mathbb{A}$, alors $\mathbb{A}$ est une sous-catégorie cogénératrice finiment régulière si et seulement si $\mathbb{A}$ est une sous-catégorie cogénératrice, c'est-à-dire si tout objet de $\mathbb{B}$ est sous-objet d'un produit d'objets de $\mathbb{A}$. Il s'en suit un théorème de représentations par sections continues de faisceaux qui contient tous les théorèmes de représentations qui utilisent habituellement des versions généralisées du théorème chinois sur les systèmes de congruences.

En appliquant les résultats à des foncteurs U oubli de structure adéquats entre catégories d'ensembles munis de structures algébriques, on obtient d'une part, de très nombreux théorèmes connus de représentation par sections continues de faisceaux dont quelques uns sont détaillés ici, et d'autre part, des nouveautés parmi lesquelles la représentation des anneaux commutatifs réguliers formellement réels par des faisceaux de corps ordonnés, celle des groupes abéliens sans torsion par des faisceaux de groupes abéliens totalement ordonnés, celle des espaces vectoriels réels par des faisceaux d'espaces vectoriels euclidiens ou par des faisceaux d'espaces vectoriels normés, celle des ensembles par des faisceaux d'ordinaux finis. Une originalité de ces dernières

représentations est que les faisceaux représentants ont en général pour bases des espaces topologiques non "spectraux" au sens de Hochster [9] car non T_o-séparés et éventuellement non quasi-compacts.
On utilise les notations et les résultats de [5], [6].

1. Foncteurs cogénérateurs finiment réguliers. On considère une catégorie localement de présentation finie $\mathbb{B}$ [7] et un foncteur $U : \mathbb{A} \to \mathbb{B}$. Un morphisme $f : B \to C$ de $\mathbb{B}$ est dit de présentation finie relative s'il est un objet de présentation finie [7] dans la catégorie $B/\mathbb{B}$ des objets de $\mathbb{B}$ au-dessous de B. Il est dit U-injectif si tout morphisme $g : B \to UA$ de B vers U se factorise à travers lui. Plus généralement, une famille $(f_i : B \to C_i)_{i\in I}$ de morphismes de même source de $\mathbb{B}$, est dite U-injective si tout morphismes $g : B \to UA$ de B vers U se factorise à travers l'un de ses membres. Les familles monomorphiques régulières de morphismes de $\mathbb{B}$, encore appelées familles monomorphiques strictes ou effectives dans [8], sont étudiées dans [5].

1.0. Définition. Le foncteur $U : \mathbb{A} \to \mathbb{B}$ est cogénérateur finiment régulier si toute famille U-injective de morphismes de présentation finie relative de $\mathbb{B}$ est monomorphique régulière.
Rappelons qu'un foncteur $U : \mathbb{A} \to \mathbb{B}$ est dit cogénérateur propre [7] ou cogénérateur au sens de Grothendieck [14] si tout morphisme $f : B \to C$ de $\mathbb{B}$ tel que l'application $\mathrm{Hom}_{\mathbb{B}}(f,UA) : \mathrm{Hom}_{\mathbb{B}}(C,UA) \to \mathrm{Hom}_{\mathbb{B}}(B,UA)$ soit bijective pour tout objet A de $\mathbb{A}$, est nécessairement isomorphique.

1.1. Proposition. Si le foncteur U est cogénérateur finiment régulier, il est cogénérateur propre donc cogénérateur.

Preuve : Soit $f : B \to C$ un morphisme de $\mathbb{B}$ tel que l'application $\mathrm{Hom}_{\mathbb{B}}(f,UA)$ soit bijective pour tout objet A de $\mathbb{A}$. La catégorie $\mathbb{B}$ étant localement de présentation finie, la catégorie $B/\mathbb{B}$ l'est aussi et le morphisme f est colimite filtrante des objets de $B/\mathbb{B}$ de présentation finie au-dessus de lui i.e. $f = \varinjlim_{i\in\mathbb{I}} f_i$ avec $f_i : B\to C_i$ de présentation finie relative. Chaque morphisme $f_i : B \to C_i$ est U-injectif donc monomorphique régulier. Le morphisme f est alors monomorphique régulier comme colimite filtrante de monomorphismes réguliers. Il reste à montrer que f est épimorphique. Soit $m,n : C \rightrightarrows D$ deux morphismes vérifiant $mf = nf$ et soit $k : D \to K$ leur conoyau. Tout morphisme $g : D \to UA$ vérifie $gmf = gnf$ donc $gm = gn$ puisque $\mathrm{Hom}_{\mathbb{B}}(f,UA)$ est bijective ; il se factorise donc à travers k ; ce qui implique que $\mathrm{Hom}_{\mathbb{B}}(k,UA)$ est bijective et donc que k est monomorphique et par suite $m = n$.
Les foncteurs qui interviennent dans les représentations par faisceaux ne possèdent jamais d'adjoint à gauche, mais ils ont nécessairement un multiadjoint à gauche [4] donné par les fibres des faisceaux représentants. On suppose donc que le foncteur U a un multiadjoint à gauche et pour chaque objet B de $\mathbb{B}$, on note $(\eta_i : B\to UA_i)_{i\in \mathrm{Spec}_U(B)}$ une famille universelle de morphismes de B vers U.

1.2. Proposition. Si le foncteur $U : \mathbb{A} \to \mathbb{B}$ a un multiadjoint à gauche, il est cogénérateur finiment régulier si et seulement si il existe une classe $\mathcal{D}$ de morphismes de $\mathbb{B}$ telle que

(1) tout morphisme diagonalement universel de B vers U est colimite filtrante de morphismes de $\mathcal{D}$ de source B et

(2) toute famille U-injective de morphismes de $\mathcal{D}$, est monomorphique régulière.

Preuve : La condition nécessaire est satisfaite en prenant pour $\mathcal{D}$ la classe des morphismes de présentation finie relative de $\mathbb{B}$. Réciproquement supposons qu'une classe $\mathcal{D}$ de morphismes de $\mathbb{B}$ satisfasse (1) et (2). Pour chaque objet B de $\mathbb{B}$, la famille universelle $(\eta_i : B \to UA_i)$ est monomorphique. En effet, si T est un objet de présentation finie de $\mathbb{B}$ et $m,n : T \rightrightarrows B$ sont deux morphismes vérifiant $\eta_i m = \eta_i n$ pour tout $i \in \mathrm{Spec}_U(B)$, alors d'après (1), pour chaque i, il existe un morphisme $d_i : B \to D_i$ de $\mathcal{D}$ au-dessus de η_i tel que $d_i m = d_i n$. La famille $(d_i)_{i\in \mathrm{Spec}_U(B)}$ est U-injective, donc monomorphique régulière d'après (2). Par suite $m = n$. Le résultat est aussi vrai pour un objet quelconque T de $\mathbb{B}$, puisque celui-ci est colimite d'objets de présentation finie de $\mathbb{B}$. Les familles (η_i) étant monomorphiques, toute famille U-injective de morphismes de $\mathbb{B}$ est aussi monomorphique puisque plus fine [5] qu'une famille (η_i). Soit $(f_k : B \to C_k)_{k\in K}$ une famille U-injective de morphismes de présentation finie relative de $\mathbb{B}$. Chaque morphisme η_i se factorise à travers un morphisme $f_{k(i)} : B \to C_{k(i)}$ avec $k(i) \in K$. D'après (1), il existe un morphisme $d_i : B \to D_i$ de $\mathcal{D}$ au-dessus de η_i, qui se factorise à travers $f_{k(i)}$. La famille $(d_i : B \to D_i)_{i\in \mathrm{Spec}_U(B)}$ ainsi obtenue est U-injective donc monomorphique régulière. Elle est moins fine que la famille $(f_k)_{k\in K}$ et même régulièrement moins fine [5] puisque toute image directe de la famille (d_i) est U-injective donc monomorphique. Il s'en suit que la famille $(f_k)_{k\in K}$ est monomorphique régulière (prop. 2.1 [5]).

Lorsque les objets de $\mathbb{A}$ sont des ensembles munis d'une structure algébrique définissable par une théorie logique du premier ordre, la catégorie $\mathbb{A}$ est à ultraproduits. Nous allons montrer que, dans ce cas, il suffit de considérer les familles finies de morphismes.

1.3. Proposition. Si le foncteur $U : \mathbb{A} \to \mathbb{B}$ a un multiadjoint à gauche et relève les ultraproduits d'objets de $\mathbb{A}$ ([6] 4.1), les assertions suivantes sont équivalentes :

(1) U est cogénérateur finiment régulier,

(2) toute famille finie U-injective de morphismes de $\mathbb{B}$, est monomorphique régulière,

(3) il existe une classe $\mathcal{D}$ de morphismes de présentation finie relative de $\mathbb{B}$ telle que

a) tout morphisme diagonalement universel de B vers U est colimite filtrante de morphismes de $\mathcal{D}$ et

b) toute famille finie U-injective de morphismes de $\mathcal{D}$ est monomorphique régulière

Preuve : (1) => (2) : Soit $(f_i : B \to C_i)_{i\in[1,n]}$ une famille finie U-injective de morphismes de $\mathbb{B}$. La catégorie $(B/\mathbb{B})^n$ étant localement de présentation finie, il existe une petite catégorie filtrante $\mathbb{K}$ et un diagramme $\left((f_{ik}:B \to C_{ik})_{i\in[1,n]}\right)_{k\in\mathbb{K}}$ d'objets de présentation finie de $(B/\mathbb{B})^n$ dont la colimite est (f_i). C'est-à-dire que les morphismes $f_{ik} : B \to C_{ik}$ sont de présentation finie relative et que pour tout $i \in [1,n]$, $f_i = \varinjlim_{k\in K} f_{ik}$. Pour chaque $k \in \mathbb{K}$, la famille $(f_{ik}:B \to C_{ik})_{i\in[1,n]}$ est U-injective donc monomorphique régulière. Notons $(f'_{ijk} : C_{ik} \to C_{ijk},\ f''_{ijk} : C_{jk} \to C_{ijk})$ la somme amalgamée de $(f_{ik} : C \to C_{ik},\ f_{jk} : C \to C_{jk})$ et $p_{ik} : \prod_{i=1}^{n} C_{ik} \to C_{ik}$ la projection canonique. Le morphisme $(f_{ik}) : B \to \prod_{i=1}^{n} C_{ik}$ est noyau des deux morphismes $(f'_{ijk}p_{ik})_{(i,j)\in[1,n]^2}$ et $(f''_{ijk}p_{jk})_{(i,j)\in[1,n]^2}$ de source $\prod_{i=1}^{n} C_{ik}$ et de but $\prod_{(i,j)=(1,1)}^{(n,n)} C_{ijk}$. Si l'on note $(f'_{ij}:C_i \to C_{ij},\ f''_{ij}:C_j \to C_{ij})$ la somme amalgamée de $(f_i : B \to C_i,\ f_j : B \to C_j)$ et $p_i : \prod_{i=1}^{n} C_i \to C_i$ la projection d'indice i, alors par passage à la colimite filtrante suivant $\mathbb{K}$, le morphisme $(f_i) : B \to \prod_{i=1}^{n} C_i$ est noyau des deux morphismes $(f'_{ij}p_i)_{(i,j)\in[1,n]^2}$ et $(f''_{ij}p_j)_{(i,j)\in[1,n]^2}$ de source $\prod_{i=1}^{n} C_i$ et de but $\prod_{(i,j)=(1,1)}^{(n,n)} C_{ij}$. Cela implique que la famille $(f_i)_{i\in[1,n]}$ est monomorphique régulière.

(2) => (3) : est satisfait en prenant pour $\mathcal{D}$ la classe des morphismes de présentation finie relative de $\mathbb{B}$.

(3) => (1) : avec la proposition 1.2, il suffit de montrer que toute famille U-injective de morphismes de $\mathcal{D}$ est monomorphique régulière. Soit $(f_k : B \to C_k)_{k\in K}$ une telle famille. Supposons qu'il n'existe aucune sous-famille finie U-injective de $(f_k)_{k\in K}$. Pour chaque partie finie K_o de K, notons $D(K_o)$ l'ensemble des $i \in \mathrm{Spec}_U(B)$ tel que η_i se factorise à travers l'un des morphismes f_k avec $k \in K_o$. Les relations $D(K_o) \neq \mathrm{Spec}_U(B)$, $D(\emptyset) = \emptyset$ et $D(K_o \cup K_1) = D(K_o) \cup D(K_1)$ montrent que les parties complémentaires des parties $D(K_o)$ dans $\mathrm{Spec}_U(B)$, quand K_o parcourt les parties finies de K, forment une base de filtre sur $\mathrm{Spec}_U(B)$. Soit F un ultrafiltre plus fin. Il existe un objet A_F de $\mathbb{A}$ tel que UA_F soit l'ultraproduit de $(UA_i)_{i\in\mathrm{Spec}_U(B)}$ suivant F. Le morphisme canoniquement défini $\eta_F : B \to UA_F$ se factorise à travers un morphisme f_k avec $k \in K$. Puisque $UA_F = \varinjlim_{I\in F} \prod_{i\in I} UA_i$ est une colimite filtrante et que le morphisme f_k est de présentation finie relative, il existe $I \in F$ tel que le morphisme $(\eta_i)_{i\in I} : B \to \prod_{i\in I} UA_i$ se factorise à travers f_k. L'inclusion $I \subset D(\{k\})$ implique alors $D(\{k\}) \in F$, ce qui est en contradiction avec le fait que le complémentaire de $D(\{k\})$ dans $\mathrm{Spec}_U(B)$ appartient à F. Il en résulte que la famille $(f_k)_{k\in K}$ possède une sous-famille finie U-injective $(f_k)_{k\in K_o}$. La sous-famille $(f_k)_{k\in K_o}$ est monomorphique régulière. La famille $(\eta_i : B \to UA_i)_{i\in\mathrm{Spec}_U(B)}$ est donc monomorphique de même que toutes les familles U-injectives. La sous-famille $(f_k)_{k\in K_o}$ est régulièrement plus fine que la famille

$(f_k)_{k\in K}$ (2 [5]) puisque toute image directe de $(f_k)_{k\in K}$ est U-injective donc monomorphique. De la proposition 2.1 [5], il résulte que la famille $(f_k)_{k\in K}$ est monomorphique régulière.

1.4. Proposition. Si le foncteur U est codense [12], il est cogénérateur finiment régulier.

Preuve : Si U est codense, tout objet B de $\mathbb{B}$ est limite de tous les objets de $\mathbb{A}$ au-dessous de lui, ce qui implique que la famille de tous les morphismes de B vers U est monomorphique régulière (prop. 5.4 [5]). Toute famille U-injective de morphismes de source B est plus fine que la famille précédente donc est monomorphique ; elle est même régulièrement plus fine puisque toute image directe d'une famille U-injective est U-injective donc monomorphique ; elle est donc monomorphique régulière (prop. 2.1 [5]).

1.5. Proposition. Le foncteur $U : \mathbb{A} \to \mathbb{B}$ est cogénérateur finiment régulier s'il existe un foncteur $V : \mathbb{K} \to \mathbb{A}$ tel que le foncteur UV soit cogénérateur finiment régulier.

Preuve : Toute famille U-injective de morphismes de présentation finie relative de même source de $\mathbb{B}$ est UV-injective donc monomorphique régulière.

2. Le critère de représentabilité.

2.0. Théorème. Soit $U : \mathbb{A} \to \mathbb{B}$ un foncteur tel que : 1) $\mathbb{B}$ est une catégorie localement de présentation finie, 2) U admet un multiadjoint à gauche, 3) tout morphisme diagonalement universel d'un objet B de $\mathbb{B}$ vers U est colimite de morphismes de présentation finie relative de source B diagonalement universels pour U, 4) U est cogénérateur finiment régulier.
Alors tout objet B de $\mathbb{B}$ détermine un faisceau F_B de base $\mathrm{Spec}_U(B)$ à valeurs dans $\mathbb{B}$ et fibres dans $\mathbb{A}$, dont l'objet des sections globales est isomorphe à B et qui est universel pour le foncteur sections globales $\Gamma : \mathbb{F}\mathrm{ais}\,\mathbb{B}\mathbb{A} \to \mathbb{B}$; c'est-à-dire que le foncteur Γ admet un adjoint à gauche pleinement fidèle. Si les conditions 1), 2), 3) sont satisfaites, la condition 4) est en fait nécessaire et suffisante pour obtenir la conclusion.

Preuve : Les conditions 1), 2), 3) sont les conditions d'applications du théorème 3.1 de [6] dont on utilise ici les notations et les résultats (cf. 3.0, 3.1, 3.3, 3.4, 3.5). Soit B un object de $\mathbb{B}$.

a) la famille universelle $(\eta_i : B \to UA_i)_{i\in \mathrm{Spec}_U(B)}$ est monomorphique puisque le foncteur U est cogénérateur. Cela implique que les morphismes de $\mathbb{B}$ diagonalement universels pour U sont épimorphiques. En effet si $\delta : B \to C$ est l'un d'eux et si $f, g : C \rightrightarrows D$ sont deux morphismes vérifiant $f\delta = g\delta$, alors pour tout $j \in \mathrm{Spec}_U(D)$, on a $\eta_j f\delta = \eta_j g\delta$ ce qui implique $\eta_j f = \eta_j g$ donc $f = g$ car la famille $(\eta_j : D \to UA_j)_{j\in \mathrm{Spec}_U(D)}$ est monomorphique.

b) Montrons que le foncteur $D : \mathbb{A}(B) \to \mathcal{D}(\mathrm{Spec}_U(B))^{op}$ est une équivalence de catégories. Il est surjectif sur les objets d'après la construction de $\mathcal{D}(\mathrm{Spec}_U(B))$ et il est fidèle puisque les morphismes de $\Delta'(B)$ étant épimorphiques, il y a au plus un morphisme entre deux objets de $\mathbb{\Delta}(B)$. Soit $\delta : B \to C$, $\delta' : B \to C'$ deux objets de $\mathbb{\Delta}(B)$ tels que $D(\delta) \subset D(\delta')$. Notons $(\delta_1 : C \to C_1, \delta'_1 : C' \to C_1)$ la somme amalgamée de (δ,δ'). La relation $D(\delta_1\delta) = D(\delta) \cap D(\delta') = D(\delta)$ implique que le morphisme diagonalement universel de présentation finie relative δ_1 est U-injectif. C'est donc un isomorphisme d'après 4). Ce qui implique que $\delta_1^{-1}\delta'_1$ est un morphisme de δ' vers δ dans la catégorie $\mathbb{\Delta}(B)$. Ainsi le foncteur D est plein.

c) Soit $(d_k : (C,\delta) \to (C_k,\delta_k))_{k\in K}$ une famille de morphismes de $\mathbb{\Delta}(B)$ telle que $D(\delta) = \bigcup_{k\in K} D(\delta_k)$. La famille $(d_k : C \to C_k)_{k\in K}$ de morphismes de présentation finie relative est U-injective. Elle est donc monomorphique régulière. Cela exprime précisément que le préfaisceau structural $P_B : \mathcal{D}(\mathrm{Spec}_U(B)^{op} \to \mathbb{B}$ défini par $P_B(D(\delta)) = C$ pour $\delta : B \to C$ dans $\mathbb{\Delta}(B)$, est un faisceau. On en déduit $F_B \simeq P_B$ et $F_B(\mathrm{Spec}_U(B)) \simeq P_B(\mathrm{Spec}_U(B)) = P_B(D(1_B)) \simeq B$.

d) Les morphismes canoniques $B \to F_B(\mathrm{Spec}_U(B))$ étant des isomorphismes, on déduit que le foncteur adjoint à gauche au foncteur $\Gamma : \mathbb{F}\mathrm{ais}\,\mathbb{B}/\mathbb{A} \to \mathbb{A}$ est pleinement fidèle (Th. 1, p.88, [12])

e) Supposons maintenant avec les conditions 1), 2), 3) que le foncteur adjoint à gauche au foncteur Γ soit pleinement fidèle. Montrons d'abord que pour $\delta : B \to C \in \Delta'(B)$, le morphisme canoniquement défini $F_B(D(\delta)) \to F_C(\mathrm{Spec}_U(C))$ est un isomorphisme. On note $(\mathrm{Spec}_U(\delta),F_\delta) : (\mathrm{Spec}_U(B),F_B) \to (\mathrm{Spec}_U(C),F_C)$ le morphisme canoniquement défini par le foncteur adjoint à gauche à $\Gamma : \mathbb{F}\mathrm{ais}\,\mathbb{B}/\mathbb{A} \to \mathbb{B}$. Alors $F_\delta : F_B \to F_C(\mathrm{Spec}_U(\delta))^*$ est un morphisme de faisceaux sur $\mathrm{Spec}_U(B)$. Montrons que le morphisme $(F_\delta)_{D(\delta)}$: $F_B(D(\delta)) \to F_C(\mathrm{Spec}_U(C))$ est un isomorphisme. Soit $j \in \mathrm{Spec}_U(C)$. Posons $i = (\mathrm{Spec}_U(\delta))(j)$ et $\alpha_j : A_i \to A_j$ l'isomorphisme défini par $(U\alpha_j)\,\eta_i = \eta_j\delta$. Puisque l'application $\mathrm{Spec}_U(\delta) : \mathrm{Spec}_U(C) \to \mathrm{Spec}_U(B)$ est un plongement homéomorphique ouvert (prop. 3.3.6 [6]) la fibre de $F_C(\mathrm{Spec}_U(\delta))^*$ au point i est la fibre de F_C en j. Le morphisme $(F_\delta)_i : (F_B)_i \to (F_C(\mathrm{Spec}_U(\delta))^*)_i$ fibre de F_δ en i est donc l'isomorphisme $U\alpha_i : UA_i \to UA_j$. On en déduit que le morphisme $(F_\delta)_{D(\delta)} : F_B(D(\delta)) \to F_C(\mathrm{Spec}_U(C))$ est un isomorphisme.

f) Montrons que le foncteur U est cogénérateur finiment régulier. Tout morphisme η_i est colimite filtrante de morphismes de $\Delta'(B)$. D'après la proposition 1.2, il suffit donc de montrer que toute famille U-injective $(\delta_k : B \to C_k)_{k\in K}$ de morphismes de $\Delta'(B)$ est monomorphique régulière. Or on a $\bigcup_{k\in K} D(\delta_k) = \mathrm{Spec}_U(B)$, donc, puisque F_B est un faisceau sur $\mathrm{Spec}_U(B)$, la famille de morphismes $(F_B(\mathrm{Spec}_U(B)) \to F_B(D(\delta_k)))_{k\in K}$ est monomorphique régulière. Compte tenu des isomorphismes $F_B(\mathrm{Spec}_U(B)) \simeq B$ et du e), la famille précédente est isomorphe à la famille (δ_k).

3. Un critère de représentabilité spécial pour les catégories arithmétiques.

Le théorème suivant contient les théorèmes de représentabilité par sections continues de faisceaux qui utilisent habituellement une version généralisée du théorème chinois sur les systèmes de congruences. Pour la définition et des exemples de catégories arithmétiques, on peut se reporter à [5] et [15].

3.0. Théorème. Soit $\mathbb{B}$ une catégorie arithmétique et $\mathbb{A}$ une sous-catégorie cogénératrice de $\mathbb{B}$ fermée pour les ultraproduits et telle que les morphismes de $\mathbb{A}$ sont exactement les monomorphismes de $\mathbb{B}$ dont le but est dans $\mathbb{A}$.
Alors $\mathbb{A}$ est une sous-catégorie multiréflexive de $\mathbb{B}$ et tout objet B de $\mathbb{B}$ détermine un faisceau F_B de base $\mathrm{Spec}_U(B)$ à valeurs dans $\mathbb{B}$ et fibres dans $\mathbb{A}$, dont l'objet des sections globales est isomorphe à B et qui est universel pour le foncteur sections globales $\Gamma : \mathbb{F}\mathrm{ais}\,\mathbb{B}\mathbb{A} \to \mathbb{B}$; c'est-à-dire que le foncteur Γ admet un adjoint à gauche pleinement fidèle.

Preuve : Soit B un objet de $\mathbb{B}$. Notons $\mathrm{Spec}_U(B)$ l'ensemble des relations d'équivalences R sur B dont l'objet quotient B/R est dans $\mathbb{A}$. Pour chaque $R \in \mathrm{Spec}_U(B)$, on note $\eta_R : B \to B/R$ le morphisme quotient. La famille $(\eta_R : B \to B/R)_{R\in \mathrm{Spec}_U(B)}$ de morphismes de B vers $\mathbb{A}$ est universelle. En effet, si $g : B \to X$ est un morphisme de B vers un objet de $\mathbb{A}$, on note R la relation d'équivalence sur B engendrée par g et $h : B/R \to X$ l'unique monomorphisme de $\mathbb{B}$ tel que $h\,\eta_R = g$; puisque X est un objet de $\mathbb{A}$, le morphisme $h : B/R \to X$ est dans $\mathbb{A}$, donc B/R est un objet de $\mathbb{A}$ et $R \in \mathrm{Spec}_U(B)$; on obtient ainsi une factorisation de g par un morphisme de la famille $(\eta_R : B \to B/R)_{R\in \mathrm{Spec}_U(B)}$. Il est immédiat qu'une telle factorisation est unique. $\mathbb{A}$ est donc une sous-catégorie multiréflexive de $\mathbb{B}$. Notons $\Delta'(B)$ l'ensemble des morphismes de $\mathbb{B}$ de la forme $\delta_R : B \to B/R$ où R est une relation d'équivalence sur B engendrée par un nombre fini d'éléments et où δ_R est le morphisme quotient. L'ensemble $\Delta'(B)$ est fermé pour les colimites finies dans $B/\mathbb{B}$. Les morphismes de $\mathbb{A}$ étant monomorphiques, les morphismes de $\Delta'(B)$ sont diagonalement universels pour $\mathbb{A}$. Ils sont aussi de présentation finie relative. Tout morphismes diagonalement universel de B vers $\mathbb{A}$ étant colimite filtrante de morphismes de $\Delta'(B)$, la condition 3) du théorème 2.0 est satisfaite.
Il reste à montrer que $\mathbb{A}$ est une sous-catégorie cogénératrice finiment régulière de $\mathbb{B}$. Pour chaque objet B de $\mathbb{B}$, la famille universelle $(\eta_R : B \to B/R)$ est monomorphique puisque $\mathbb{A}$ cogénératrice dans $\mathbb{B}$. Par suite toute famille U-injective de morphismes de $\mathbb{B}$ est monomorphique. De la proposition 7.11 [5], il résulte que toute famille finie U-injective de morphismes de $\Delta'(B)$ est monomorphique régulière. Le résultat découle alors de la proposition 1.3 en prenant comme morphismes de classe $\mathcal{D}$ de source B, les morphismes de $\Delta'(B)$.

4. Applications. Le théorème 2.0 permet de retrouver de très nombreux théorèmes connus de représentation par sections continues de faisceaux. Il suffit de l'appliquer à des foncteurs $U : \mathbb{A} \to \mathbb{B}$ oubli de structure adéquats. De nombreux exemples de foncteurs U pour lesquels les hypothèses 1), 2), 3) du théorème sont satisfaites sont donnés dans [6]. Il reste au lecteur à déterminer dans quels cas, le foncteur U est cogénérateur finiment régulier. Nous en étudions quelques uns. Pour les catégories arithmé-

tiques (4.3 à 4.7), on utilise plutôt le théorème 3.0. avec lequel on est ramené à montrer que le foncteur inclusion $U : \mathbb{A} \to \mathbb{B}$ est cogénérateur, c'est-à-dire que tout objet de $\mathbb{B}$ est sous-objet d'un produit d'objets de $\mathbb{A}$. Or c'est une propriété souvent bien connue dont la preuve repose essentiellement sur le lemme de Zorn. Les représentations (4.8 à 4.12) sont nouvelles. Les faisceaux représentants possèdent l'originalité d'avoir pour bases des espaces topologiques non "spectraux" au sens de Hochster [9] car non T_o-séparés et éventuellement non quasi-compacts.

4.0. Représentation d'un anneau commutatif par un faisceau d'anneaux locaux [2], [3]. Compte tenu de 7.0 [6], il suffit de montrer que le foncteur $U : \mathbb{L}occ \to \mathbb{A}nc$ est cogénérateur finiment régulier. Un ultraproduit d'anneaux locaux étant un anneau local, le foncteur U relève les ultraproduits. Utilisons la proposition 1.3 avec la classe $\mathcal{D}$ des morphismes de la forme $A \to A[a^{-1}]$. Soit $(A \to A[a_i^{-1}])_{i\in[1,n]}$ une famille U-injective. Pour chaque $P \in \mathrm{Spec}_U(A)$, le morphisme $A \to A_P$ se factorise à travers un morphisme $A \to A[a_{i(P)}^{-1}]$ avec $i(P) \in [1,n]$; alors $a_{i(P)} \notin P$. L'idéal de A engendré par l'ensemble des éléments a_i pour $i \in [1,n]$ n'étant contenu dans aucun idéal premier de A, est égal à A. La suite $a_1,\dots,a_n$ engendre donc le A-module A. La famille $(A \to A[a_i^{-1}])_{i\in[1,n]}$ est donc monomorphique régulière (8.0 [5]). On obtient ainsi la représentation classique de A par son faisceau structural $\tilde{A}$.

4.1. Représentation d'un anneau commutatif par un faisceau d'anneaux indécomposables [3], [13].

Soit $U : \mathbb{A}ncInd \to \mathbb{A}nc$ le foncteur inclusion (7.5 [6]). Pour un anneau $A \in \mathbb{A}nc$, la famille des anneaux quotients $(A \to A/PA)$ où P décrit l'ensemble des idéaux premiers de l'anneau des idempotents de A, est une famille universelle de morphismes de A vers U. Chaque morphisme $A \to A/PA$ est colimite filtrante de morphismes diagonalement universels de présentation finie relative de la forme $A \to A/Ae$ où e est un idempotent de A. Montrons que le foncteur U est cogénérateur finiment régulier. Un ultraproduits d'anneaux indécomposables étant indécomposable, le foncteur U relève les ultraproduits. Utilisons la proposition 1.3 avec la classe $\mathcal{D}$ des morphismes de la forme $A \to A/Ae$. Soit $(A \to A/Ae_i)_{i\in[1,n]}$ une famille finie U-injective. Pour chaque $P \in \mathrm{Spec}_U(A)$, il existe $i_{(P)} \in [1,n]$ tel que le morphisme $A \to A/PA$ **se** factorise à travers $A \to A/Ae_{i(P)}$ i.e tel que $e_{i(P)} \in P$. Par suite l'idempotent $\prod_{i=1} e_i$ appartient à tous les idéaux premiers d'idempotents de A ; il est donc nul. La famille $(A \to A/Ae_i)_{i\in[1,n]}$ est donc monomorphique régulière (8.2 [5]).

4.2. Représentation d'un treillis distributif par un faisceau de treillis locaux [1], [3]. Compte tenu de 7.9 [6], il suffit de montrer que le foncteur $U : \mathbb{T}rDLoc \to \mathbb{T}rD$ est cogénérateur finiment régulier. Ce foncteur relève les ultraproduits puisqu'un ultraproduit de treillis locaux est un treillis local. Utilisons la proposition 1.3 avec la classe $\mathcal{D}$ des morphismes quotients de la forme $E \to E/(a)$ où (a) est le filtre principal de E engendré par a. Soit $(E \to E/(a_i))_{i\in[1,n]}$ une famille finie U-injective. Pour chaque $\Phi \in \mathrm{Spec}_U(E)$, le morphisme $E \to E/\Phi$ factorise à travers un morphisme $E \to E/(a_{i(\Phi)})$ avec $i(\Phi) \in [1,n]$, et par suite $a_{i(\Phi)} \in \Phi$. L'élément $a_1 \vee \ldots \vee a_n$ appartient alors à tous les filtres premiers de E ; il est égal à 1. Donc $E/(a_1) \cap \ldots \cap (a_n) = E/(a_1 \vee \ldots \vee a_n) = E$. La famille $(E \to E/(a_i))_{i \in [1,n]}$ est monomorphique régulière d'après 7.0.1, 8.5.1, 8.5.2 [5].

4.3. Représentation d'un anneau commutatif régulier par un faisceau de corps commutatifs [3], [10].

Le foncteur inclusion $U : \mathbb{K}c \to \mathbb{A}ncReg$ (7.3 [6]) satisfait les hypothèses du théorème 3.0. En effet la catégorie $\mathbb{A}ncReg$ est arithmétique (8.5 [5]), $\mathbb{K}c$ est une sous-catégorie cogénératrice de $\mathbb{A}ncReg$ puisque l'intersection des idéaux maximaux d'un anneau régulier $A \in \mathbb{A}ncReg$ est réduite à zéro et donc la famille des anneaux quotients $(A \to A/P)_{P \in Spec(A)}$ est monomorphique, $\mathbb{K}c$ est fermée pour les ultraproduits et tout sous-anneau régulier d'un corps commutatif est un corps.

4.4. Représentation d'un anneau fortement régulier par un faisceau de corps [0].

Le foncteur inclusion $U : \mathbb{K} \to \mathbb{A}nForReg$ (7.4 [6]) satisfait les hypothèses du théorème 3.0, la catégorie $\mathbb{A}nForReg$ étant arithmétique (8.5 [5]) et tout anneau $A \in \mathbb{A}nForReg$ étant un sous-anneau d'un produit de corps.

4.5. Représentation d'un groupe abélien réticulé par un faisceau de groupes abéliens totalement ordonnés [11].

Le foncteur inclusion $U : \mathbb{A}bTotOrd \to \mathbb{A}bRet$ (7.10 [6]) satisfait les hypothèses du théorème 3.0, la catégorie AbRet étant arithmétique (8.5 [5]).

4.6. Représentation d'un anneau commutatif fortement réticulé par un faisceau d'anneaux totalement ordonnés [11].

Le foncteur inclusion $U : \mathbb{A}ncTotOrd \to \mathbb{A}ncForRet$ (7.11 [6]) satisfait les hypothèses du théorème 3.0, la catégorie AncForRet étant arithmétique (8.5 [5]).

4.7. Représentation d'un anneau commutatif régulier fortement réticulé par un faisceau de corps commutatifs ordonnés [11].

Le foncteur inclusion $\mathbb{K}cOrd \to \mathbb{A}ncRegForRet$ (7.12 [6]) satisfait les hypothèses du théorème 3.0.

4.8. Représentation d'un anneau commutatif régulier formellement réel par l'anneau des sections globales d'un faisceau de corps commutatifs ordonnés.

On considère la catégorie $\mathbb{A}ncRegFormRl$ des anneaux commutatifs unitaires réguliers formellement réels i.e. qui vérifient l'axiome : $\forall x_1,\ldots,x_n$, $1+x_1^2 + \ldots + x_n^2$ inversible, et des homomorphismes d'anneaux et le foncteur oubli de structure $U : \mathbb{K}cOrd \to \mathbb{A}ncRegFormRl$. Le foncteur U admet un multiadjoint à gauche. Le spectre de $A \in \mathbb{A}ncRegFormRl$ relativement à U est l'ensemble des parties P de A telles que (0) $-1 \notin P$ (1) $P + P \subset P$ (2) $PP \subset P$ (3) $P \cup (-P) = A$ (4) $\forall x \in A, \forall y \in A(xy \in -P \Rightarrow (x \in P$ ou $y \in P))$ (cf. 7.27 [6]). La famille des anneaux quotients $(A \to A/_{P \cap (-P)})$ munis des ordres quotients de P, est une famille universelle de morphismes de A vers U. La topologie spectrale est engendrée par les parties $D(a_1,\ldots,a_n) = \{P : -(a_1^2 + \ldots + a_n^2) \in P\}$ où $a_1,\ldots,a_n \in A$. Elle n'est pas T_0-séparée car toutes les parties P qui définissent un ordre sur A i.e. qui vérifient en plus (5) $P \cap (-P) = 0$, sont des points denses de $\mathrm{Spec}_U(A)$. Les hypothèses (1), (2), (3) du théorème 2.0 sont satisfaites. Montrons que U est cogénérateur finiment régulier. Les idéaux maximaux M de A sont réels, donc les corps quotients A/M sont formellement réels et par suite ordonnables. La famille des anneaux quotients de A par ses idéaux maximaux est donc une famille de morphismes de A vers U. Or c'est une famille monomorphique, donc le foncteur U est cogénérateur. Le foncteur U relève les ultraproduits. Chaque morphisme $A \to A/_{P \cap (-P)}$ est colimite filtrante de morphismes quotients $A \to A/I$ où I est un idéal de type fini de A. Si $(A \to A/_{I_1}, \ldots, A \to A/_{I_n})$ est une famille finie U-injective, elle est monomorphique donc monomorphique régulière puisque la catégorie AncRegFormReel est arithmétique (prop. 7.0.1. [5]). Il s'ensuit que le foncteur U est cogénérateur finiment régulier.

4.9. <u>Représentation d'un groupe abélien sans torsion par le groupe des sections globales d'un faisceau de groupes abéliens totalement ordonnés.</u>

Le foncteur U : $\mathbb{A}$bTotOrd $\to$ $\mathbb{A}$bSTor (7.30 et 7.31 [6]) est cogénérateur finiment régulier car il est surjectif sur les objets puisque tout groupe abélien sans torsion est totalement ordonnable.

4.10. <u>Représentation d'un espace vectoriel réel par l'espace vectoriel des sections globales d'un faisceau d'espaces vectoriels euclidiens.</u>

Montrons que le foncteur U : $\mathbb{E}$ucl $\to$ $\mathbb{V}$ec($\mathbb{R}$) est cogénérateur finiment régulier (7.32 [6]), en utilisant la proposition 1.2 avec la classe $\mathcal{D}$ des morphismes de la forme $E \to E/X$ où X est un sous-espace vectoriel de dimension finie de E. Soit $(E \to E/X_i)_{i \in I}$ une famille U-injective de morphismes de $\mathcal{D}$. Soit $i(1), \ldots, i(n) \in I$. L'espace vectoriel $X_{i(1)} + \ldots + X_{i(n)}$ est de dimension finie donc possède une base $e_1, \ldots, e_p$ et un supplémentaire X' de codimension finie. Tout élément x de E étant de la forme $x = x_1 e_1 + \ldots + x_p e_p + x'$ avec $x_1, \ldots, x_p \in \mathbb{R}$ et $x' \in X'$, on définit la forme quadratique positive q sur E par $q(x) = x_1^2 + \ldots + x_p^2$. Alors $\mathrm{ISO}(q) = X'$, donc $q \in \mathrm{Spec}_U(E)$. Il existe $i(n+1) \in I$ tel que le morphisme quotient $E \to E/_{X_{i(n+1)}}$ factorise $E \to E/_{\mathrm{ISO}(q)}$, c'est-à-dire tel que $X_{i(n+1)} \subset \mathrm{ISO}(q)$ donc tel que $(X_{i(1)} + \ldots + X_{i(n)}) \cap X_{i(n+1)} = \{0\}$. Il s'ensuit que la famille des morphismes quotients $(E \to E/X_{i(1)}, \ldots, E \to E/X_{i(n+1)})$ est monomorphique régulière (8.4 [5]). D'après la prop. 6.0 [5], la famille $(E \to E/X_i)_{i \in I}$ est monomorphique régulière.

4.11. <u>Représentation d'un espace vectoriel réel par l'espace vectoriel des sections globales d'un faisceau d'espaces vectoriels normés.</u>

Le foncteur U : $\mathbb{N}$orm($\mathbb{R}$) $\to$ $\mathbb{V}$ect($\mathbb{R}$) (7.34 [6]) est cogénérateur finiment régulier puisqu'il factorise le foncteur cogénérateur finiment régulier $\mathbb{E}$ucl $\to$ $\mathbb{V}$ect($\mathbb{R}$) (prop.1.5)

4.12. <u>Représentation d'un ensemble par l'ensemble des sections globales d'un faisceau d'ordinaux finis.</u>

Montrons que le foncteur U : $\mathbb{O}$rdfin $\to$ $\mathbb{E}$ns (7.36 [6]) est cogénérateur finiment régulier en utilisant la proposition 1.2 avec la classe $\mathcal{D}$ des morphismes de la forme $E \to E/R$ où R est une relation d'équivalence sur E engendrée par un ensemble fini.

Soit $(E \to E/R_i)_{i \in I}$ une famille U-injective de morphismes de $\mathcal{D}$. Soit $i(1),\ldots,i(n) \in I$. Soit R la relation d'équivalence sur E engendrée par $R_{i(1)} \cup \ldots \cup R_{i(n)}$. Elle est finiment engendrée. La relation d'équivalence $\sim$ sur E définie par $x \sim y$ si et seulement si (x = y ou (les classes d'équivalences de x et de y suivant R sont des singletons)) possède un ensemble fini de classes d'équivalence. L'ensemble quotient $E/\sim$ peut être muni d'une structure d'ordre total et est donc en bijection avec un ordinal fini. Il existe alors $i(n+1) \in I$ tel que l'application quotient $E \to E/R_{i(n+1)}$ factorise l'application quotient $E \to E/\sim$, c'est-à-dire tel que $R_{i(n+1)} \subset \sim$. Or les deux relations d'équivalence R et $\sim$ sont premières entre-elles (8.3 [5]). Donc les relations d'équivalences $R_{i(1)},\ldots,R_{i(n)}$ sont premières avec $R_{i(n+1)}$. Il s'ensuit que la famille $(E \to E/R_{i(1)},\ldots,E \to E/R_{i(n+1)})$ est monomorphique régulière (8.3. [5]). D'après la proposition 6.0 de [5] la famille $(E \to E/R_i)_{i \in I}$ est monomorphique régulière.

4.13. Quelques contre-exemples. On montre facilement que les foncteurs $\mathbb{D}$om $\to$ $\mathbb{A}$nc, $\mathbb{K}$c $\to$ $\mathbb{A}$nc, $\mathbb{A}$ncDifLoc $\to$ $\mathbb{A}$ncDif, $\mathbb{L}$oc $\to$ $\mathbb{A}$n [6] ne sont pas cogénérateurs propres, donc ne sont pas cogénérateurs finiment réguliers et par suite ne donnent pas de théorèmes de représentations.

REFERENCES

[0] R.F. ARENS et J. KAPLANSKY. Topological representation of algebras, Trans. Amer. Math. Soc. 63, pp. 457-481, 1948.

[1] A. BREZULEANU et R. DIACONESCU. Sur la duale de la catégorie des treillis, Rev. Roumaine. Math. Pures et Appl. 14, pp. 331-323, 1969.

[2] J.C. COLE. The bicategory of topoï and Spectra, preprint.

[3] M. COSTE. Localisation, spectra and sheaf representation, Lecture Notes in Math. 753, Springer-Verlag. Berlin-New-York, 1979.

[4] Y. DIERS. Familles universelles de morphismes, Ann. Soc. Sci. Bruxelles, 93, III, pp. 175-195, 1979.

[5] Y. DIERS. Sur les familles monomorphiques régulières de morphismes, Cahiers de Top Geom Diff, XXI-4, pp. 441-425, 1980.

[6] Y. DIERS. Une construction universelle des spectres, topologies spectrales et faisceaux structuraux, Archiv der Math, à paraître.

[7] P. GABRIEL et F. ULMER. Lokal präsentierbare Kategorien, Lecture Notes in Math. 221, Springer-Verlag, Berlin-New-York, 1971.

[8] A. GROTHENDIECK, M. ARTIN, J.L. VERDIER. Théorie des topos et cohomologie étale des schémas, Lecture Notes in Math 269, Springer-Verlag, Berlin Heideberg New-York, 1972.

[9] M. HOCHSTER. Prime ideal structure in commutative rings, Trans. Amer. Math. Soc. 142, pp. 43-60, 1969.

[10] P.T. JOHNSTONE. Rings, Fields, and Spectra. Jour. Alg. 49, pp. 238-260, 1977.

[11] K. KEIMEL. The representation of lattice-ordered groups and rings by sections in sheaves. Lecture Notes in Math. 248, Springer-Verlag, Berlin-New-York, 1971.

[12] S. MACLANE. Categories for the working Mathematician, Springer-Verlag, New-York-Heidelberg-Berlin, 1971.

[13] R.S. PIERCE. Modules over commutative regular rings. Mem. Amer. Math. Soc. 70, 1967.

[14] H. SCHUBERT. Categories, Springer-verlag, Berlin-Heidelberg-New-York, 1972.

[15] A. WOLF. Sheaf representations of Arithmetical Algebras. Mem. Amer. Math. Soc. 148, pp. 87-93, 1974.

Kan extensions and systems of imprimitivity.

Armin Frei *)

Given a diagram $P \xrightarrow{K} T \xrightarrow{M'} A$ of functors we consider the following problems: When is M' a right Kan extension $\mathrm{Ran}_K M$ of some functor $M: P \longrightarrow A$? Can the functors M satisfying $M' \cong \mathrm{Ran}_K M$ be classified in a similar way as the representations inducing a given one are classified by Mackey's systems of imprimitivity? The first question has been answered in [F,K]; the sole reason for treating it again briefly is that it fits naturally in the discussion of the second. At the end of the paper we apply the general theory to a special situation in module theory.

All concepts used are V-concepts, where V is a bicomplete closed category and all right Kan extensions are pointwise, given by the Kan formula as in (2).

In the diagram

$$(1) \qquad P \xrightarrow{K} T \xrightarrow{D} S_K \xrightarrow{E} [P,V]^{op}$$

P denotes a small category, K any functor and S_K the <u>shape category of K</u>, given by $|S_K| = |T|$ and by $S_K(X,Y) = \mathrm{Nat}_{\cdot}(T(Y,K\cdot),T(X,K\cdot))$. D is the obvious extension of the identiy on objects to a functor and E is the embedding, identifying an object X in S_K with the functor $T(X,K\cdot)$.

Let A be a complete category (it actually suffices that A contain $\mathrm{Ran}_K F$ for all F in $[P,A]$) and $F: P \longrightarrow A$ a functor. By the definition of the formal Hom-functor (see [A], or [B,K] where that notion is called indexed limit) and the Yoneda lemma, the right Kan extension $F^K = \mathrm{Ran}_K F$ is given by

*) Supported by the Fonds National Suisse

(2) $F^K(-) = Hom_P(T(-,K),F)$.

F^K admits a canonical extension $F^K = \bar{F}D$ where

$$\bar{F}(\#) = Hom_P(E\#,F).$$

We recall that a functor with domain T which factors over D is said to be shape invariant. In turn, $\bar{F}$ admits a canonical extension $\bar{F} = \hat{F}E$ where

$$\hat{F}(*) = Hom_P(*,F)$$

which is clearly continuous.

The operations $(\ )^K$, $(\bar{\ })$ and $(\hat{\ })$ extend to functors in obvious ways and $(\hat{\ })$ is just the functor Ran_Y , where Y is the Yoneda embedding $Y: P \longrightarrow [P,V]^{op}$. Indeed

$$Ran_Y F(*) = Hom_P([P,V]^{op}(*,Y),F) = Hom_P(*,F) = \hat{F}(*).$$

As Y is fully faithful one has that $Y*(\hat{\ }) \cong Id[P,A]$.

On the other hand, for any continuous functor $M''' : [P,V]^{op} \longrightarrow A$ one has $M''' \cong M''' Ran_Y Y \cong Ran_Y(M'' Y)$ as Y is codense; thus $(\hat{\ })Y^* \cong Id(Cont[[P,V]^{op},A]$, where $Cont[[P,V]^{op},A]$ denotes the full subcategory of $[[P,V]^{op},A]$ consisting of continuous functors.

We also observe that $((ED)^*\hat{F})(-) = \hat{F}ED(-) = Hom_P(ED(-),F)$ $= Hom_P(\ (-,K),F) = F^K(-)$.

Summarizing we have

Theorem 1. Let A be complete. Then, with the notations above, the dia gramm

$$\begin{array}{ccc}
[P,A] & \xrightarrow{\quad \widehat{(\;)} \quad} & \mathrm{Cont}[[P,V]^{op},A] \\
\Vert & \searrow^{(\;)^K} \quad [T,A] \quad \swarrow_{(ED)^*} & \Vert \\
[P,A] & \xleftarrow[\quad Y^* \quad]{} & \mathrm{Cont}[[P,V]^{op},A]
\end{array}$$

commutes up to natural isomorphisms.

<u>Corollary 2.</u> A functor $M': T \longrightarrow A$ is of the form $M' \cong M^K$ for some $M: P \longrightarrow A$ if and only if it is of the form $M' \cong M''' \, ED$ with M''' continuous; furthermore for a given $M' = M''' \, ED$ with M''' continuous, the functor $M = M''' \, Y$ is, up to isomorphism, the unique one satisfying $M^K \cong M'$ and $M''' \cong \hat{M}$.

<u>Remark.</u> From the proof of the theorem we have:
If a functor $M''': [P,V]^{op} \longrightarrow A$ satisfies $M''' \, \mathrm{Ran}_Y Y \cong \mathrm{Ran}_Y M''' \, Y$ then $M'''(*) \cong \mathrm{Hom}_P(*, M''' \, Y)$. By the definition of Hom_P we have an isomorphism

$$A(A, M'''(*)) \cong A(A, \mathrm{Hom}_P(*, M''' \, Y)) \cong [P,V](*, A(A, M''' \, Y)) \cong [P,V]^{op}(A(A, M''' Y), *)$$

that is, M''' has a left adjoint, and hence preserves all limits that exist in $[P,V]^{op}$. When $A = V$, then $M'''(*) \cong \mathrm{Nat}(M''' \, Y, *)$ is representable. Hence the functors in $\mathrm{Cont}[[P,V]^{op},A]$ preserve all limits.

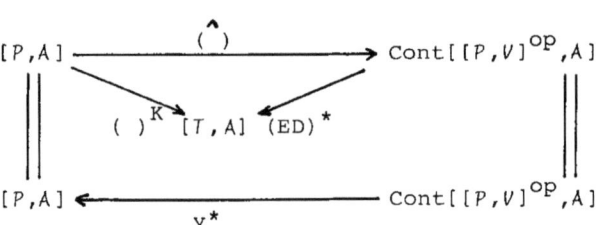

According to Theorem 1 the continuous extensions M''' of M' classify the candidates M satisfying $M' \cong M^K$. As $[P,V]^{op}$ is a rather large category it is preferable, where possible, to classify the candidates M by functors having domain S_K . We next investigate that possibility.

We call a functor M": $S_K \longrightarrow A$ a <u>system of imprimitivity</u> if M" = M"'E with M"' continuous and denote by Imps$[S_K,A]$ the full subcategory of $[S_K,A]$ consisting of systems of imprimitivity. The functor $(\bar{\ })$ clearly takes values in Imps$[S_K,A]$; we use the same symbol $(\bar{\ })$ for the corresponding functor with codomain Imps$[S_K,A]$. The functor E^*, restricted to Cont$[[P,V]^{op},A]$ takes values in Imps$[S_K,A]$, and we use E^* for the corresponding functor Cont$[[P,V]^{op},A] \longrightarrow$ Imps$[S_K,A]$. With this notation we have:

<u>Theorem 3.</u> In the situation of diagram (1) assume T, and hence S_K to be small. Then the following statements are equivalent:

(i) E^*: Cont$[[P,V]^{op},V] \longrightarrow$ Imps$[S_K,V]$ is an equivalence.

(ii) E^*: Cont$[[P,V]^{op},A] \longrightarrow$ Imps$[S_K,A]$ is an equivalence for all complete A.

(iii) $(\bar{\ })$: $[P,V] \longrightarrow$ Imps$[S_K,V]$ is an equivalence.

(iv) $(\bar{\ })$: $[P,A] \longrightarrow$ Imps$[S_K,A]$ is an equivalence for all complete A.

(v) E is codense.

Proof. (ii) implies (i) and (iv) implies (iii) trivially.
We have that $(E^* \circ (\hat{?}))(\#) = \mathrm{Hom}_P(E\#,?) = (\bar{?})(\#)$, that is, $E^*(\hat{\ }) = (\bar{\ })$. Since $(\hat{\ })$ is an equivalence, this entails that (i) is equivalent to (iii) and (ii) is equivalent to (iv). Next we show that (iii) implies (v):

For any $F,G\colon \mathcal{P} \longrightarrow \mathcal{V}$ we have an isomorphism

$\psi\colon [\mathcal{P},\mathcal{V}](F,G) \longrightarrow [\mathcal{P},\mathcal{V}](\mathrm{Ran}_E E(F),G)$ given by

$[\mathcal{P},\mathcal{V}](F,G)$

$\cong \mathrm{Imps}[\mathcal{S}_K,\mathcal{V}](\bar{F},\bar{G})$ via $(\bar{\ })$,

$\cong \mathrm{Imps}[\mathcal{S}_K,\mathcal{V}](\mathrm{Hom}_{\mathcal{P}}(E\#,F),\mathrm{Hom}_{\mathcal{P}}(E\#,G))$, by the definition of $(\bar{\ })$,

$\cong \mathrm{Ran}_E([\mathcal{P},\mathcal{V}]^{op}(G,E\#))(F)$, by the Kan formula,

$\cong [\mathcal{P},\mathcal{V}]^{op}(G,\mathrm{Ran}_E E(F))$ as representables preserve $\mathrm{Ran}_E E$,

$\cong [\mathcal{P},\mathcal{V}](\mathrm{Ran}_E E(F),G)$,

natural in F and G, which, by the Yoneda lemma is induced by a natural isomorphism $\varphi\colon \mathrm{Ran}_E E \longrightarrow \mathrm{Id}$. The counit of $\mathrm{Ran}_E E$ is φE, again an isomorphism, hence E is codense.

It remains to show that (v) implies (ii). If E is codense, that is, the counit of $\mathrm{Ran}_E E$ is an isomorphism, we may and do choose $\mathrm{Ran}_E E = \mathrm{Id}$ with counit $1\colon \mathrm{Id}\, E \longrightarrow E$.

For B in $\mathrm{Cont}[[\mathcal{P},\mathcal{V}]^{op},\mathcal{A}]$ we have that $B(1)$ is the unit for $B = \mathrm{Ran}_E(BE)$ and for any $A\colon [\mathcal{P},\mathcal{V}]^{op} \longrightarrow \mathcal{A}$

$$\mathrm{Nat}(A,B) \xrightarrow{E^*} \mathrm{Nat}(AE,BE) \xrightarrow{\mathrm{Nat}(AE,B(1))} \mathrm{Nat}(AE,BE)$$

is the isomorphism of the universal property of $\mathrm{Ran}_E(BE)$. This holds a fortiori if A is continuous, and, as the second arrow is an isomorphism so is E^*. It is clear from the definitions that E^* hits all objects of $\mathrm{Imps}[\mathcal{S}_K,\mathcal{A}]$, hence it is an equivalence.

If $M'\colon \mathcal{T} \longrightarrow \mathcal{A}$ is of the form $M' = M''D$ with M'' in $\mathrm{Imps}[\mathcal{S}_K,\mathcal{A}]$ we say that M' is <u>imprimitive</u> and that M" is a system of imprimitivity <u>associated</u> with M'.

From Theorems 1 and 3 we then have

<u>Corollary 4.</u> A functor $M'\colon \mathcal{T} \longrightarrow \mathcal{A}$ is of the form $M' \cong M^K$ for some $M\colon \mathcal{P} \longrightarrow \mathcal{A}$ if and only if it is imprimitive.

<u>Corollary 5.</u> The functor $(\bar{\ })$ is an equivalence if and only if E is codense; in this case for a given imprimitive M' with associated system of imprimitivity M" there is an M, unique up to isomorphism, with $M^K \cong M'$ and $\bar{M} \cong M''$.

The terminology used comes from the theory of representations of groups: Taking for $K\colon \mathcal{P} \longrightarrow \mathcal{T}$ the embedding $kH \longrightarrow kG$ of the group-algebras of finite groups $H \subset G$ considered as one-object categories, and taking $\mathcal{V}$ = k-Mod, the functor $(\ )^K$ corresponds to coinducting representations of H, considered as functors $kH \longrightarrow$ k-Mod. Our $\mathrm{Imps}[\mathcal{S}_K, \mathcal{A}]$ is then isomorphic to the category of systems of imprimitivity in the sense of Mackey. For details see [F], [K] and [M].

Theorem 3 has the following application in module theory. Let $K\colon P \longrightarrow T$ be a ringhomomorphism, interpreted as a functor between one-object categories. Here $\mathcal{V} = \mathcal{Ab}$. The shape category of K is the ring $S^{op} = (\mathrm{End}_P T)^{op}$ again taken as a one-object category and D becomes the ringhomomorphism which takes t in T to the left multiplication by t. Interpreting the P-modules as functors from P to $\mathcal{Ab}$, the category $[P\mathcal{V}]^{op}$ becomes $(P\text{-Mod})^{op}$ and the functor $E\colon S^{op} \longrightarrow (P\text{-Mod})^{op}$ takes the unique object of S^{op} to the P-module T. The functor $(\bar{\ })\colon$ P-Mod $\longrightarrow S^{op}$-Mod is given by $(\bar{-}) = \mathrm{Hom}_P(T,-)$. A Morita theorem (see for instance [P]) says that if T is a progenerator (that is a finitely generated projective generator) in P-Mod, then $(\bar{\ })$ is an equivalence. By Theorem 3 the functor E is then codense, hence the P-module T is dense in P-Mod. We thus have

Theorem 6. Let P be a ring and T a P-algebra. If T is a progenerator it is dense in P-Mod.

Remark. The discussion leading to Theorem 3 points out that when $(\bar{\ }): [P,A] \longrightarrow \mathrm{Imps}[S_K,A]$ is an equivalence it generalizes a Morita equivalence.

Bibliography

[A] C. Auderset, Adjonctions et monades au niveau des 2-catégories, Cahiers de Topol. et Géom. Diff., Vol. XV,1(1974).

[B,K] F. Borceux and G.M. Kelly, A notion of limit for enriched categories, Bull. Austral. Math. Soc. 12 (1975).

[F] A. Frei, Shape and induced representations, to appear in Quaestiones Mathematicae.

[F,K] A. Frei and H. Kleisli, A question in categorical shape theory: when is a shape-invariant functor a Kan extension?, Springer L.N. in Math. 719 (1979).

[K] H. Kleisli, Coshape-invariant functors and Mackey's induced representation theorem, Cahiers de Topol. et Géom.Diff., Vol. XXII-1 (1981).

[M] G.W. Mackey, Induced representations of groups and quantum mechanics, Benjamin-Boringhieri (1968).

[P] B. Pareigis, Kategorien und Funktoren, Math. Leitfäden, Teubner, (1969).

SMOOTH STRUCTURES

by

Alfred Frölicher

A smooth structure on a set S consists of a set $C \subset S^{\mathbb{R}}$ of curves and a set $F \subset \mathbb{R}^S$ of functions such that C and F determine each other by the condition that $F \circ C \subset C^\infty(\mathbb{R}, \mathbb{R})$. The sets with smooth structures yield a category $\mathcal{S}$ if we take as morphisms those maps which behave good with respect to the curves or, equivalently, with respect to the functions.

The main results presented here go in two directions : first we draw the attention to the excellent categorical properties of $\mathcal{S}$; and then we will show how classical calculus of C^∞-differentiable manifolds (not necessarily finite dimensional!) fits in this set-up. Within $\mathcal{S}$ one finds also objects with singularities, but we shall not at all go here in this direction.

Among the categorical properties there are those which we call the elementary ones, e.g. completeness and cocompleteness. They are easily obtained and are valid for each category which is generated in a similar way, replacing $\mathbb{R}$ by any fixed set B and $C^\infty(\mathbb{R}, \mathbb{R})$ by any submonoid of B^B. For other examples, cf [3] or [4].

For certain monoids the category is cartesian closed. A necessary and sufficient condition was given in [3]. The verification of this condition for the monoid $C^\infty(\mathbb{R}, \mathbb{R})$ is difficult and was first proved by Lawvere, Schanuel and Zame [12]. The functor yielding cartesian closedness will be constructed explicitly by describing the smooth structure of the function-spaces.

In §3 it will be shown that for any Fréchet-space and for many C^∞-Fréchet-manifolds the smooth curves together with the smooth functions form a smooth structure and that the C^∞-maps are exactly the $\mathcal{S}$-morphisms. The basic result in order to obtain this is a caracterization of smooth maps between Fréchet spaces which generalizes theorems of Boman [2], of Bochnak and Siciak [1] and of Hain [7]. The proof is facilitated by using a minimal caracterization of C^1-maps between locally convex spaces which does for instance not require the linearity of the differential at a point but nevertheless yields the usual results of calculus and in particular the usual notion of C^∞-maps.

In recent years several authors used with advantage cartesian closed categories in order to develop calculus for non-normed vector spaces : convergence structures were used in [6], compactly generated spaces in [14], arc-determined spaces in [11]. Which cartesian closed category is the natural one for calculus ? If one wants to study C^∞-maps, the answer seems clear : vector spaces with a compatible smooth structure. Some ideas and results in this direction are given in the last section.

The results presented here are very easy to formulate and to understand; however several proofs require hard analysis and thus can only be indicated in this expository article. All proofs were carried out in detail in a seminar on smooth functions at the University of Geneva; in particular another proof of the theorem of Lawvere, Schanuel and Zame due to H. Joris and the author was presented. I wish to express my gratitude for a very active participation and substantial contributions at this seminar in particular to Gonzalo Arzabe, Henri Joris and Oscar Pino-Ortiz.

§1 THE CATEGORY $\mathcal{S}$ OF SMOOTH SPACES AND ITS ELEMENTARY PROPERTIES

A smooth structure on a set S is a couple (C,F) where $C \subset S^{\mathbb{R}}$ and $F \subset \mathbb{R}^S$ such that the "duality" $C = D_*F$ and $F = D^*C$ holds, with

$$D_*F = \{c : \mathbb{R} \to S;\ f \circ c \in C^\infty(\mathbb{R},\mathbb{R}) \text{ for all } f \in F\};$$
$$D^*C = \{f : S \to \mathbb{R};\ f \circ c \in C^\infty(\mathbb{R},\mathbb{R}) \text{ for all } c \in C\}.$$

A smooth space is a triple (S,C,F) where S is a set and (C,F) is a smooth structure on it.

The smooth spaces form a category $\mathcal{S}$ for which, by definition, the morphisms from (S,C,F) so (S',C',F') are those maps $\alpha : S \to S'$ which satisfy $\alpha_*(C) \subset C'$ or, equivalently $\alpha^*(F') \subset F$.

The set of smooth structures on a fixed set S is ordered in the usual manner : (C,F) is called finer than (C',F') if the identity map of S is a morphism from (S,C,F) to (S,C',F'). For any set $C_0 \subset S^{\mathbb{R}}$ there is a finest structure (C,F) on S such that $C_0 \subset C$; it is called the structure generated by C_0 and is obtained as follows : $F = D^*C_0$; $C = D_*F$. Similarly one has the structure (C,F) generated by any set $F_0 \subset \mathbb{R}^S$: it is the coarsest structure with $F_0 \subset F$ and is obtained as $C = D_*F_0$; $F = D^*C$. It follows that the smooth structures of a fixed set S form a complete lattice, that the forgetful functor from $\mathcal{S}$ to sets has a left and a right adjoint, and that

$\mathcal{S}$ is complete and cocomplete, limits or colimits being obtained (as in the category of topological spaces) by taking them in the category of sets and then putting the initial resp. final structure on them. We note in particular that the product of smooth spaces (S_i, C_i, F_i), $i \in I$, is the object (S,C,F) with $S = \underset{i\in I}{\times} S_i$ and C consisting of those curves $c : \mathbb{R} \to S$ whose component $c_i : \mathbb{R} \to S_i$ belong for all $i \in I$, to C_i.

An object (S,C,F) of $\mathcal{S}$ is called <u>separated</u> if for all $a \neq b \in S$ there exists $f \in F$ with $f(a) \neq f(b)$. $\mathcal{S}_{sep}$ denotes the respective full subcategory of $\mathcal{S}$. The inclusion functor has an obvious left adjoint and it follows that $\mathcal{S}_{sep}$ is also complete and cocomplete. The forgetful functor from $\mathcal{S}_{sep}$ to sets still commutes with limits, but not with all colimits. The one point set with its unique smooth structure is obviously a final object of $\mathcal{S}$, and it also yields a representation of the forgetful functor from $\mathcal{S}$ to sets.

Another important object is the triple $(\mathbb{R}, C^{\infty}(\mathbb{R},\mathbb{R}), C^{\infty}(\mathbb{R},\mathbb{R}))$. It will be denoted simply by $\mathbb{R}$ and is generator and cogenerator of $\mathcal{S}_{sep}$. For any object $X = (S_X, C_X, F_X)$ of $\mathcal{S}$ we have :

$$C_X = \mathcal{S}(\mathbb{R}, X) \ ; \qquad F_X = \mathcal{S}(X, \mathbb{R}) .$$

The results of this section do not depend on the nature of the monoid $C^{\infty}(\mathbb{R},\mathbb{R})$. As it was shown in [3], they hold for the category $\mathcal{K}$ generated analogously by any monoid M of maps of any set B to itself (i.e. M a submonoid of B^B). Cartesian closedness of $\mathcal{K}$ however depends on M; in [3] a necessary and sufficient condition was given. In the following section we discuss this condition and its verification for the case $B = R$, $M = C^{\infty}(\mathbb{R},\mathbb{R})$.

§2 <u>CARTESIAN CLOSEDNESS OF $\mathcal{S}$ AND $\mathcal{S}_{sep}$</u>

From the mentioned properties of the one-point object it follows that if there is a functor $H : \mathcal{S}^{op} \times \mathcal{S} \to \mathcal{S}$ yielding cartesian closedness, then it can be chosen such that the underlying set of $H(Y,Z)$ is the function space $\mathcal{S}(Y,Z)$. In particular one must get on $\mathcal{S}(\mathbb{R},\mathbb{R}) = C^{\infty}(\mathbb{R},\mathbb{R})$ a smooth structure (Γ,Φ) such that, with $\tilde{\gamma}(x,y) := \gamma(x)(y)$,

$$\Gamma = \{ \gamma : \mathbb{R} \to C^{\infty}(\mathbb{R},\mathbb{R}) \ ; \ \tilde{\gamma} : \mathbb{R}\,\pi\,\mathbb{R} \to \mathbb{R} \text{ a morphism } \}$$

and $\Phi = D^*\Gamma$. Since trivially $\Gamma \subset D_*(D^*\Gamma)$ the couple (Γ,Φ) will be a smooth structure iff $D_*(D^*\Gamma) \subset \Gamma$. Let us discuss the meaning of this condition. According to the discription of products in $\mathcal{S}$, a map $G : \mathbb{R}^2 \to \mathbb{R}$ is a morphism

$\mathbb{R}\pi\mathbb{R} \to \mathbb{R}$ iff for all $\sigma,\tau \in C^\infty(\mathbb{R},\mathbb{R})$ one has $G\circ(\sigma,\tau) \in C^\infty(\mathbb{R},\mathbb{R})$, i.e. iff G is smooth along all smooth curves (σ,τ) of $\mathbb{R}^2$. According to a remarkable theorem of Boman [2] this is equivalent to $G \in C^\infty(\mathbb{R}^2,\mathbb{R})$ and hence $\gamma \mapsto \tilde{\gamma}$ yields a bijection $\Gamma \to C^\infty(\mathbb{R}^2,\mathbb{R})$. From this and the definition of ϕ as $D^*\Gamma$ we get

$$\Phi = \{\varphi : C^\infty(\mathbb{R},\mathbb{R}) \longrightarrow \mathbb{R};\ x \mapsto \varphi(G(x,-)) \text{ is in } C^\infty(\mathbb{R},\mathbb{R}) \text{ for all } G \in C^\infty(\mathbb{R}^2,\mathbb{R})\}$$

We call the elements of Φ smooth functionals. One does not know all of them explicitly; however for the linear ones one has

Proposition 1. The linear smooth functionals φ have compact support (i.e. for each φ there exists a compact K of $\mathbb{R}$ with the property $f_1 | K = f_2 | K \Rightarrow \varphi(f_1) = \varphi(f_2)$) and satisfy

$$\varphi(\lim_{n\to\infty} f_n) = \lim_{n\to\infty} \varphi(f_n)$$

if $f_1, f_2, \ldots$ is a sequence in $C^\infty(\mathbb{R},\mathbb{R})$ such that for all $k \geq 0$ the derivatives $f_n^{(k)}$ converge locally uniformly for $n \to \infty$. This means that the linear smooth functionals are exactly the distributions of compact support.

This result is due to van Que and Reyes [13] ; it can be proved by constructing for convenient subsequences $f_{n_1}, f_{n_2}, \ldots$ a function $G \in C^\infty(\mathbb{R}^2,\mathbb{R})$ such that $G(1/k,y) = f_{n_k}(y)$ and $G(0,y) = \lim_{n\to\infty} f_n(y)$.

Proposition 2. If $G : \mathbb{R}^2 \to \mathbb{R}$ satisfies

I For all $x \in \mathbb{R}$, $G(x,-) \in C^\infty(\mathbb{R},\mathbb{R})$,

II For all linear smooth functionals φ, $x \mapsto \varphi\, G(x,-)$ is in $C^\infty(\mathbb{R},\mathbb{R})$,

then $G \in C^\infty(\mathbb{R}^2,\mathbb{R})$.

This important result was proved by Lawvere, Schanuel and Zame [12]. It can be proved by showing first the continuity of G and its first order partial derivatives $\partial_1 G$ and $\partial_2 G$. This is quite delicate for $\partial_1 G$ and we found it useful to show first that $\partial_1 G$ is partially continuous in the second variable and $\partial_1\partial_1 G$ is locally bounded. Once one has obtained the continuous differentiability of G, the proof is completed by showing that $\partial_1 G$ and $\partial_2 G$ satisfy the same conditions I and II ; for $\partial_2 G$ this is easy, and for $\partial_1 G$ one makes use of proposition 1.

From proposition 2 is follows immediately that (Γ,Φ) is a smooth structure on $C^\infty(\mathbb{R},\mathbb{R})$. We remark that for this it would be enough to prove proposition 2 under the asumption that I holds and II holds for all smooth functionals φ. However, if one would allow non-linear ones in the proof that G or $\partial_1 G$ or $\partial_2 G$ are continuous it would be hard to get further, because one does not have the analogue of proposition 1 for non-linear smooth functionals

Theorem. The category $\mathcal{S}$ of smooth spaces is cartesian closed. The same holds for $\mathcal{S}_{sep}$.

The function-space structure can be described explicitly In fact, for objects Y,Z of $\mathcal{S}$ one defines on the function-space $\mathcal{S}(Y,Z)$ a structure (C,F) by

$$C = \{d : \mathbb{R} \to \mathcal{S}(Y,Z);\ \tilde{d} : \mathbb{R}\,\pi\, Y \to Z \text{ a morphism}\}$$
$$F = D^*C.$$

Using that (Γ,Φ) is a smooth structure on $C^\infty(\mathbb{R},\mathbb{R})$ it is easy to show that (C,F) is actually a smooth structure. Denoting the smooth space formed by the set $\mathcal{S}(Y,Z)$ with this structure by H(Y,Z) it is straightforward to show that one has the universal property

$$\alpha : X \to H(Y,Z) \text{ a morphism} \iff \tilde{\alpha} : X\,\pi\, Y \to Z \text{ a morphism}$$

and this yields functoriality of H and cartesian closedness of $\mathcal{S}$.

Since Z separated implies H(Y,Z) separated, cartesian closedness of $\mathcal{S}_{sep}$ is obtained by restriction of the functor H.

Because in proposition 2 only linear smooth functionals are used it follows easily that the structure of H(Y,Z) is generated by the functions $\mathcal{S}(Y,Z) \to \mathbb{R}$ of the form $\alpha \mapsto \varphi(f\circ\alpha\circ c)$ where φ is a distribution of compact support, $c \in C_Y = \mathcal{S}(\mathbb{R},Y)$ and $f \in F_Z = \mathcal{S}(Z,\mathbb{R})$.

§3 THE SMOOTH STRUCTURE OF FRECHET SPACES AND MANIFOLDS.

For $\mathbb{R}^n$ the couple $\left(C^\infty(\mathbb{R},\mathbb{R}^n),\ C^\infty(\mathbb{R}^n,\mathbb{R})\right)$ is a smooth structure. This is not at all trivial, but it is equivalent to Bomans theorem already quoted [2], which says that a function on $\mathbb{R}^n$ is smooth if it is smooth along all smooth curves. Using partitions of unity one gets a more general result : for any finite dimensional paracompact C^∞-manifold V, $\left(C^\infty(\mathbb{R},V),\ C^\infty(V,\mathbb{R})\right)$ is a smooth structure on V. So every such manifold can be considered as a smooth space, and the C^∞-maps between them are exactly the $\mathcal{S}$-morphisms. In order to get the same results for a greater class of vector spaces and manifolds we need first of all a theorem which generalizes Boman's result ;

this theorem will at the same time generalize results of Bochnak-Siciak [1] and Hain [7].

The following set-up for calculus between locally convex spaces is useful (E,F will in this section always denote separated locally convex spaces).

Definition. A map $f : E \to F$ between locally convex spaces is called of class C^1 if for all $x,h \in E$.

$$df(x,h) := \underset{\lambda\to 0}{\text{w-lim}}\ 1/\lambda \cdot \big(f(x+\lambda h)-f(x)\big)$$

exists and yields a continuous map $df : E\times E \to F$.

By w-lim we mean the (unique) limit in F with respect to the weak topology.

Hence $df(x,h)$ is caracterized by

$$\lim_{\lambda\to 0} \frac{1}{\lambda}\cdot \big((l\circ f)(x+\lambda h) - (l\circ f)(x)\big) = l\big(df(x,h)\big)$$

for all $l \in F'$, F' being the topological dual of F.

We require so little because it is enough to get the usual properties, in particular the linearity of $df(x,-)$. In fact one has :

Proposition 1. If $f : E \to F$ is of class C^1, then

$$\lim_{\substack{\lambda\to 0\\ x\to x_0\\ h\to h_0}} 1/\lambda\cdot \big(f(x+\lambda h) - f(x)\big) = df(x_0,h_0)$$

We remark that here the limit is with respect to the topology of F (and not only with respect to the weak topology!), and the fact that this simultaneous limit exists shows that we are close to what is sometimes called strict differentiability. As easy consequences we have :

Corollary. If f is of class C^1, then f is continuous and $df(x,-)$ is linear.

In order to prove proposition 1 the following "mean value theorem" is useful :

Proposition 2. Let $A \subset E$ be convex and closed; $c,d : I \to E$ maps of an open interval $I \subset \mathbb{R}$ into E related by $(l\circ c)^{\cdot}(\lambda) = l(d(\lambda))$ for all $\lambda \in I$ and $l \in E'$ (special case : $d = c^{\cdot}$). Then for any $\alpha<\beta$ of I one has :

$$d(\lambda) \in A \text{ for } \alpha<\lambda<\beta \Rightarrow c(\beta) - c(\alpha) \in (\beta-\alpha)\cdot A\ .$$

The proof is a simple application of the Hahn-Banach theorem.

For a map $f : E_1\times\ldots\times E_n \to F$ the notion "partially of class C^1" is defined in the usual way; i.e. the partial differentials $d_i f: E_1 \times \ldots \times E_n \times E_i \to F$ have to exist and have to be continuous in all variables It follows as usual that f is of class C^1 if and only if it is partially of class C^1.

Inductively we define now maps of class C^n :

Definition. $f : E \to F$ is of class C^{n+1} if it is of class C^1 and df is of class C^n. f is of class C^∞ if it is of class C^n for all $n \in N$.

For d^n there is a chain rule. It is complicated and for this the operator T behaves much better; it is defined as

$$Tf : E \times E \to F \times F$$
$$(x,h) \mapsto (f(x), df(x,h))$$

The chain rule then says : if $f : E \to F$ and $g : F \to G$ are of class C^n, then also $g \circ f$, and $T^n(g \circ f) = T^n g \circ T^n f$.

Since for f of class C^1 the map df is linear (and continuous) in the second variable, only the first partial differential is of interest and yields a map $D^2 f : E \times E \times E \to F$. $D^2 f$ exists and is continuous iff f is of class C^2, and then $D^2 f(x,-,-)$ is bilinear and symmetric. Recursively on defines $D^1 f$ as df and D^{n+1} as the first partial differential of $D^n f$. One shows that f is of class C^n if and only if $D^1 f, \ldots, D^n f$ exist and are continuous, and then $D^n f(x,-,\ldots,-)$ is n-multilinear and symmetric.

If we want to show that a map f of class C^{n-1} is even of class C^n, we must verify existence and continuity of the first partial differential of $D^{n-1} f$. If we suppose that for all maps $g : \mathbb{R}^n \to E$ of the form

$$(\lambda_1, \ldots, \lambda_n) \mapsto a + \lambda_1 \cdot h_1 + \ldots + \lambda_n \cdot h_n$$

the composite map $f \circ g$ is of class C^n, we get easily the existence of $D^n f(x,-,\ldots,-)$ and its multilinearity and symmetry. It is then enough to show that the map $(x,h) \to D^n f(x,h,\ldots,h)$ is continuous. Under the asumption that E admits a denumerable basis for the zero-neighborhoods (i.e.if E is metrizable) we can show this continuity if we suppose that $f \circ g$ is of class C^n for all $g : \mathbb{R}^2 \to E$ of class C^∞, using the following lemma (cf. [7]) :

Lemma. Let E be metrizable and $a_0 = \lim_{n\to\infty} a_n$, $b_0 = \lim_{n\to\infty} b_n$ be limits of convergent sequences in E. Then there exist a strictly increasing function $\sigma : \mathbb{N} \to \mathbb{N}$, a sequence of reals λ_n with a limit $\lambda_0 = \lim_{n\to\infty} \lambda_n$, and a function

$g : \mathbb{R}^2 \to E$ of class C^∞ such that

$$g(\lambda_n, \mu) = a_{\sigma(n)} + \mu \cdot b_{\sigma(n)} \qquad \text{for all } n \in \mathbb{N} \cup \{0\}.$$

This also explains why in the following proposition we assume $n \geq 2$ (in fact it is false for $n = 1$).

Proposition 3. Let $f : E \to F$ be a map between locally convex spaces; E metrizable ; $n \in \mathbb{N}$, $n \geq 2$. Then f is of class C^n if and only if $f \circ g : \mathbb{R}^n \to F$ is of class C^n for all $g : \mathbb{R}^n \to E$ of class C^∞

This proposition is better than the respective result of Hain [7]. In order to show that f is of class C^n, Hain (who restricts to Banach spaces) supposes that $f \circ g$ is of class C^{n+1} for all $g : \mathbb{R}^{n+1} \to E$ of class C^∞. He could not get the above result because he uses the classical notion of "map of class C^n " , which we shall call here "Fréchet-C^n " .

There is only a slight difference between our notion C^n and Fréchet-C^n. If $f : E \to F$ is of class C^n, then the map

$$f^{(n)} : E \to L_n(E;F)$$

has the property that

$$\widetilde{f^{(n)}} = D^n f : E \times \ldots \times E \to F$$

is continuous, and this does not imply the continuity of $f^{(n)}$ with respect to the norms. However, if f is of class C^n and $f^{(n)}$ <u>is</u> norm-continuous, then f is Fréchet-C^n; and this is in particular true if f is of class C^{n+1}. Hence :

$$C^{n+1} \Rightarrow \text{Fréchet-}C^n \Rightarrow C^n$$

This shows that for Banach spaces our C^∞-notion coincides with the classical notion of Fréchet-C^∞. For not-normable spaces many different notions "of class C^n" exist; but almost all of them yield the same notion "of class C^∞" as the one we use (cf. [10]).

Proposition 4. Let $g : \mathbb{R}^n \to F$ be such that $l \circ g$ is of class C^{n+1} for all $l \in F'$ (F' the topological dual of F) and suppose that F is complete (in fact it is enough to assume that the Mackey-topology of F is sequentially complete, or, still weaker, that F is locally complete, (cf [9]). Then g is of class C^n.

Using our set-up the proof is almost the same as that given by Bochnak and Siciak who gave this proposition in the case $n = 1$, cf. [1].

Combining Boman's theorem with propositions 3 and 4 one gets easily the following theorem, announced in [5] :

Theorem 1 . Let $f : E \to F$ be a map between locally convex spaces and suppose that E is metrizable and F (locally) complete. Then the following conditions are equivalent :

1) f is of class C^∞

2) $f_*(C^\infty(\mathbb{R},E)) \subset C^\infty(\mathbb{R},F)$

3) $f^*(C^\infty(F,\mathbb{R})) \subset C^\infty(E,\mathbb{R})$

4) $f^*(F') \subset C^\infty(E,\mathbb{R})$

Corollary. For any Fréchet space E, the couple $(C^\infty(\mathbb{R},E)\ ,\ C^\infty(E,\mathbb{R}))$ is a smooth structure on E. Hence Fréchet spaces can be considered as smooth spaces and the $\mathcal{S}$-morphisms between them are exactly the maps of class C^∞.

If we want to get objects of $\mathcal{S}$ from Fréchet manifolds we must make sure that partitions of unity exist.

Theorem 2. Let V be a paracompact Fréchet manifold modelled over a Fréchet space E which has the property that to each neighborhood V of zero there exists a C^∞-function $f : E \to \mathbb{R}$ with $f(0) = 1$ and $f(x) = 0$ for $x \notin V$. Then $(C^\infty(\mathbb{R},V),\ C^\infty(V,\mathbb{R}))$ is a smooth structure on V. The $\mathcal{S}$-morphisms between such spaces are exactly the maps of class C^∞.

Remark. According to the cartesian closedness of $\mathcal{S}$ we get for any such Fréchet-manifolds V,W the natural smooth structure on the function space $C^\infty(V,W)$. This structure can be described explicitly in a simple way and has the universal property that for any such manifold X a map $f : X \to C^\infty(V,W)$ is of class C^∞ iff $\tilde{f} : X \sqcap V \to W$ is of class C^∞. One can then ask : when is $C^\infty(V,W)$ again a manifold ? Of course Fréchet manifolds are not the natural things to look at in this set-up. More natural are manifolds modelled over vector spaces equipped with a compatible smooth structure. The last section will give some ideas in this direction.

§4 CALCULUS FOR SMOOTH VECTOR SPACES

A smooth vector space is a vector space with a compatible smooth structure, i.e. such that the vector space operations are $\mathcal{S}$-morphisms. For an arbitrary object $X = (S,C,F)$ of $\mathcal{S}$ the function set F, being identified with $H(X, \mathbb{R})$, is a smooth vector space. If we define, for $p \in S$, in the usual way an equivalence relation "$\sim_p$" on F by

$$f \sim_p g \iff (f\circ c)^{\cdot}(0) = (g\circ c)^{\cdot}(0) \text{ for all } c \in C \text{ with } c(0) = p$$

then the quotient is, due to cartesian closedness of $\mathcal{S}$, also a smooth vector space, called the cotangent space of X at p. Similarly we can introduce the tangent space of X at p as a quotient of the space $\{c \in C;\ c(0) = p\}$; it is not, in general, a vector space, but can be imbedded as a subspace into the smooth vector space formed by the derivations of F with respect to the evaluation at p. We do not go further in this direction here; but these remarks show that smooth vector spaces come in a natural way. It does not seem very useful to put a topology on them.

The question we want to study is the following : are the $\mathcal{S}$-morphisms between smooth vector spaces indefinitely differentiable in the usual sense ? In order to obtain such a result, some restrictions on the spaces seem useful.

Let E be a smooth vector space, (C_E, F_E) its smooth structure. We put

$$E' := E^* \cap F_E$$

where E^* notes the algebraic dual of E. So E' is the vector space of the real-valued linear smooth functions on E.

Definition. The smooth vector space E is called convenient if E' separates points, generates the smooth structure and yields a complete bornology on E; cf [8].

This completeness condition is equivalent to the condition that any locally convex topology on E yielding E' as topological dual is locally complete; cf [9]. According to the results of §3 we see that for any Fréchet space E the natural smooth structure $\left(C^{\infty}(\mathbb{R},E),\ C^{\infty}(E,\mathbb{R})\right)$ is convenient.

If $c : \mathbb{R} \to E$ is a $\mathcal{S}$-morphism from $\mathbb{R}$ to a convenient smooth vector space E (i.e. $c \in C_E$) one deduces from the separation and the completeness hypothesis that there exists a unique map, denoted by $c^{\cdot}$, from $\mathbb{R}$ to E such that

$$l \circ c^{\cdot} = (l \circ c)^{\cdot} \quad \text{for all } l \in E'.$$

From the other asumption (that E' generates the smooth structure of E) it follows then immediately that $c^{\cdot} : \mathbb{R} \to E$ is also a $\mathcal{S}$-morphism. Inductively we obtain $\mathcal{S}$-morphisms $c^{(n)} : \mathbb{R} \to E$ for $n \in \mathbb{N}$ and we see that c is indefinitely differentiable in the usual sense with respect to any locally convex topology on E yielding E' as topological dual. Moreover one verifies that the (linear) map $H(\mathbb{R},E) \to H(\mathbb{R},E)$ sending c into $c^{\cdot}$ is a $\mathcal{S}$-morphism.

Using this we get similar results for the general case :

Theorem 1. Let $\alpha : E_1 \to E_2$ be a $\mathcal{S}$-morphism between convenient smooth vector spaces. Then the map

$$d\alpha : E_1 \pi E_1 \to E_2$$

defined by $d\alpha(a,h) = (\alpha \circ c_{a,h})^{\cdot}(0)$ where $c_{a,h}(\lambda) = a + \lambda h$ is also a $\mathcal{S}$-morphism. For any $a \in E_1$ the map $d\alpha(a,-)$ is linear. The (linear) map $H(E_1,E_2) \to H(E_1 \pi E_1, E_2)$ sending α into $d\alpha$ is also a $\mathcal{S}$-morphism.

If E_2' separates points of E_2, then obviously $\big(H(E_1,E_2)\big)'$ separates points of $H(E_1,E_2)$. And if E_2' generates the smooth structure of E_2, then (cf the remark at the end of §2) the functions $H(E_1,E_2) \to \mathbb{R}$ of the form $\alpha \mapsto \varphi(l \circ \alpha \circ c)$ for $l \in E_2'$, $c \in C_{E_1}$ and φ a distribution of compact support generate the smooth structure of $H(E_1, E_2)$; since these functions are linear $\big(H(E_1,E_2)\big)'$ certainly generates the structure of $H(E_1,E_2)$. By showing in addition that $H(E_1,E_2)$ satisfies the completeness condition provided E_2 does, one sees that for E_2 convenient also $H(E_1, E_2)$ is convenient. Hence we have :

Theorem 2. The category formed by the convenient smooth vector spaces with the $\mathcal{S}$-morphisms is, by restriction of the functor H, cartesian closed.

Other properties of that category as well as the category formed by the same objects but with only the linear $\mathcal{S}$-morphisms are being studied; in particular duality and reflexivity questions.

By introducing the spaces $L_n(E_1,E_2)$ of n-multilinear $\mathcal{S}$-morphisms $E_1 \pi \ldots \pi E_1 \to E_2$ one can of course introduce for a $\mathcal{S}$-morphism $\alpha : E_1 \to E_2$ between convenient smooth vector spaces the maps $\alpha^{(n)} : E_1 \to L_n(E_1,E_2)$ which are also $\mathcal{S}$-morphisms, and one has the usual relations between the higher derivatives $\alpha^{(n)}$ and the higher differentials $d^n\alpha$.

Added in proof.

The convenient smooth vector spaces can be identified with the spaces considered by A. Kriegl ("Die richtigen Räume für Analysis im unendlich-dimensionalen", preprint, Vienna 1981, to appear in Monatshefte für Mathematik), namely the separated locally convex spaces which are bornological and locally complete.

R E F E R E N C E S

[1] J. Bochnak and J. Siciak : "Analytic functions in topological vector spaces" Studia Math. 39, 1971, p. 77-112

2 J. Boman : "Differentiability of a function and of its compositions with functions of one variable", Math. Scand. 20, 1967, p. 249-268.

3 A. Frölicher : "Durch Monoide erzeugte kartesisch abgeschlossene Kategorien", Seminarberichte aus dem Fachber. Math. der Fernuniversität Hagen 5, 1979, p. 7-48 .

4 A. Frölicher : "Catégories carésiennement fermées engendrées par des monoîdes", Cahiers de Top. et Géom. diff. XXI/4, 1980, p. 367-375.

5 A. Frölicher : "Applications lisses entre espaces et variétés de Fréchet C.R. Ac. Sci. Paris 293, 1981, p. 125-127.

6 A. Frölicher and W. Bucher : "Calculus in Vector Spaces without Norm", Lecture Notes in Math. 30, Springer 1966.

7 R.M. Hain : "A characterization of smooth functions defined on a Banach space", Proc. Am. Math. Soc. 77, 1979, p. 63-67.

8 H. Hogbe-Nlend : "Bornologies and functional analysis", Mathematics Studies 26, North-Holland 1977.

9 H. Jarchow : "Locally convex Spaces", Teubner 1981.

10 H.H. Keller :"Differential Calculus in locally convex spaces", Lecture Notes in Math. 417, Springer 1974.

11 A. Kriegl : "Eine Theorie glatter Mannigfaltigkeiten und Vekorbündel", Dissertation, Wien 1980.

12 F.W. Lawvere, S.H. Schanuel and W.R. Zame : "On C^∞ Function Spaces", Preprint 1981.

13 N. Van Que and G. Reyes : "Théorie des distributions et théorèmes d'extension de Whitney", Exposé 8, Géom. diff. synth. fasc. 2, Rapport de Recherches DMS 80-12, Université de Montréal 1980.

14 U. Seip : "A convenient Settino for Smooth Manifolds". J. of pure and appl. Algebra 21, 1981, p. 279-305.

Section de Mathématiques
Université de Genève
2-4, rue du Lièvre

CH-1211 GENEVE 24

Enriched algebras, spectra and homotopy limits

John W. Gray

0. Introduction. The purpose of this paper is the same as that of [5]; to show how certain properties of homotopy limits are consequences of what either are or should be standard facts about categories enriched in a closed category. The property to be explained here is as follows: in [16], Thomason shows that the degreewise homotopy limit of a diagram of pointed simplicial spectra is a pointed simplicial spectrum. He calls this the homotopy limit in the category of such spectra. It is eminantly reasonable to suppose that, in fact, it is the homotopy limit in this category, but two things have to be proved. i). The category Spec $\underline{K}_*$ of pointed simplicial spectra is a complete simplicial category, since only such categories have homotopy limits. ii). The component projections pr_n : Spec $\underline{K}_* \to \underline{K}_*$, for each degree n, have simplicially enriched left adjoints, and hence preserve homotopy limits.

The required tools are mostly at hand for ordinary categories in the form of known properties of the category Dyn S of algebras for an endofunctor S of a category $\underline{A}$. In Section 1 these tools are sharpened and extended to the case of enriched categories. In Section 2 Spec $\underline{A}$ is described for an arbitrary complete $\underline{V}$-category ($\underline{V}$ a closed category) and a pair of $\underline{V}$-adjoint functors $\Sigma \dashv \Omega$. Finally, in Section 3, these results are specialized to pointed simplicial spectra. Note that the spectra treated here are those for which $X_n \to \Omega X_{n+1}$ is an isomorphism.

1. $\underline{V}$-categories. Throughout this section $\underline{V}$ denotes a complete, cocomplete, symmetric, monoidal closed category. $\underline{V}$-cat denotes the category of $\underline{V}$-enriched categories and functors, regarded both as a symmetric, monoidal closed category itself and as a 2-category in which the 2-cells are $\underline{V}$-natural transformations; i.e., natural transformations $t : F \Rightarrow G : \underline{A} \to \underline{B}$ between $\underline{V}$-functors such that for all A and B, the diagrams

$$\begin{array}{ccc}
\underline{A}(A,B) & \xrightarrow{F_{A,B}} & \underline{B}(FA,FB) \\
{\scriptstyle G_{A,B}}\downarrow & & \downarrow{\scriptstyle (1,t_B)} \\
\underline{B}(GA,GB) & \xrightarrow[(t_A,1)]{} & \underline{B}(FA,GB)
\end{array}$$

commute. For basic information, see [5], [8] and references therein.

1.1. Proposition. $\underline{V}$-cat is a complete 2-category.

Proof: It is well known that $\underline{V}$-cat has limits. Thus it is sufficient to show that it has cotensors with the arrow category $\underline{2}$ (cf. [21]). If $\underline{A} \in \underline{V}$-cat then $\underline{2} \pitchfork \underline{A}$ is the $\underline{V}$-category whose objects are morphisms in $\underline{V}$. If $f : A \to B$ and $f' : A' \to B'$ are two such, then $\underline{2} \pitchfork \underline{A}(f,f') \in \underline{V}$ is the pullback

$$\begin{array}{ccc}
\underline{2} \pitchfork \underline{A}(f,f') & \xrightarrow{d_1} & \underline{A}(B,B') \\
{\scriptstyle d_0}\downarrow & & \downarrow{\scriptstyle (f,1)} \\
\underline{A}(A,A') & \xrightarrow[(1,f')]{} & \underline{A}(A,B')
\end{array}$$

It is easily checked that there is a natural bijection between V-natural transformations $t : F \Rightarrow G : \underline{A} \to \underline{B}$ and $\underline{V}$-functors $\bar{t} : \underline{A} \to \underline{2} \pitchfork \underline{B}$ such that $\theta\bar{t} = t$ where $\theta : d_0 \to d_1$ in the $\underline{V}$-natural transformation such that $\theta_f = f$.

1.2. Proposition. Let

$$\begin{array}{ccc}
\underline{B} & \xrightarrow{K} & \underline{B}' \\
{\scriptstyle F}\uparrow\downarrow{\scriptstyle U} & & {\scriptstyle F'}\uparrow\downarrow{\scriptstyle U'} \\
\underline{A} & \xrightarrow[H]{} & \underline{A}'
\end{array}$$

be a diagram of $\underline{V}$-functors such that there are $\underline{V}$-adjunctions $F \dashv U$, $F' \dashv U'$. Then there is a natural bijection between $\underline{V}$-natural transformation $\theta : HU \Rightarrow U'K$ and $\theta^{\#} : F'H \Rightarrow KF$ such that for all A and B the diagrams

$$D : \quad \begin{array}{ccccc}
\underline{A}(A,UB) & \xrightarrow{H} & \underline{A}'(HA,HUB) & \xrightarrow{(1,\theta_B)} & \underline{A}'(HA,U'KB) \\
\Big\| & & & & \Big\| \\
\underline{B}(FA,B) & \xrightarrow{K} & \underline{B}'(KFA,KB) & \xrightarrow[(\theta^{\#}_A,1)]{} & \underline{B}'(F'HA,KF)
\end{array}$$

commute (cf. [19]).

Proof: Given $\theta : HU \to U'K$, then $\theta^{\#} = \varepsilon' KF \circ F'\theta F \circ F'H\eta$, where $\eta : \underline{A} \Rightarrow UF$, $\varepsilon : FU \Rightarrow \underline{B}$, $\eta' : \underline{A}' \Rightarrow U'F'$, and $\varepsilon' : F'U' \to \underline{B}'$ are the adjunction natural transformations. See [4], I, 6.6 for the proof that this establishes a bijection as indicated. The diagram D commutes because of the commutativity of Figure 1 in which the sides, top and bottom are the sides, top and bottom of D. The regions labeled * commute by the definition of $\underline{V}$-naturality. The other regions commute trivially.

1.3. **Definition**: 1) Let $S : \underline{A} \to \underline{A}$ be a $\underline{V}$-endofunctor. The comma category $S \downarrow \underline{A}$ is the $\underline{V}$-category whose objects are morphisms $\varphi : SA \to B$. If $\varphi' : SA' \to B'$ then $S \downarrow \underline{A}(\varphi,\varphi') \in \underline{V}$ is the pullback

$$\begin{array}{ccc} S \downarrow \underline{A}(\varphi,\varphi') & \xrightarrow{p_1} & \underline{2} \pitchfork \underline{A}(\varphi,\varphi') \\ p_0 \downarrow & & \downarrow d_0 \\ \underline{A}(A,A') & \xrightarrow{S} & \underline{A}(SA,SA') \end{array}$$

Dually $\underline{A} \downarrow S$ has as objects morphisms $\varphi : A \to SB$ and $\underline{A} \downarrow S(\varphi,\varphi')$ is the pullback of the diagram

$$\begin{array}{ccc} \underline{A} \downarrow S(\varphi,\varphi') & \xrightarrow{p_1} & \underline{A}(B,B') \\ p_0 \downarrow & & \downarrow S \\ \underline{2} \pitchfork \underline{A}(\varphi,\varphi') & \xrightarrow{d_1} & \underline{A}(SB,SB') \end{array}$$

ii) Dyn S denotes the $\underline{V}$-category of S-algebras in $\underline{A}$. Its objects are morphisms $\varphi : SA \to A$ and Dyn $S(\varphi,\varphi') \in \underline{V}$ is the equalizer of the two morphisms

$$S \downarrow \underline{A}(\varphi,\varphi') \begin{array}{c} \xrightarrow{d_1 p_1} \\ \xrightarrow[d_0]{} \end{array} \underline{A}(A,A')$$

Dually, coDyn S denotes the $\underline{V}$-category of S-coalgebras in $\underline{A}$. Its objects are morphisms $\varphi : A \to SA$ and coDyn $S(\varphi,\varphi') \in \underline{V}$ is the equalizer of the two morphisms

$$\underline{A} \downarrow S(\varphi,\varphi') \begin{array}{c} \xrightarrow{p_1} \\ \xrightarrow[d_0 p_0]{} \end{array} \underline{A}(A,A')$$

(cf. [6] and [10]).

1.4. **Remarks**: There are diagrams

$$\begin{array}{ccc} & S \downarrow \underline{A} & \\ U_1 \swarrow & \Downarrow \theta & \searrow U_2 \\ \underline{A} & \xrightarrow{S} & \underline{A} \end{array} \qquad \begin{array}{ccc} & \underline{A} \downarrow S & \\ U_2 \swarrow & \Downarrow \psi & \searrow U_1 \\ \underline{A} & \xrightarrow{S} & \underline{A} \end{array}$$

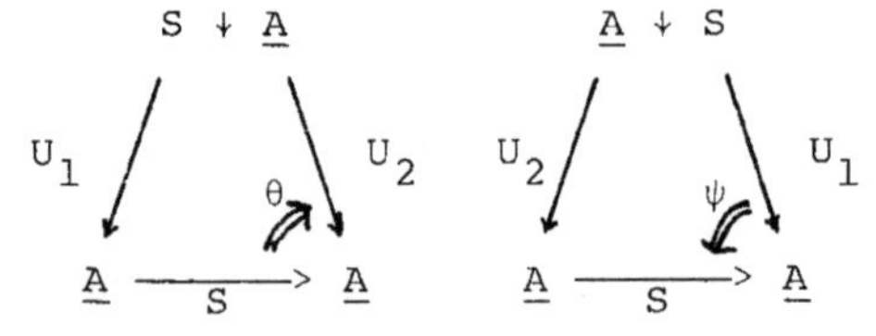

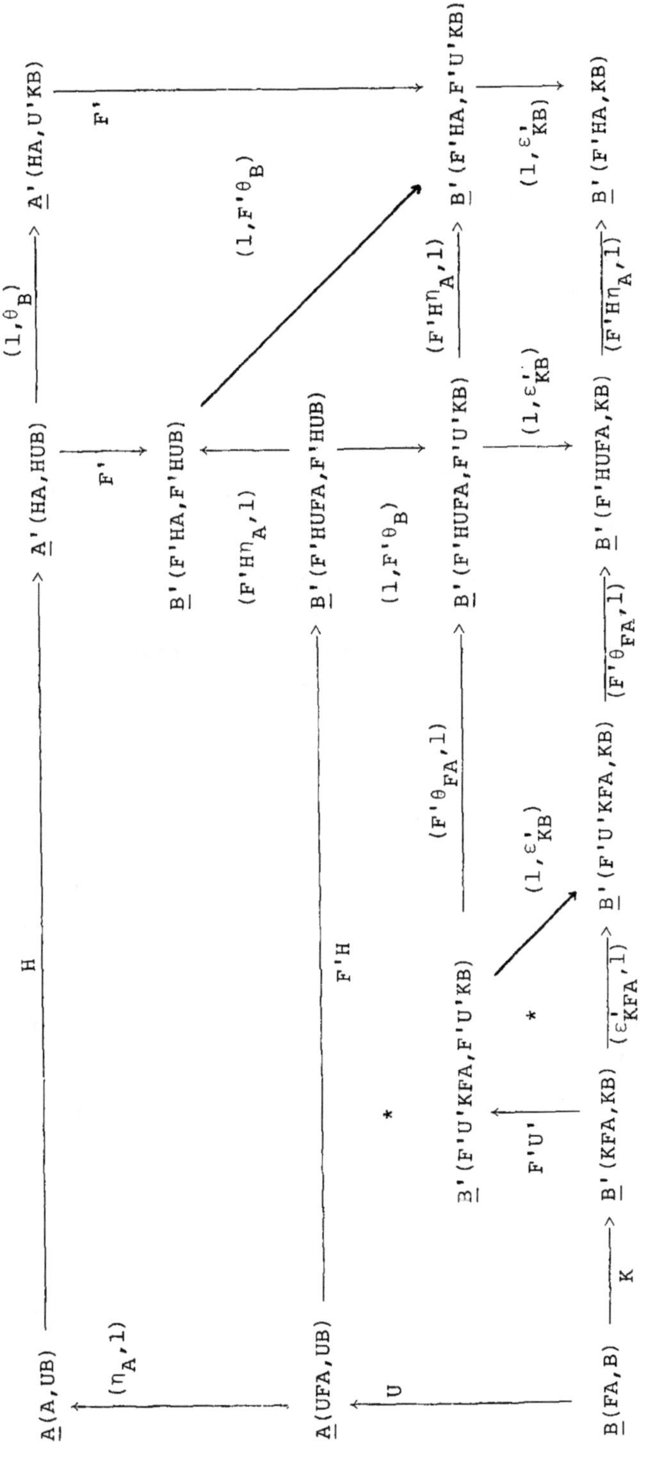
A(A,UB)
A'(HA,HUB)
A'(HA,U'KB)
H
(1,θ_B)
(η_A,1)
F'
B'(F'HA,F'HUB)
(1,F'θ_B)
A(UFA,UB)
B'(F'HUFA,F'HUB)
B'(F'HA,F'U'KB)
F'H
(F'Hη_A,1)
U
(F'θ_FA,1)
B'(F'HUFA,F'U'KB)
B'(F'U'KFA,F'U'KB)
*
(1,ε'_KB)
B(FA,B)
K
B'(KFA,KB)
F'U'
(ε'_KFA,1)
B'(F'U'KFA,KB)
B'(F'HUFA,KB)
B'(F'HA,KB)

Figure 1

which are universal for pairs of V-functors $F_1, F_2 : \underline{X} \to \underline{A}$ and a $\underline{V}$-natural transformation $SF_1 \Rightarrow F_2$ (resp. $F_1 \Rightarrow SF_2$). Similarly there are diagrams

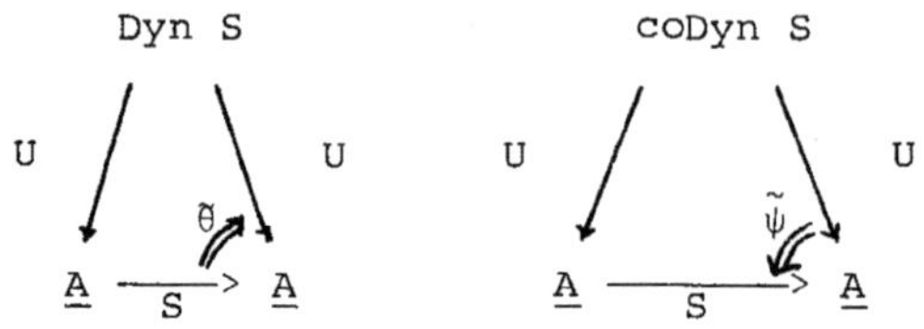

which are universal for a single $\underline{V}$-functor $F : \underline{X} \to \underline{A}$ and a $\underline{V}$-natural transformation $SF \Rightarrow F$ (resp., $F \Rightarrow SF$) (cf. [6] 6.4)

1.5. <u>Proposition</u>. Given a $\underline{V}$-adjunction $S \dashv T$ then there are $\underline{V}$-isomorphisms $S \downarrow \underline{A} \simeq \underline{A} \downarrow T$ over $\underline{A} \times \underline{A}$ and Dyn S $\simeq$ coDyn T over $\underline{A}$.

<u>Proof</u>. The natural transformation $\eta : A \Rightarrow TS$ determines a unique $\underline{V}$-functor $\bar{\eta} : S \downarrow \underline{A} \to \underline{A} \downarrow T$ such that $\psi\bar{\eta} = T\theta \cdot \eta U_1$. Similarly, $\varepsilon : ST \Rightarrow \underline{A}$ determines $\bar{\varepsilon} : \underline{A} \downarrow T \to S \downarrow \underline{A}$ such that $\theta\bar{\varepsilon} = \varepsilon U_2 \cdot S\psi$. The adjunction equations are equivalent to $\bar{\eta} = \bar{\varepsilon}^{-1}$ since, for instance

$$\begin{aligned}\theta\bar{\varepsilon}\,\bar{\eta} &= (\varepsilon U_2 \cdot S\psi)\bar{\eta} = \varepsilon U_2\bar{\eta} \cdot S\psi\bar{\eta}\\ &= \varepsilon U_2 \cdot S(T\theta \cdot \eta U_1) = \varepsilon U_2 \cdot ST\theta \cdot S\eta U_1\\ &= \theta \cdot \varepsilon S U_1 \cdot S\eta U_1 = \theta \cdot (\varepsilon S \cdot S\eta)U_1 = \theta\end{aligned}$$

so $\bar{\varepsilon}\,\bar{\eta} = \mathrm{id}$. Clearly $\bar{\varepsilon}$ and $\bar{\eta}$ restrict to give the second isomorphism.

1.6. <u>Proposition</u>. If $\underline{A}$ is a complete $\underline{V}$-category then Dyn S is a complete $\underline{V}$-category.

<u>Proof</u>. It is well known that Dyn S has ordinary limits which are created by $U : \mathrm{Dyn}\ S \to \underline{A}$ (cf. [1]). We show here that $\underline{A}$ has indexed limits. Recall that if $G : \underline{I} \to \underline{V}$ and $\underline{F} : \underline{I} \to \underline{A}$ are $\underline{V}$-functors then the indexed limit $\{G,F\} \in \underline{A}$ satisfies

$$\underline{A}(A,\{G,F\}) \simeq [\underline{I},\underline{V}](G,\underline{A}(A,F))$$

For the moment, suppose for clarity that $S : \underline{A} \to \underline{B}$ is a $\underline{V}$-functor. Then there is a diagram like the one in 1.2,

$$\begin{array}{ccc}
\underline{A}^{op} & \xrightarrow{S^{op}} & \underline{B}^{op} \\
\{-,F\}\uparrow\ \downarrow \underline{A}(-,F) & & \{-,SF\}\uparrow\ \downarrow \underline{B}(-,SF) \\
[\underline{I},\underline{V}] & \xrightarrow[id]{} & [\underline{I},\underline{V}]
\end{array}$$

and hence a bijection between $\underline{V}$-natural transformations $\underline{A}(-,F) \Rightarrow \underline{B}(-,SF) \circ S$ and $\{-,SF\} \Rightarrow S \circ \{-,F\}$ Let $t : \underline{A}(-,F) \Rightarrow \underline{B}(-,SF) \circ S$ be the transformation whose components $t_A : \underline{A}(A,F) \to \underline{B}(SA,SF)$ have components $t_{A,i} = S_{A,Fi} : \underline{A}(A,Fi) \to \underline{B}(SA,SFi)$ Then t corresponds to a natural transformation $c : \{-,SF\} \Rightarrow S \circ \{-,F\}$ between functors with codomain $\underline{B}^{op}$ whose components are morphisms $c_G : S(\{G,F\}) \to \{G,SF\}$ in $\underline{B}$

Now we return to an endofunctor $S : \underline{A} \to \underline{A}$. A $\underline{V}$-functor $\bar{\varphi} : \underline{I} \to \text{Dyn } S$ corresponds to a functor $F : \underline{I} \to \underline{A}$ and a $\underline{V}$-natural transformation $\varphi : SF \Rightarrow F$. Let

$$\lambda = \{G,\varphi\} \circ c_G : S\{G,F\} \to \{G,F\}$$

We claim that $\lambda = \{G,\bar{\varphi}\} \in \text{Dyn } S$. Let $\psi : SA \to A$ be an object of Dyn S. Then there is an equalizer diagram

$$\begin{array}{ccccc}
\text{Dyn } S(\psi,\lambda) & \longrightarrow & \underline{A}(A,\{G,F\}) & \longrightarrow & \underline{A}(SA,S\{G,F\}) \\
 & & & & \downarrow (1,c_G) \\
 & & & & \underline{A}(SA,\{G,SF\}) \\
 & & & (\psi,1) \searrow & \downarrow (1,\{1,\varphi\}) \\
 & & & & \underline{A}(SA,\{G,F\})
\end{array}$$

which is isomorphic to an equalizer diagram

$$\begin{array}{ccccc}
E & \longrightarrow & [\underline{I},\underline{V}](G,\underline{A}(A,F) & \xrightarrow{(1,t_A)} & [\underline{I},\underline{V}](G,\underline{A}(SA,SF)) \\
 & & & (1,(\Psi,1)) \searrow & \downarrow (1,(1,\varphi)) \\
 & & & & [\underline{I},\underline{V}](G,\underline{A}(SA,F))
\end{array}$$

the crucial step being given by the commutative diagram in 1.2. But in $[\underline{I},\underline{V}]$ one has an equalizer diagram

$$\begin{array}{ccccc}
A(\psi,\varphi) & \longrightarrow & \underline{A}(A,F) & \xrightarrow{t_A} & \underline{A}(SA,SF) \\
 & & (\psi,1) & \searrow & \downarrow (1,\varphi) \\
 & & & & \underline{A}(SA,F)
\end{array}$$

and $[\underline{I},\underline{V}](G,-)$ preserves equalizes so

$$\text{Dyn } S(\psi,\lambda) \simeq E \simeq [\underline{I},\underline{V}](G,\underline{A}(\psi,\varphi))$$

Hence $\lambda = \{G,\bar{\varphi}\}$.

1.7. Proposition. If $S : \underline{A} \to \underline{A}$ preserves coproducts then S generates a free $\underline{V}$-monad.

Proof. See [1], [3], [9], [12] and [16] for discussions of the free monad (triple) problem. This is the simplest possible case in which it is obvious that one gets a $\underline{V}$-monad. We sketch the details in order to have them in the main example. Let $\hat{S} : \underline{A} \to \underline{A}$ be the $\underline{V}$-functor whose value on objects is $\hat{S}(A) = \coprod_{n=0}^{\infty} S^n(A)$, where $S^0(A) = A$, and whose $\underline{V}$-structure is given by the composition

$$\underline{A}(A,B) \xrightarrow{<S^m_{A,B}>} \prod_m \underline{A}(S^m A, S^m B) \xrightarrow{\prod_m (1,in_m)} \prod_m \underline{A}(S^m A, \coprod_n S^n B)$$

$$\prod_m \underline{A}(S^m A, \coprod_n S^n B) \cong \underline{A}(\coprod_m S^m A, \coprod_n S^n B) \cong \underline{A}(\hat{S}A,\hat{S}B)$$

$$\hat{S}_{A,B} : \underline{A}(A,B) \longrightarrow \underline{A}(\hat{S}A,\hat{S}B)$$

where $in_m : S^m B \Rightarrow \hat{S}B$ in the m'th summand. Let $\eta = in_0 : Id \Rightarrow \hat{S}$ and $s = in_1 : S \Rightarrow \hat{S}$. Define $\mu : \hat{S}\,\hat{S} \Rightarrow \hat{S}$ to be the transformation whose components are the maps

$$\mu_A : \coprod_{m=0}^{\infty} \coprod_{n=0}^{\infty} S^m S^n A \longrightarrow \coprod_{p=0}^{\infty} S^p A$$

such that $\mu_A \circ in_{m,n} = in_{m+n}$. Then η,μ and s are $\underline{V}$-natural transformations and $\hat{S} = (\hat{S},\mu,\eta)$ is a $\underline{V}$-monad. If $\lambda = <\lambda_p> : \coprod_p S^p A \to A$ is an $\hat{S}$-algebra, then the algebra equations are easily seen to be equivalent to the recursive conditions: $\lambda_0 = id$, $\lambda_1 : SA \to A$ arbitrary, and $\lambda_{n+1} = \lambda_1 \circ S\lambda_n$ (actually, $\lambda_{m+n} = \lambda_m \circ S^m \lambda_n$). Hence, $s : S \Rightarrow \hat{S}$ induces (by composition) a $\underline{V}$-isomorphism $\bar{s} : \underline{A}^{\hat{S}} \to \text{Dyn } S$.

1.8. Corollary. If $S : \underline{A} \to \underline{A}$ preserves coproducts then $U : \text{Dyn } S \to \underline{A}$ has a $\underline{V}$-left adjoint given by

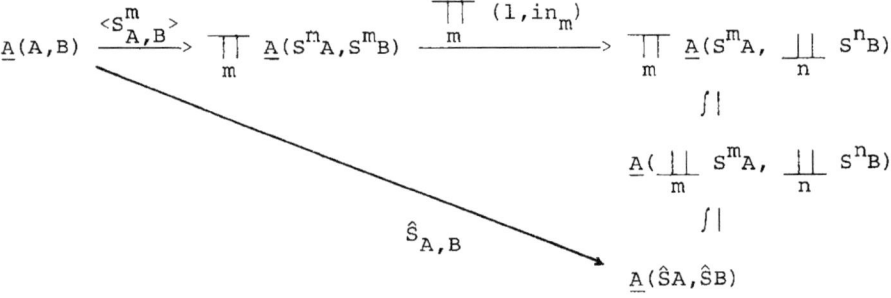

$$F(A) = (f_A : \coprod_{n=1}^{\infty} S^n A \longrightarrow \coprod_{n=0}^{\infty} S^n A)$$

Proof: $\underline{A}^{\hat{S}} \to \underline{A}$ has a $\underline{V}$-left adjoint given by $F_S(A) = \mu_A : \hat{S}\hat{S}A \to \hat{S}A$. Composing with s gives the map $f_A : S\hat{S}A \to \hat{S}A$ in which

$$f_A : \coprod_{n=0}^{\infty} SS^n A \longrightarrow \coprod_{n=0}^{\infty} S^n A$$

satisfies $f_A \circ in_{1,n} = in_{n+1}$. Rewritting $\coprod_{n=0}^{\infty} SS^n A = \coprod_{n=1}^{\infty} S^n A$, then $f_A \circ in_n = in_n$ for $n \geq 1$.

1.9. Definition. Let $\text{coDyn}_i S$ denote the $\underline{V}$-full subcategory of $\text{coDyn}\ S$ determined by isomorphisms $\varphi : A \xrightarrow{\sim} SA$.

1.10. Proposition. If S preserves sequential colimits, then $\text{coDyn}_i S$ is a $\underline{V}$-reflective subcategory of $\text{coDyn}\ S$.

Proof: The inclusion functor has a $\underline{V}$-left adjoint K whose value on an object $\varphi : A \to SA$ in $\text{coDyn}\ S$ is the isomorphism.

$$\lambda = K(\varphi) : \varinjlim_n S^n A \xrightarrow{\sim} \varinjlim_n S^{n+1} A \simeq S \varinjlim_n S^n A$$

where the maps in the sequential colimit are given by $S^n\varphi : S^n A \to S^{n+1} A$. Let $\tilde{S}A = \varinjlim_n A$ and let $q_n : S^n A \to \tilde{S}A$ be the map to the colimit, so $q_{n+1} \circ S^n\varphi = q_n$. The maps that induce the two isomorphisms in λ are given by

$$\begin{array}{ccccc}
S^{n+1}A & = & S^{n+1}A & = & SS^nA \\
\downarrow q_{n+1} & & \downarrow q'_{n+1} & & \downarrow Sq_n \\
\varinjlim S^pA & \xleftarrow{\sim} & \varinjlim S^{p+1}A & \xleftarrow{\sim} & S \varinjlim S^pA
\end{array}$$

Hence, $\lambda \circ q_{n+1} = Sq_n$. Finally, if $h : \tilde{S}A \to B$ is any map, let $h_n = hq_n : S^n A \to B$. A diagram

$$(\infty) \qquad \begin{array}{ccc}
\tilde{S}A & \xrightarrow{h} & B \\
\downarrow \lambda & & \downarrow \psi \\
S\tilde{S}A & \xrightarrow{Sh} & SB
\end{array}$$

commutes iff $\psi \circ h \circ q_{n+1} = Sh \circ \lambda \circ q_{n+1}$ iff $\psi \circ h_{n+1} = Sh \circ Sq_n = Sh_n$.

Step 1. To show that K is left adjoint to the inclusion, consider also the square

$$(0) \qquad \begin{array}{ccc} A & \xrightarrow{\;h_0\;} & B \\ \varphi\downarrow & & \downarrow\psi \\ SA & \xrightarrow{\;Sh_0\;} & SB \end{array}$$

If ψ is an isomorphism then square (∞) commutes iff square (0) commutes. For suppose square (∞) commutes. Then $h_{n+1} = \psi^{-1} \circ Sh_n$. In particular, $h_1 = \psi^{-1} \circ Sh_0$. Since one also has $h_1 \circ \varphi = h_0$, it follows that $Sh_0 \circ \varphi = \psi \circ h_0$; i.e., that square (0) commutes. Conversely, given h_0 making square (0) commute, define h_{n+1} recursively by $h_{n+1} = \psi^{-1} \circ Sh_n$. Then

$$\begin{aligned} h_{n+1} \circ S^n\varphi &= \psi^{-1} \circ Sh_n \circ S^n\varphi = \psi^{-1} \circ S(h_n \circ S^{n-1}\varphi) \\ &= \psi^{-1} \circ S(h_{n-1}) = h_n . \end{aligned}$$

Hence $h = \langle h_n \rangle : \tilde{S}A \to B$ makes square (∞) commute (cf. [3], 4.6).

Step 2. Showing K is $\underline{V}$-left adjoint to the inclusion is harder because of the recursive step. We must also consider squares

$$(n) \qquad \begin{array}{ccc} S^nA & \dashrightarrow & B \\ S^n\varphi\downarrow & & \downarrow\psi \\ S^{n+1}A & \dashrightarrow & SB \end{array}$$

If ψ is an isomorphism, then the $\underline{V}$-object of such squares (i.e., $\mathrm{coDyn}\ S(S^n\varphi,\psi) \in \underline{V}$) can be taken to be the equalizer

$$\begin{array}{ccccc} & & \underline{A}(S^{n+1}A,SB) & \xrightarrow{(1,\psi^{-1})} & \underline{A}(S^{n+1}A,B) \\ & S \nearrow & & & \searrow (S^n\varphi,1) \\ E_n \to \underline{A}(S^nA,B) & & \xrightarrow{\qquad id \qquad} & & \underline{A}(S^nA,B) \end{array}$$

For the square (∞) there is the equalizer (where id is omitted)

$$\begin{array}{ccccccc} E_\infty \to \underline{A}(\tilde{S}A,B) & \xrightarrow{S} & \underline{A}(S\tilde{S}A,SB) & \xrightarrow{(1,\psi^{-1})} & \underline{A}(S\tilde{S}A,B) & \xrightarrow{(\lambda,1)} & \underline{A}(\tilde{S}A,B) \\ \| & & \| & & \| & & \| \\ \varprojlim \underline{A}(S^nA,B) & \xrightarrow{S} & \varprojlim \underline{A}(S^{n+1}A,SB) & \xrightarrow{(1,\psi^{-1})} & \varprojlim \underline{A}(S^{n+1}A,B) & \xleftarrow{\lim(S^n\varphi,1)} & \varprojlim \underline{A}(S^nA,B) \end{array}$$

The maps $q_n : S^nA \to \tilde{S}A$ induce projection maps $p_n : E_\infty \to E_n$ and $E_\infty = \varprojlim E_n$ (since equalizers commute with inverse limits). The maps in this inverse system are induced by the natural transformations illustrated in the first two rows of Figure 2. In this diagram E_i is the equalizer of the i'th row and the identity map. The third and fourth rows show that there is also an induced map $E_n \to E_{n+1}$. Inspection shows that $E_n \to E_{n+1} \to E_n$ is the identity map, by the definition of E_n. The composition $E_{n+1} \to E_n \to E_{n+1}$ is also the identity because there is a commutative diagram

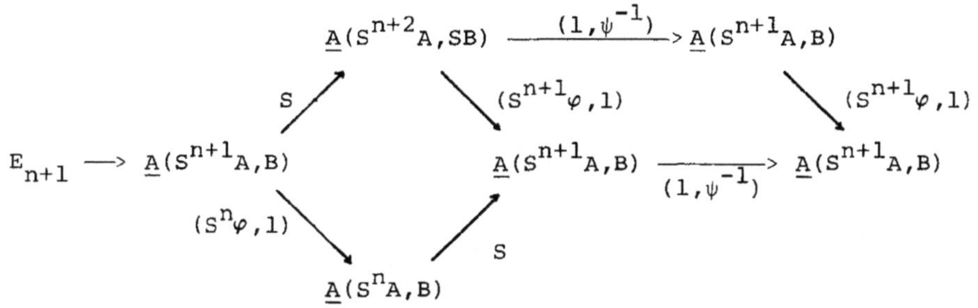

Hence, all of the maps in the inverse system are isomorphisms, so $E_\infty \simeq E_0$; i.e., K is $\underline{V}$-left adjoint to the inclusion.

1.11. Proposition. If S preserves indexed limits then coDyn S and $\text{coDyn}_i S$ are complete.

Proof: Given $\underline{V}$-function $G : \underline{I} \to \underline{V}$ and $\bar{\varphi} : \underline{I} \to$ coDyn S where $\bar{\varphi}$ corresponds to a $\underline{V}$-natural transformation $\varphi : F \Rightarrow SF$, consider $\{G,\varphi\} : \{G,F\} \to \{G,SF\} \simeq S\{G,F\}$ in coDyn S. A construction very similar to that of 1.6 shows that this is the indexed limit of G and $\bar{\varphi}$ in coDyn S. If $\bar{\varphi}$ lies in $\text{coDyn}_i S$ then φ is an isomorphism and hence so is $\{G,\varphi\}$.

2. $\underline{V}$-spectra. Let $\underline{A}$ be a complete $\underline{V}$-category and let $\Sigma \dashv \Omega$ be a pair of $\underline{V}$-adjoint endofunctors of $\underline{A}$. (The notation is chosen to reflect the topological situation.) Let $\underline{A}^\infty = \prod_{n=0}^{\infty} \underline{A}$ with projection $\underline{V}$-functors $pr_n : \underline{A}^\infty \to \underline{A}$ for $n \geq 0$. Write $A_n = pr_n(A)$ and, sometimes, $A = \langle A_0, A_1, \ldots, \rangle$. Let $(-)^+ : \underline{A}^\infty \to \underline{A}^\infty$ be the unique $\underline{V}$-functor such that $pr_n(-)^+ = pr_{n-1}$ for $n \geq 1$ and $pr_0(-)^+ = \phi$ (the initial object of $\underline{A}$); i.e.,

$$\langle A_0, A_1, \ldots \rangle^+ = \langle \phi, A_0, A_1, \ldots \rangle$$

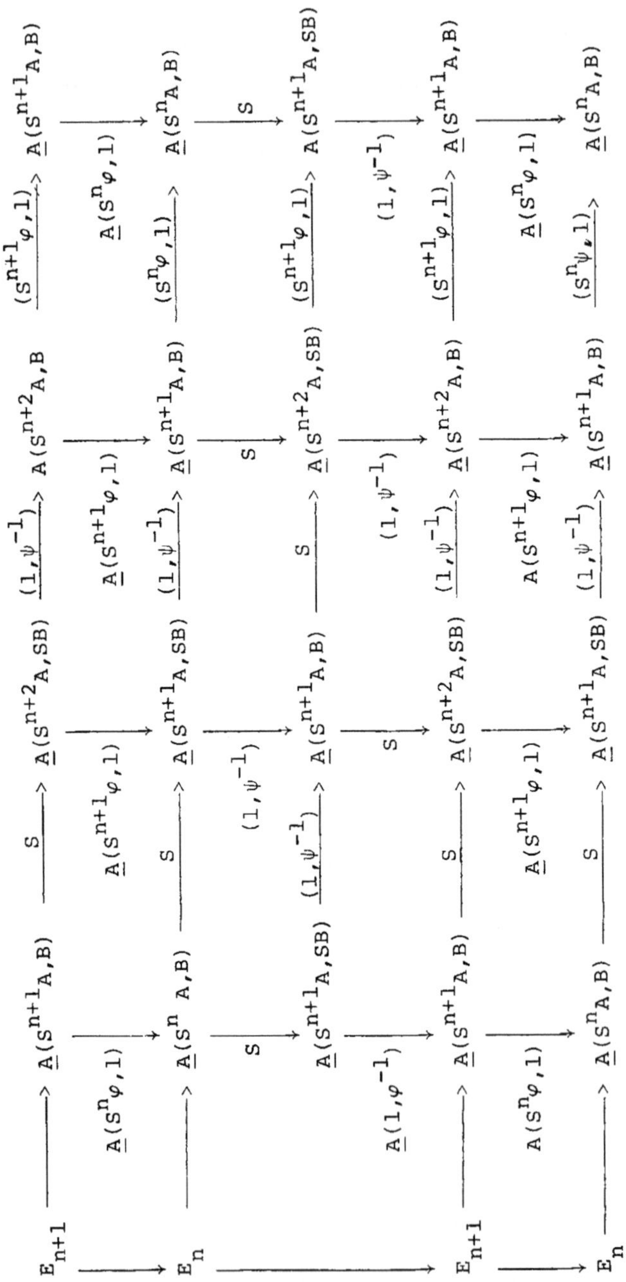

Figure 2

Similarly $(-)^- : \underline{A}^\infty \to \underline{A}^\infty$ satisfies $pr_n(-)^- = pr_{n+1}$ so

$$\langle A_0, A_1, \ldots\rangle^- = \langle A_1, A_2, \ldots\rangle$$

Clearly $(-)^+ \dashv (-)^-$ is a $\underline{V}$-adjunction. Now Σ and Ω also induce endofunctors of $\underline{A}^\infty$, given by $\Sigma^\infty = \prod_{n=0}^{\infty} \Sigma$ and $\Omega^\infty = \prod_{n=0}^{\infty} \Omega$, which are $\underline{V}$-adjoint. Hence $\Sigma^+ = (-)^+ \circ \Sigma^\infty$ and $\Omega^- = \Omega^\infty \circ (-)^-$ are $\underline{V}$-adjoint. Here

$$\Sigma^+\langle A_0, A_1, \ldots\rangle = \langle \phi, \Sigma A_0, \Sigma A_1, \ldots\rangle$$

$$\Omega^-\langle B_0, B_1, \ldots\rangle = \langle \Omega B_1, \Omega B_2, \ldots\rangle$$

2.1. Definitions. i) Prespec $\underline{A}$ = Dyn Σ^+. ii) Spec $\underline{A}$ is the $\underline{V}$-full subcategory of Prespec $\underline{A}$ corresponding to $\mathrm{coDyn}_i\Omega^-$ under the isomorphism Dyn $\Sigma^+ \cong$ coDyn Ω^- of 1.5.

2.2. Remark. An object of Prespec $\underline{A}$ is an object $A = \langle A_0, A_1, \ldots\rangle$ of $\underline{A}^\infty$ together with a map $\varphi : \Sigma^+ A \to A$ whose components are maps $\varphi_{n+1} : \Sigma A_n \to A_{n+1}$. It belongs to Spec $\underline{A}$ if the transpose maps $\varphi_n^\# : A_n \to \Omega A_{n+1}$ are isomorphisms for $n \geq 0$.

2.3. Theorem. If Ω preserves sequential colimits then Spec $\underline{A}$ and Prespec $\underline{A}$ are complete $\underline{V}$-categories and there are pairs of $\underline{V}$-adjoint functors

$$\text{Spec } \underline{A} \underset{I}{\overset{K}{\leftrightarrows}} \text{Prespec } \underline{A} \underset{U}{\overset{F}{\leftrightarrows}} \underline{A}^\infty \underset{pr_n}{\overset{L_n}{\leftrightarrows}} \underline{A}$$

Proof: i) $\underline{A}^\infty$ is clearly complete. The left adjoint L_n is the $\underline{V}$-full and faithful functor such that $(L_n(A))_p = A$ if $p = n$ and ϕ otherwise.

ii) By 1.6, Prespec $\underline{A}$ is complete. Since Σ has a right adjoint it preserves coproducts so the constructions of 1.7 and 1.8 apply, giving a left adjoint F such that $UF(A) = \coprod_{p=0}^{\infty} (\Sigma^+)^p(A)$. Thus $UF(A)_n = \coprod_{p=0}^{n} \Sigma^p A_{n-p}$ and the structure map $\Sigma UF(A) \to UF(A)$ has as components the maps

$$\Sigma\Big(\coprod_{p=0}^{n} \Sigma^p A_{n-p}\Big) = \coprod_{p=0}^{n} \Sigma^{p+1} A_{n-p} = \coprod_{q=1}^{n+1} \Sigma^q A_{n+1-q} \to \coprod_{q=0}^{n+1} \Sigma^q A_{n+1-q}$$

which omit the first summand.

ii) Since Ω^- has a $\underline{V}$-left adjoint, it preserves indexed limits so Spec $\underline{A} \cong \mathrm{coDyn}_i\Omega^-$ is complete, by 1.11. By hypothesis,

Ω preserves sequential colimits, so 1.10 applies, giving a left adjoint

$$K : \text{Prespec } \underline{A} = \text{Dyn } \Sigma^+ \simeq \text{coDyn } \Omega^- \to \text{coDyn}_i \Omega^- \simeq \text{Spec } \underline{A}$$

A prespectrum $\varphi: \Sigma^+ A \to A$ corresponds to a coalgebra $\varphi^\#: A \to \Omega^- A$ with components $\varphi_n^\# : A_n \to \Omega A_{n+1}$. The reflection into $\text{coDyn}_i \Omega^-$ has as underlying $\underline{A}$-object the colimit $\varinjlim_n (\Omega^-)^n (A)$. Thus $UK(A)_n = \varinjlim_j \Omega^j A_{n+j}$. The structure map $\Sigma UK(A) \to UK(A)$ is the transpose of the isomorphism $\varinjlim(\Omega^-)^n(A) \xrightarrow{\sim} \Omega^- \varinjlim(\Omega^-)^n(A)$ whose components are the isomorphisms

$$UK(A)_n = \varinjlim \Omega^j A_{n+j} \simeq \varinjlim \Omega^{j+1} A_{n+j+1} \simeq \Omega \varinjlim \Omega^j A_{n+1+j} = \Omega\, UK(A)_{n+1}$$

2.4. Remarks. i) The composition $Q_\infty = KFL_0 : \underline{A} \to \text{Spec } \underline{A}$, which is of interest in homotopy theory (cf. [11]), is given as follows: $FL_0(A)_n = \Sigma^n A$ so

$$Q_\infty(A)_n = \varinjlim_j \Omega^j \Sigma^{n+j} A$$

In particular, $Q(A) = Q_\infty(A)_0 = \varinjlim \Omega^j \Sigma^j A$ is the stabilizing functor.

ii) Let $p_n = pr_n UI : \text{Spec } \underline{A} \to \underline{A}$. Then p_n preserves indexed limits since all three functors have $\underline{V}$-left adjoints. If $H : \underline{I} \to \text{Spec } \underline{A}$, let $H_n = p_n H$. Then for any $G : \underline{I} \to \underline{V}$, one has $\{G,F\}_n \simeq \{G,F_n\}$.

3. Pointed simplicial spectra. Let $\underline{K} = [\Delta^{op}, \text{Sets}]$ denote the cartesian closed category of simplicial sets, and let $\underline{K}_*$ denote the category of pointed simplicial sets; i.e., objects X of $\underline{K}$ together with a map $x : 1 \to X$.

3.1. $\underline{K}_*$ as a $\underline{K}$-category. Let $1 : \underline{K} \to \underline{K}$ denote the constant $\underline{K}$-functor such that $1(X) = 1$ (the terminal object of $\underline{K}$). Then $\underline{K}_* = \text{Dyn } 1$. By 1.6, $\underline{K}_*$ is complete as a $\underline{K}$-category. The underlying functor $U : \underline{K}_* \to \underline{K}$ has a left adjoint F given by $F(X) = X \sqcup 1$, (although this does not follow from the construction of 1.8 since $1(-)$ does not preserve coproducts). From the general description of Dyn S, one has the following:

i) If $Y_*, Z_* \in \underline{K}_*$, then $\underline{K}_*(Y_*, Z_*) \in K$ is the pullback

$$\begin{array}{ccccc} \underline{K}_*(Y_*,Z_*) & \longrightarrow & 1 & = & \underline{K}(1,1) \\ \downarrow & & \downarrow z & & \downarrow (1,z) \\ \underline{K}(Y,Z) & \xrightarrow[(y,1)]{} & Z & = & K(1,Z) \end{array}$$

ii) $X \in \underline{K}$, then $X \pitchfork Y_* \in K_*$ is given by

$$1 = \underline{K}(X,1) \xrightarrow[(1,y)]{} \underline{K}(X,Y)$$

iii) $\underline{K}_*$ is also tensored over $\underline{K}$, $X \mathbin{\overline{\otimes}} Y_*$ being the pushout

$$\begin{array}{ccc} X \times 1 & \longrightarrow & 1 \\ {\scriptstyle X \times y}\downarrow & & \downarrow \\ X \times Y & \longrightarrow & X \mathbin{\overline{\otimes}} Y_* \end{array}$$

iv) One has the following formulas

a) $\underline{K}_*(X \mathbin{\overline{\otimes}} Y_*,Z_*) \simeq \underline{K}(X,\underline{K}_*(Y_*,Z_*)) \simeq \underline{K}_*(Y_*,X \pitchfork Z_*)$

b) $\underline{K}_*(X_*,Y_*)_n = \underline{K}_*(\Delta[n] \mathbin{\overline{\otimes}} Y_*,Z_*)_0$

c) $\underline{K}_*(X \amalg 1,Y_*) \simeq \underline{K}(X,Y)$.

3.2. $\underline{K}_*$ <u>as a closed category</u>. If $X_*,Y_* \in \underline{K}_*$, Let $1 \to \underline{K}_*(X_*,Y_*)$ correspond by adjointness to $X_* \to 1 \to 1 \pitchfork Y_*$ (from iv) a) and ii). Further, let $X_* \wedge Y_*$ denote the pushout

$$\begin{array}{ccc} (X \times 1) \amalg (1 \times Y) & \longrightarrow & 1 \\ {\scriptstyle \langle X \times y, x \times Y \rangle}\downarrow & & \downarrow \\ X \times Y & \longrightarrow & X_* \wedge Y_* \end{array}$$

Then $\underline{K}_*$ is a symmetric monoidal closed category with internal hom given by $1 \to \underline{K}_*(X_*,Y_*)$ and $\otimes$-product given by $1 \to X_* \wedge Y_*$. (We suppress the base points from the notation hereafter.) The underlying functor $U : \underline{K}_* \to \underline{K}$ comes equipped with canonical maps $U\underline{K}_*(X_*,Y_*) \to \underline{K}(X,Y)$ and $UX_* \times UY_* \to U(X_* \times Y_*)$ making U into a normal closed functor. The left adjoint $F = (-) \amalg 1$ to U preserves products (i.e., $(X \amalg 1) \wedge (Y \amalg 1) \simeq (X \times Y) \amalg 1)$ so it is also a closed functor. Hence

d) $-\amalg 1$ preserves tensors; i.e., $(X \times Y) \amalg 1 = X \mathbin{\overline{\otimes}} (Y \amalg 1)$

e) U preserves cotensors; i.e., $U(X \pitchfork Y_*) = \underline{K}(X,Y)$.

3.3. Definitions. i) Let S^1 be the coequalizer (in $\underline{K}$) of the two maps

$$\Delta[0] \underset{d^1}{\overset{d^0}{\rightrightarrows}} \Delta[1] \xrightarrow{q} S^1$$

and let S^1_* have the base point $qd^0 : 1 = \Delta[0] \to S^1$.

ii) If $\underline{A}$ is a complete and cocomplete $\underline{K}_*$-category, then $\Sigma_{\underline{A}} = S^1_* \tilde{\otimes} - : \underline{A} \to \underline{A}$ and $\Omega_{\underline{A}} = S^1_* \pitchfork - : \underline{A} \to \underline{A}$ are defined and are $\underline{K}_*$-adjoint.

iii) An object $X \in \underline{A}$ is called sequentially $\underline{K}_*$-small if $\underline{A}(X,-)$ perserves sequential colimits.

3.4. Proposition. If $\underline{A}$ has a strongly generating family of sequentially $\underline{K}_*$-small objects, then $\Omega_{\underline{A}}$ preserves sequential colimits.

Proof: S^1_* is clearly sequentially $\underline{K}_*$-small in $\underline{K}_*$. Hence, if $\{A_\alpha\}$ is the family in $\underline{A}$, then

$$\begin{aligned}\underline{A}(A_\alpha, S^1_* \pitchfork \varinjlim B_i) &\simeq \underline{K}_*(S^1_*, \underline{A}(A_\alpha, \varinjlim B_i)\\ &\simeq \varinjlim \underline{K}_*(S^1_*, \underline{A}(A_\alpha, B_i)) \simeq \varinjlim \underline{A}(A_\alpha, S^1_* \pitchfork B_i)\\ &\simeq \underline{A}(A_\alpha, \varinjlim S^1_* \pitchfork B_i)\end{aligned}$$

so $S^1_* \pitchfork \varinjlim B_i \simeq \varinjlim S^1_* \pitchfork B_i$.

3.5. Example: $\{\Delta[n] \amalg 1\}$ is a strongly generating family of sequentially $\underline{K}_*$small objects in $\underline{K}_*$. However, $\Omega_{\underline{K}_*}(-) = \underline{K}_*(S^1_*,-)$ preserves sequential colimits anyway.

3.6. Definition. If $\underline{A}$ is a $\underline{K}_*$ -category, then $U_*\underline{A}$ is the $\underline{K}$-category with the same objects and $U_*\underline{A}(A,B) = U(\underline{A}(A,B))$.

By [5], 2.2.3, if $\underline{A}$ is a complete and/or cocomplete $\underline{K}_*$-category, then $U_*\underline{A}$ is a complete and/or cocomplete $\underline{K}$-category. Furthermore, by [5], 2.4.3, there is a close relationship between indexed limits in $\underline{A}$ and in $U_*\underline{A}$. Namely, if $H : \underline{I} \to U_*\underline{A}$ and $G : \underline{I} \to \underline{K}$ are given then, regarding the left adjoint F to U as a $\underline{K}$-functor $F : \underline{K} \to U_*\underline{K}_*$, the $\underline{K}$-functors $H : \underline{I} \to H_*\underline{A}$ and $FG : \underline{I} \to U_*\underline{K}_*$ correspond to $\underline{K}_*$-functors $H^\# : F_*\underline{I} \to \underline{A}$ and $(FG)^\# : F_*\underline{I} \to \underline{K}_*$ and $\{G,H\}_{U_*\underline{A}} = \{(FG)^\#, H^\#\}_{\underline{A}}$.

If $\underline{I}$ is any $\underline{K}$-category, then an indexing functor $Z_{\underline{I}} : \underline{I} \to \underline{K}$ is defined in [5], 4.5.1, which reduces to $N(\underline{I}/-)$ in case $\underline{I}$ is the free $\underline{K}$-category on an ordinary category, such that indexed limits over

$Z_{\underline{I}}$ are homotopy limits. Hence for a complete $\underline{K}_*$-category $\underline{A}$, homotopy limits in $U_*\underline{A}$ are given by either of the formulas

$$\text{i)} \quad \operatorname{holim} H = \{Z_{\underline{I}}, H\}_{U_*\underline{A}} \left(= \int_i N(\underline{I}/i) \pitchfork_{U_*\underline{A}} H(i)\right)$$

$$\text{ii)} \quad \operatorname{holim} H = \{(Z_{\underline{I}} \amalg 1)^{\#}, H^{\#}\}_{\underline{A}} \left(= \int_i (N(\underline{I}/i) \amalg 1) \pitchfork_{\underline{A}} H^{\#}(i)\right)$$

where the cotensore are actually the same by the construction of cotensors in $U_*\underline{A}$ (cf., [5], 2.2.3).

3.7. Theorem. If $\underline{A}$ is a complete $\underline{K}_*$-category which has a strongly generating family of sequentially $\underline{K}_*$-small objects, then Spec $\underline{A}$ is a complete $\underline{K}_*$-category. U_* Spec $\underline{A}$ has homotopy limits that are preserved by the projection functors $p_n : U_* \text{ Spec } \underline{A} \to U_*A$, $n \geq 0$.

Proof: Immediate, by 2.3, 2.4, ii), and 3.4.

3.8. Example: The category k-sp of k-spaces is a complete and cocomplete $\underline{K}$-category. If $|-| : \underline{K} \to$ k-sp denotes geometric realization and S : k-sp $\to \underline{K}$ denotes the singular functor, then this structure is given as follows: if $X, Y \in$ k-sp and $Z \in \underline{K}$, then k-sp$(X,Y) = S(Y^X)$, $Z \bar{\otimes} X = |Z| \times X$ and $Z \pitchfork X = X^{|Z|}$. Similarly, the category of pointed k-spaces is a complete and cocomplete $\underline{K}_*$category.

Bibliography

[1] M. Barr, Coequalizers and free triples, Math. Z. 116(1970), 307-322.

[2] A. K. Bousfield and D. M. Kan, Homotopy limits, Completions and Localizations, Lecture Notes in Mathematics 304, Springer-Verlag, New York, 1972.

[3] G. Gierz, K. H. Hofmann, K. Keimel, J. D. Lawson, M. Mislove and D. S. Scott, A Compendium of Continuous Lattices, Springer-Verlag, New York, 1980.

[4] J. W. Gray, Formal Category Theory: Adjointness for 2-categories, Lecture Notes in Mathematics 391, Springer-Verlag, New York, 1974.

[5] J. W. Gray, Closed categories, lax limits, and homotopy limits, J. Pure and Applied Alg. 19(1980), 127-158.

[6] J. W. Gray, The existence and construction of lax limits, Cahiers de Top. et Géom. Diff. 21(1980), 277-304.

[7] J. W. Gray, Two results on homotopy limits, (to appear).

[8] G. M. Kelly, The Basic Concepts of Enriched Category Theory, (to appear).

[9] J. Lambek and B. A. Rattray, Localization and sheaf reflectors, Trans. Am. Math. Soc. 210(1975), 279-293.

[10] E. Manes, Algebraic Theories, Graduate Texts in Mathematics, Springer-Verlag, 1976.

[11] J. P. May, Infinite loop space theory, Bull. Am. Math. Soc. 83(1977), 456-494.

[12] D. Scott, The lattice of flow diagrams, Symposium on Semantics of Algorithmic Languages, E. Engeler, ed., Lecture Notes in Mathematics 188, Springer-Verlag, New York (1971), 311-366.

[13] D. S. Scott, Continuous lattices, in Toposes, Algebraic Geometry and Logic, Lecture Notes in Mathematics 274, Springer-Verlag, New York, 1972, 97-136.

[14] M. B. Smyth and G. D. Plotkin, The category-theoretic solution of recursive domain equations, D. A. I. Research Report No. 60, Edinburgh, 1978.

[15] R. W. Thomason, Homotopy colimits in the category of small categories, Math. Proc. Camb. Phil. Soc. 85(1979), 91-109.

[16] R. W. Thomason, Algebraic K-theory and etale cohomology, Preprint 1980.

[17] H. Wolff, Free monads and the orthogonal subcategory problem, J. Pure and Applied Alg. 13(1978), 233-242.

[18] M. Tierney, Categorical Constructions in Stable Homotopy Theory. A seminar given at the ETH. Zürich, in 1967. Lecture Notes in Mathematics 87, Springer-Verlag, New York, 1969.

[19] G. M. Kelly, Adjunction for enriched categories, Reports of the Midwest Category Seminar III, Lecture Notes in Mathematics 106, Springer-Verlag, New York, 1969, 166-177.

[20] L. G. Lewis, The stable category and generalized Thom spectra, Dissertation, University of Chicago, 1978.

[21] R. Street, Fibrations in bicategories, Cahiers de Top. et Géom. Diff. 21 (1980), 111-160.

GENERAL CONSTRUCTION OF MONOIDAL CLOSED STRUCTURES IN TOPOLOGICAL , UNIFORM AND NEARNESS SPACES

Georg Greve

Abstract/Introduction: In the following paper we consider topological structures on function spaces and cartesian products being connected by an exponential law of the form $C(X\circledast Y,Z) \cong C(X,C(Y,Z))$. Topological categories provided with such a "monoidal closed" structure are suitable base categories for topological algebra, algebraic topology, automata- or duality theory, in particular if $\circledast$ is symmetric or the usual direct product. We start from a purely categorical point of view proving an extension theorem which later turns out to be very convenient for the construction of monoidal closed structures in concrete categories, namely in topological spaces, uniform spaces, merotopic spaces and nearness spaces. Enlarging a theorem of Booth and Tillotson [2] it is shown that there are arbitrary many (non symmetric) monoidal closed structures in these categories, hence there is a great difference to the symmetric case, where closed structures seem to be unique (cp. Činčura [3], Isbell [8]). A further application of the extension theorem is a criterion for monoidal- resp. cartesian closedness of MacNeille completions. Of course a symmetric monoidal closed structure is uniquely determined by its values on a finally and initially dense subcategory, but also the converse statement is true, i.e. monoidal closed structures can be obtained by extending a suitable structure from a subcategory to its MacNeille completion.

0. PRELIMINARIES

Throughout this paper $T : C \to Set$ denotes a topological functor, so we have T-initial liftings of (class indexed) sources (cp. [13],[7]), i.e. for every source $(x_i : X \to TA_i)_{i\in I}$ there is a source $(f_i : A \to A_i)_{i\in I}$, $Tf_i = x_i$, such that every map $y : TB \to TA$ with $Tf_i y = Tg_i$, $g_i : B \to A_i$, can be lifted to a morphism $l : B \to A$. A is said to carry the initial structure with respect to all x_i. We assume that there is a unique object $1\in ObC$ with a one element underlying set and for technical reasons T is supposed to be amnestic ($Ti\in ObSet$ for an isomorphism implies $i\in ObC$). Topological functors defined in this way are faithful, there are T-final liftings of sinks, all constant maps are morphisms in C. 1 is a representing object for T, hence we can identify elements $a \in TA$ with morphisms $a : 1 \to A$. A (full) subcategory $E < C$ is called finally dense, iff the sink $(f : E \to A)_{E \in ObE, f \in C(E,A)}$ is final for every $A \in ObC$. The dual notion is "initially dense". If $E < C$ is finally and initially dense, then C is the MacNeille completion of E, resp. of T/E (cp. [1]). A Tensorproduct in C is a unitary and associative functor $\otimes : C \times C \to C$ with $T(\square \otimes \square) = T\square \times T\square$, i.e. there are natural isomorphisms $A \otimes 1 \cong A \cong 1 \otimes A$ and $A \otimes (B \otimes C) \cong (A \otimes B) \otimes C$ which (automatically) have canonical underlying bijections. $\otimes$ is called a monoidal closed structure if $\otimes$ has a right adjoint $H : C^{op} \times C \to C$. Obviously we get $C(A,B) \cong C(1 \otimes A, B) \cong C(1, H(A,B)) \cong TH(A,B)$, so H can always chosen to be an inner homfunctor in C, i.e. $T \circ H = C(\square, \square)$. $\otimes$ is called symmetric if $A \otimes B \cong B \otimes A$ holds (in a natural way). Note that for $a \in TA$, $b \in TB$ there are lifted sections ${}_a s^{AB} : B \to A \otimes B$, $s_b^{AB} : A \to A \otimes B$ with $T({}_a s^{AB})(b) = T(s_b^{AB})(a) = (a,b)$ and lifted projections $p_A : A \otimes B \to A$, $p_B : A \otimes B \to B$ with $Tp_A(a,b) = a$ resp. $Tp_B(a,b) = b$. The definition of a monoidal closed structure used here coincides (up to natural isomorphisms) with the classical notion of Eilenberg and Kelly ([4], cp. [3] Thm 2.1).

<u>0.1 Remark</u> (cp. also [15])
For a tensorproduct $\otimes : C \times C \to C$ the following statements are equivalent:

(i) $\otimes$ is a monoidal closed structure.
(ii) For all $A \in ObC$ $\square \otimes A$ preserves colimits.
(iii) For all $A \in ObC$ $\square \otimes A$ preserves final epi-sinks.

<u>0.2 Remark</u>
Take $\otimes$ to be a tensorproduct in C such that for all $A \in ObC$ the diagonal $\delta_A : TA \to TA \times TA$, $\delta_A(a) = (a,a)$, can be lifted to a morphism $d_A : A \to A \otimes A$.
Then $\otimes$ is the usual product in C.

<u>Proof:</u> The universal property of $\otimes$ is shown by the following diagram:

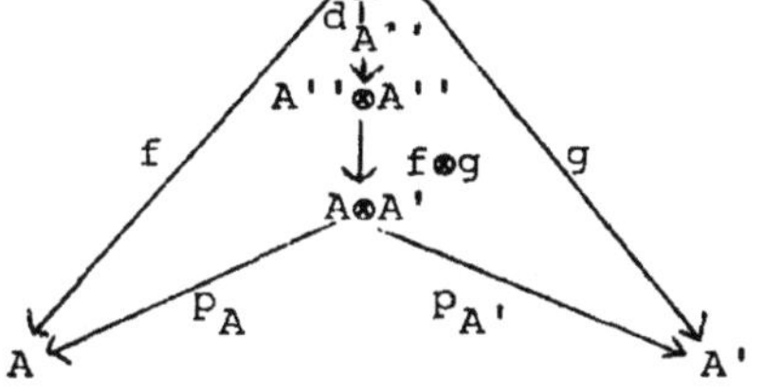

□

1. The extension theorem

In the following section we describe how to construct a monoidal closed structure from a partially given structure on $\mathcal{C}$. For simplicity we introduce the following

1.1 Definition: Take $\mathcal{B}<\mathcal{A}<\mathcal{C}$, $\mathcal{E}<\mathcal{C}$ to be full subcategories and $\otimes_p : \mathcal{A}\times\mathcal{B} \to \mathcal{C}, H_p : \mathcal{B}^{op}\times\mathcal{E} \to \mathcal{C}$ to be functors with $T(A\otimes_p B) = TA\times TB$ and $TH_p(B,E) = \mathcal{C}(B,E)$ in a natural way. Then $(\otimes_p, H_p)$ is called a partial monoidal closed structure with respect to $\mathcal{A}$, $\mathcal{B}$ and $\mathcal{E}$ if the following conditions hold:

(i) $1\in\mathcal{B}$ and there are natural isomorphisms $A\otimes_p 1 \cong A$, $1\otimes_p B \cong B$ for all $A\in\mathrm{Ob}\mathcal{A}$, $B\in\mathrm{Ob}\mathcal{B}$.

(ii) $\mathcal{B}\otimes_p\mathcal{B}\subset\mathcal{B}$, $\mathcal{A}\otimes_p\mathcal{B}\subset\mathcal{A}$ and there is a natural isomorphism $A\otimes_p(B\otimes_p B') \cong (A\otimes_p B)\otimes_p B'$.

(iii) There is a natural isomorphism $\gamma_{ABE}: \mathcal{C}(A\otimes_p B,E) \cong \mathcal{C}(A,H_p(B,E))$.

$(\otimes_p, H_p)$ is called symmetric if $\mathcal{A} = \mathcal{B}$ and there is a natural isomorphism $A\otimes_p B \cong B\otimes_p A$.

Note again that the underlying maps of the transformations appearing in 1.1 (i), (ii) and (iii) are canonical. For example we have $\gamma_{ABE}(f)(a)(b) = f(a,b)$ (the functor T is omitted) for all $a\in A$, $b\in B$. The following statement is the main result of this section:

1.2 Extension theorem:

Take $(\otimes_p, H_p)$ to be a partial monoidal closed structure with respect to $\mathcal{A},\mathcal{B},\mathcal{E}$ (cp. 1.1, $\mathcal{B}<\mathcal{A}<\mathcal{C}$, $\mathcal{E}<\mathcal{C}$)

(a) If $\mathcal{A}$ is finally dense, $\mathcal{E}$ initially dense and if for all $B\in\mathrm{Ob}\mathcal{B}$ the diagonal $\delta_B : TB \to TB\times TB$ can be lifted to a morphism $d_B : B \to B\otimes_p B$ then there is a monoidal closed structure $\otimes$ extending $\otimes_p$, i.e. $\otimes/\mathcal{A}\times\mathcal{B} = \otimes_p$.

(b) If $\mathcal{A} = \mathcal{B}$, $\mathcal{A}$ finally dense, $\mathcal{E}$ initially dense and $(\otimes_p, H_p)$ is symmetric, then $(\otimes_p, H_p)$ can be extended to a symmetric monoidal closed structure on $\mathcal{C}$.

Proof: For $M,N\in\mathrm{Ob}\mathcal{C}$ consider the source

(1.2.1) $\quad (\mathcal{C}(f,g) : \mathcal{C}(M,N) \to TH_p(B,E))_{B\in\mathrm{Ob}\mathcal{B},E\in\mathrm{Ob}\mathcal{E},f\in\mathcal{C}(B,M),g\in\mathcal{C}(N,E)}$.

Providing $\mathcal{C}(M,N)$ with the initial structure with respect to this source we get an inner homfunctor $H : \mathcal{C}^{op}\times\mathcal{C} \to \mathcal{C}$ with $H/\mathcal{B}^{op}\times\mathcal{E} = H_p$. Furthermore consider the sink consisting of all morphisms

(1.2.2) $Tl \times Tk : T(A \otimes_p B) \to TM \times TN$, $A \in Ob\mathcal{A}$, $B \in Ob\mathcal{B}$, $l : A \to M$, $k : B \to N$, and all sections $\sigma_m : TN \to TM \times TN$, $\sigma(n) = (m,n)$.

Providing $TM \times TN$ with the final structure with respect to this sink we get a functor $\otimes : \mathcal{C} \times \mathcal{C} \to \mathcal{C}$ with $\otimes / \mathcal{A} \times \mathcal{B} = \otimes_p$.

Let us first show that $\otimes$ is unitary: We have the following diagram

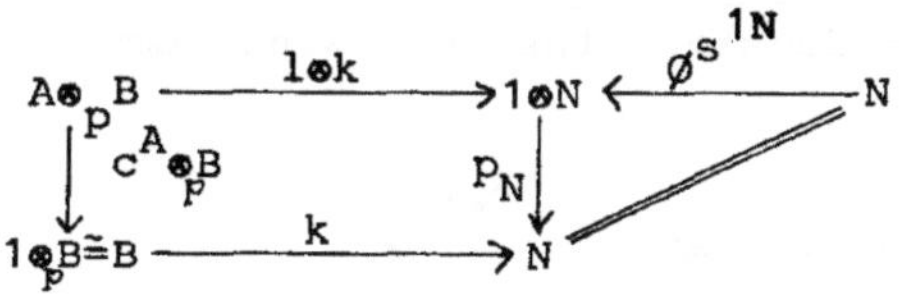

($c^A : A \to 1$ denotes the canonical morphism, ${}_{\emptyset}s^{1N}$ is the lifted section).

The sink on the roof of the diagram is final, hence we have a lifted projection p_N with $p_N \, {}_{\emptyset}s^{1N} = N$, i.e. p_N is a bijective retraction, hence an isomorphism. Furthermore consider the following diagram:

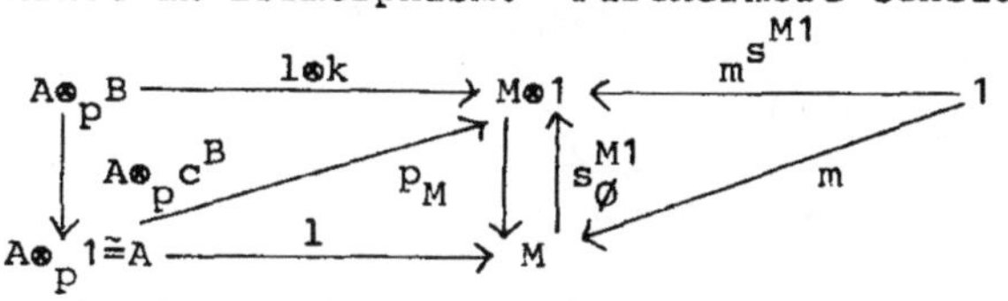

($c^B : B \to 1$ canonical, ${}_m s^{M1}$ lifted section).

The same argument as before yields a lifted projection p_M. Moreover $\mathcal{A}$ is finally dense, hence the sink on the bottom of the diagram is final, hence there is a lifted section $s_{\emptyset}^{M1}$, so we have $M \otimes 1 \cong M$.

Now let us prove that $\otimes$ is left adjoint to H, i.e. we have to find a natural isomorphism $\xi_{M,N,L} : \mathcal{C}(M \otimes N, L) \to \mathcal{C}(M, H(N,L))$. Take $t : M \otimes N \to L$ and define $t' : TM \to \mathcal{C}(N,L)$ by $t'(m) := t \, {}_m s^{MN}$, where ${}_m s^{MN} : N \to M \otimes N$ is a lifted section. Consider the following diagram:

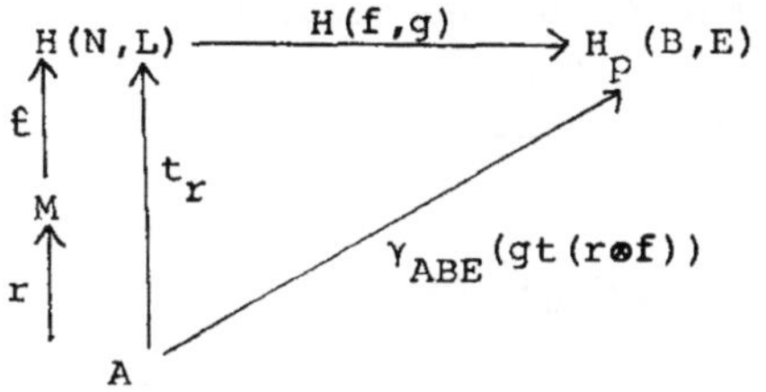

(cp. 1.1) The source $(H(f,g) : H(N,L) \to H_p(B,E))_{B \in Ob\mathcal{B}, E \in Ob\mathcal{E}}$ is initial, so for every $r : A \to M$ there is a morphism $t_r : A \to H(N,L)$ with $Tt_r = t'Tr$. Furthermore the sink of all morphisms $r : A \to M$ is final, hence there is a morphism $\tilde{t} : M \to H(N,L)$ with $T\tilde{t} = t'$.

Now given a morphism $u : M \to H(N,L)$ and define $u' : T(M \otimes N) \to TL$ by $u'(m,n) := T(Tu(m))(n)$. This yields the following commutative diagram:

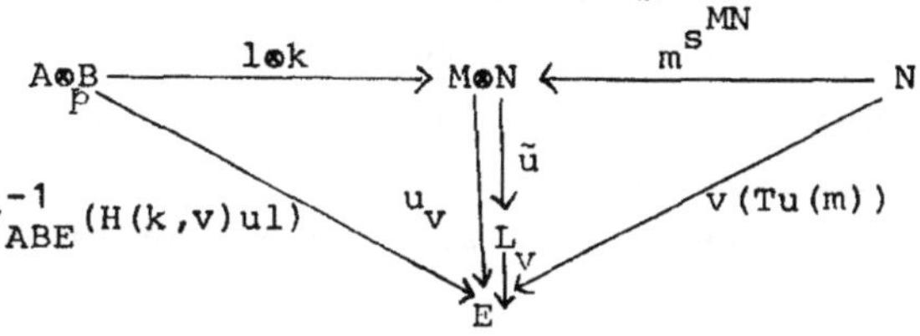

The sink of all $1\otimes k : A\otimes_p B \to M\otimes N$, $A\in \mathrm{Ob}\mathcal{A}$, $B\in\mathrm{Ob}\mathcal{B}$, and all sections ${}_m s^{MN} : N \to M\otimes N$ is final, hence for each $v : L \to E$ there is a morphism $u_v : M\otimes N \to E$ with $Tu_v = Tv\, u'$. The source of all $v : L \to E$, $E\in\mathrm{Ob}\mathcal{E}$, is initial, so u' can be lifted to a morphism $\breve{u} : M\otimes N \to L$. Defining ξ_{MNL} by $\xi_{MNL}(t) := \hat{t}$ ξ is a natural bijection because of $\tilde{\hat{t}} = t$. It remains to show that $\otimes$ is associative: Consider the following commutative diagram (1.2.3):

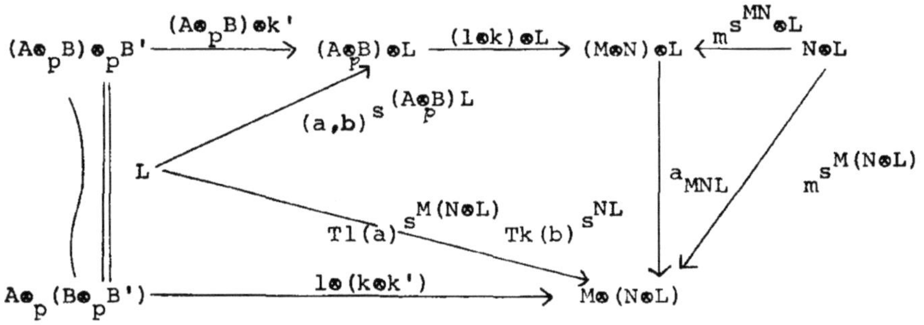

$\square\otimes L$ has a right adjoint, hence preserves final epi-sinks (cp. 0.1), i.e. the sink consisting of all $(1\otimes k)\otimes L : (A\otimes_p B)\otimes L \to (M\otimes N)\otimes L$, $A\in\mathcal{A}$, $B\in\mathcal{B}$, and all ${}_m s^{MN}\otimes L$ is final. Furthermore because of $A\otimes_p \mathcal{B}\subset\mathcal{A}$ the sink of all $(A\otimes_p B)\otimes k' : (A\otimes_p B)\otimes_p B' \to (A\otimes_p B)\otimes L$, $B'\in\mathcal{B}$ and all ${}_{(a,b)}s^{(A\otimes B)L} : L \to (A\otimes_p B)\otimes L$ is final, so the whole sink appearing on the roof of the diagram is final. The canonical associativity $\alpha : (TM\times TN)\times TL \to TM\times(TN\times TL)$ therefore can be lifted to a morphism $a_{MNL} : (M\otimes N)\otimes L \to M\otimes(N\otimes L)$. Now consider the following diagram:

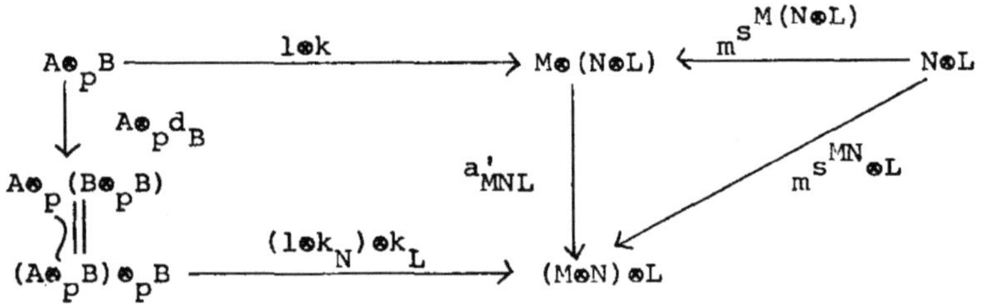

d_B denotes the lifted diagonal and $k_N := p_N k$ resp. $k_L := p_L k$ are the components of k, which are morphisms in $\mathcal{C}$ because $\otimes$ is unitary. Again the sink on the roof of the diagram is final, so the associativity $\alpha^{-1} : TM\times(TN\times TL) \to (TM\times TN)\times TL$ in *Set* can be lifted to a morphism $a'_{MNL} : M\otimes(N\otimes L) \to (M\otimes N)\otimes L$. Of course we have $a_{MNL}a'_{MNL} = M\otimes(N\otimes L)$ and $a'_{MNL}a_{MNL} = (M\otimes N)\otimes L$, i.e. $\otimes$ is associative.

(b) We define $\otimes$ and H as in (1.2.1) resp. (1.2.2). Then the proof of

(a) shows that $\otimes$ is unitary and left adjoint to H. Also the morphism a_{MNL} is still availble (cp. (1.2.3)). The finality of the sinks appearing in the following diagram now proves that the canonical symmetry $TM \times TN \cong TN \times TM$ in *Set* can be lifted to an isomorphism $sym^M_N : M\otimes N \to N\otimes M$:

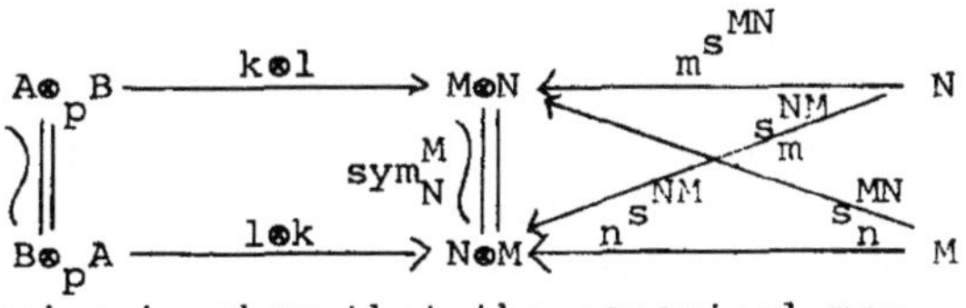

It remains to show that the canonical map $\alpha^{-1} : TM\times(TN\times TL) \to (TM\times TN)\times TL$ can be lifted.
This is done by the following morphism:

$$M\otimes(N\otimes L) \xrightarrow{M\otimes sym^N_L} M\otimes(L\otimes N) \xrightarrow{sym^M_{L\otimes N}} (L\otimes N)\otimes M \xrightarrow{a_{LNM}} L\otimes(N\otimes M) \xrightarrow{sym^L_{N\otimes M}} (N\otimes M)\otimes L \xrightarrow{sym^N_M\otimes L} (M\otimes N)\otimes L\ .$$

Note that the extended structure $\otimes$ above is something like the "coarsest" monoidal closed structure extending $(\otimes_p, H_p)$, i.e. given another monoidal closed structure $\otimes'$ on C with $\otimes'/A\times B = \otimes_p$ there is a natural transformation $\beta_{MN} : M\otimes N \to M\otimes' N$ such that the underlying morphism of β_{MN} is the identity.
We get the following corollary, which of course admits further generalization:

1.3 Corollary: Take $B\subset C$ to be a class of cartesian objects, i.e. for every $B\in B$ $\square\pi B$ has a right adjoint. Then there is a monoidal closed structure $\otimes$ on C with $M\otimes B = M\pi B$ for every $M\in ObC$, $B\in ObB$.

Proof: The productfunctor $\pi : C\times B \to C$ has a right adjoint $H_p : B^{op}\times C \to C$ and applying the usual trick we can assume $TH_p(B,M) = C(B,M)$ in a natural way (we identify B with the full subcategory generated by B). For $B,B'\in B$, $M\in ObC$ we get $TH_p(B,H_p(B',M)) = C(B,H_p(B',M)) \cong C(B\pi B',M)$, thus providing $C(B\pi B',M)$ with the final structure with respect to the isomorphism above we get an object $H_p(B\pi B',M)$ which is isomorphic to $H_p(B,H_p(B',M))$. Of course $H_p(B\pi B',\square)$ is a functor and we get for $M,N\in ObC$: $C(M\pi B\pi B',N) \cong C(M\pi B,H_p(B',N)) \cong C(M,H_p(B,H_p(B',N))) \cong C(M,H_p(B\pi B',N))$. That means: Finite products of cartesian objects are cartesian. Now take B^e to be the closure of B under finite products, then $\pi : C\times B^e \to C$ has a right adjoint $H_p^e : B^{e^{op}}\times C \to C$ and obviously (π, H_p) is a partial monoidal closed structure fullfilling the assumptions of the extension theorem.

In particular the subclass of all cartesian objects yields a monoidal closed structure which of course is a good approximation for cartesian closedness. Furthermore taking C to be the category *Top* of topological spaces, it is well known that for every compact Hausdorff space the functor $\square\pi B$ has a right adjoint which is given by the compact open topology. Hence every class of compact Hausdorff spaces yields a monoidal closed structure. So corollary 1.3 can be understood as a generalization of a corresponding theorem of Booth and Tillotson (cp. [2], Theorem 2.6). Moreover it is shown that the condition of regularity in that theorem is superfluous.

<u>1.4 Corollary:</u> Take C to be the MacNeille completion of $A<C$.

(a) C has a symmetric monoidal closed structure if there is a partial symmetric monoidal closed structure with respect to A (i.e. with respect to A,A,A).

(b) C is cartesian closed iff there is a functor $H_p : A^{op}\times A \to C$ and a natural isomorphism $C(A\pi A', A'') \cong C(A, H_p(A', A''))$ for all $A, A', A''\in \mathrm{Ob}A$, where π is the product in C.

<u>Proof:</u> We have to prove (b): (1.2.1) and (1.2.2) with $\otimes_p = \pi$ yield a pair of adjoint functors $(\otimes, H)$. Because of remark 0.2 is suffices to show that for each $M\in \mathrm{Ob}C$ the diagonal $\delta_M : TM \to TM\times TM$ is a morphism in C. Consider the following diagram:

$$\begin{array}{ccc} A & \xrightarrow{\quad f \quad} & M \\ {\scriptstyle d_A}\big\downarrow & & \big\downarrow{\scriptstyle d_M} \\ A\pi A & \xrightarrow{\quad f\otimes f \quad} & M\otimes M \end{array}$$

The finality of the sink $f : A \to M$, $A\in \mathrm{Ob}A$, yields the assumption.

Note that because of 0.2 it is useless to apply 1.2 (a) to MacNeille completions, we would get again 1.4 (b). Especially if A in 1.4 is already symmetric monoidal closed, its structure can be extended to the MacNeille completion. In [13] it is shown, that every semitopological functor has a MacNeille completion, so 1.4 can be applied to categories which have a symmetric monoidal closed structure with respect to a semitopological functor.

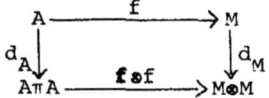

Finally it should be mentioned that R. Börger independently found 1.4(b) using quite different methods (not published). He got this statement as a generalization of a theorem of A. Frölicher , who considered MacNeille completions of one object categories via so called "cartesian monoids" (cp. [5]).

2. Applications to *Top*, *Near*, *Mero* and *Unif*.

In the following we shall apply the results of section 1 to the categories of topological spaces (*Top*), nearness spaces (*Near*), merotopic spaces (*Mero*) and uniform spaces (*Unif*). Note that a class *A* of compact Hausdorff spaces is called rigid (or "strongly rigid" cp. [9]) iff every continuous map between two of them is either constant or an identity. Rigid classes of compact spaces are considered explicitely for example in [9] or [14], here they are used to get different monoidal closed structures in the categories mentioned above. In order to state our main theorem on monoidal closed structures in *Top* we have to consider the following simple lemmata:

2.1 Lemma: Take X to be a compact Hausdorff space and $x_i \in X$, $y_i \in \mathbb{R}$ (= real numbers), $x_i \neq x_j$, $1 \leq i,j \leq n$. Then there is a continuous function $f : X \to \mathbb{R}$ with $f(x_i) = y_i$.

Proof: $M := \bigcup_{i=1}^{n} \{x_i\}$ is a discrete and closed subspace of X, hence $g : M \to \mathbb{R}$, $g(x_i) = y_i$ is continuous, hence there is a continuous map $f : X \to \mathbb{R}$ with $f(x_i) = y_i$ because X is T_4.

2.2 Lemma: Take X to be a compact Hausdorff space and denote by $C_{co}(X)$ resp. $C_{pw}(X)$ the set of all continuous functions from X to $\mathbb{R}$ equipped with the compact open topology resp. the topology of pointwise convergence. Then $C_{co}(X) = C_{pw}(X)$ iff X is finite.

Proof: For finite X the equation is obvious. We assume $C_{co}(X) = C_{pw}(X)$. A subbase of the topology of $C_{co}(X)$ consists of sets $U(K,O) = \{f : X \to \mathbb{R} \mid f(K) \subset O\}$ where $K \subset X$ is compact and $O \subset \mathbb{R}$ open. A subbase of the topology of $C_{pw}(X)$ consists of $U(\{x\},O)$, $x \subset X$, $O \subset \mathbb{R}$ open. Now suppose that X is not finite, take $O :=]0,1[$ and assume that there are open sets $\emptyset \neq O_i \subset \mathbb{R}$ and $x_i \in X$, $1 \leq i \leq n-1$, with

$$(2.2.1) \qquad U(X,O) = \bigcap_{i=1}^{n-1} U(\{x_i\},O_i).$$

Because of $U(\{x\},O) \cap U(\{x\},O') = U(\{x\},O \cap O')$ for $x \in X$, $O,O' \subset \mathbb{R}$ open, we can assume $x_i \neq x_j$ for $1 \le i,j \le n-1$. Now there is an element $x_n \in X$, $x_n \neq x_i$, $1 \le i \le n-1$, and given $y_i \in O_i$, $1 \le i \le n-1$, there is a continuous function $f : X \to \mathbb{R}$ with $f(x_i) = y_i$, $1 \le i \le n-1$, and $f(x_n) = 1$ (cp. lemma 2.1). Hence $f \in \bigcap_{i=1}^{n-1} U(\{x_i\},O_i)$ but $f \notin U(X,O)$, so (2.2.1) cannot hold. Of course also the equation $U(X,O) = \bigcup_{j\in J} \bigcap_{i=1}^{n} U(\{x_{ij}\},O_{ij})$ for points $x_{ij} \in X$ and open sets $O_{ij} \subset \mathbb{R}$ is not true, thus we have a contradiction to $C_{co}(X) = C_{pw}(X)$.

<u>2.3 Lemma:</u> Take $M \subset \mathrm{Ob}\,\mathit{Top}$ to be a rigid class of compact Hausdorff spaces and denote by M^e the class of all finite products of spaces of M. Then all continuous mappings between the spaces of M^e are either constant or canonical, i.e. every morphism $f : \prod_{i=1}^{n} M_i \to M$, $M_i \in M$ is constant or a projection.

<u>Proof:</u> We prove the assumption for $n = 2$, then for arbitrary n the proof is obvious. So consider spaces $X,Y,Z \in M$, $Y \neq Z$. For a continuous mapping $f : X\pi Y \to Z$ define $\hat{f} : X \to C_{co}(Y,Z)$ by $\hat{f}(x)(y) := f(x,y)$. ($C_{co}(Y,Z)$ is the set of all continuous maps from Y to Z with the compact open topology). The exponential law yields the continuity of $\hat{f}$. Moreover we have $C_{co}(Y,Z) \cong Z$, hence $\hat{f}$ is constant or an isomorphism, so we obtain that f is either constant or a projection. Now assume $Y = Z$. If $X \neq Z$ we get the statement above by interchanging X and Y. It remains to consider the case $X = Y = Z$, which is already mentioned by Herrlich [6], p. 134.

<u>2.4 Theorem:</u> Take M to be a rigid class of non finite compact Hausdorff spaces. Then there are as many different monoidal closed structures in *Top* as there are non-empty subclasses of M.

<u>Proof:</u> Take M to be a rigid class of non finite compact Hausdorff spaces, $\emptyset \neq N \subset M$, $\emptyset \neq P \subset M$, $N \neq P$, say $N \in N$, $N \notin P$. Then according to corollary 1.3 we get monoidal closed structures $\otimes^P$ and $\otimes^N$ with right adjoint H^P resp. H^N. $H^P(N,\mathbb{R})$ is the set $\mathit{Top}(N,\mathbb{R})$ provided with the initial structure with respect to the source of all maps $H^P(N,\mathbb{R}) \xrightarrow{H^P(f,\mathbb{R})} C_{co}(P,\mathbb{R})$, $P \in P^e$, $f : P \to N$, where P^e is the class of all finite products of spaces of P. Of course all $f : P \to N$ must be constant, so for $K \subset P$ compact and $O \subset R$ open we get $H^P(f,\mathbb{R})^{-1}(U(K,O)=U(f(K),O)$ hence $H^P(N,\mathbb{R})$ carries the topology of the pointwise convergence, which

by lemma 2.2 is different from $H^{N}(N,R) = C_{co}(N,R)$.

Next let us consider the category *Unif* of uniform spaces. For uniform spaces X and Y we have the "semi-uniform" product $X \circledast Y$, which is defined to be the set $X \times Y$ provided with the initial uniformity induced by all semi-uniform functions from $X \times Y$ to spaces Z (cp. Isbell [8], p. 44). The semi-uniform product is left adjoint to the structure of uniform convergence on function spaces and yields a monoidal closed structure on *Unif* (cp. [8], p. 46). The forgetful functor $F : \mathit{Unif} \to \mathit{Top}$ is known to preserve initial sources. For compact Hausdorff spaces K,K' we have $K \circledast K' = K \pi K'$, (cp. [8], p.46), so it is easy to set up the following

<u>2.5 Theorem:</u> Take M to be a rigid class of non finite compact Hausdorff spaces. Then there are as many different monoidal closed structures in *Unif* as there are non-empty subclasses of M.

<u>Proof:</u> Take $\emptyset \neq K \subset M$ and denote by K^{e} the class of all finite products of spaces of K. Then $\circledast/(\mathit{Unif} \times K^{e})$ yields a partial monoidal closed structure $(\circledast_p, H_p)$ with respect to C, K^{e}, C, which is extendable by 1.2(a) to a monoidal closed structure $\circledast^{K}$ on *Unif*. Take H^{K} to be right adjoint to $\circledast^{K}$, then for uniform spaces X,Y the function space $H^{K}(X,Y)$ is constructed by requiring the source of all $H^{K}(X,Y) \xrightarrow{H^{K}(f,Y)} C_u(K,Y)$ to be initial, where $C_u(K,Y)$ is the set of all uniform continuous functions from K to Y provided with the structure of uniform convergence, $K \in K^{e}$, $f : K \to X$. It follows that the source $FH^{K}(X,Y) \xrightarrow{FH^{K}(f,Y)} FC_u(K,Y)$ is initial in *Top*. Moreover we have $FC_u(K,Y) = C_{co}(K,FY)$, i.e. for a compact Hausdorff space X the function space $FH^{K}(X,\mathbb{R})$ carries the initial topology with respect to all $FH^{K}(X,\mathbb{R}) \xrightarrow{FH^{K}(f,\mathbb{R})} C_{co}(K,\mathbb{R})$, $K \in K^{e}$, $f : K \to X$ continuous, i.e. $FH^{K}(X,\mathbb{R})$ coincides with the corresponding function space constructed in the proof of 2.4, hence distinct non empty subclasses of M yield different monoidal closed structures in *Unif*.

Now let us have a look at the category *Mero* of merotopic and at the category *Near* of nearness spaces (cp. for example [7]). We shortly repeat the relevant facts on these categories: A merotopic space is a pair (X,μ) such that X is a set and μ is a set of covers of X satisfying the following conditions:

(i) If $\mathcal{A}\in\mu$ and $\mathcal{A}<\mathcal{B}$, i.e. for each $A\in\mathcal{A}$ there is an $B\in\mathcal{B}$ with $A\subset B$, then $\mathcal{B}\in\mu$

(ii) $\emptyset\notin\mu$ and $\{X\}\in\mu$

(iii) $\mathcal{A}\in\mu$, $\mathcal{B}\in\mu$ implies $\mathcal{A}\wedge\mathcal{B} = \{A\cap B \mid A\in\mathcal{A}, B\in\mathcal{B}\}\in\mu$.

The members of μ are called uniform covers, a uniform continuous map $f : (X,\mu) \to (Y,\mu)$ is a set map $f : X \to Y$ satisfying $f^{-1}(\mathcal{A})\in\mu$ for all $\mathcal{A}\in\mu$. f is final iff the following condition holds: A cover $\mathcal{A}$ of Y is in ν iff $f^{-1}(\mathcal{A})\in\mu$. For a set I and merotopic spaces (X_i,μ_i) (X_i disjoint) $\coprod_{i\in I} X_i := (\bigcup_{i\in I} X_i, \coprod_{i\in I}\mu_i)$, where $\coprod_{i\in I}\mu_i$ is the set of all covers of $\bigcup_{i\in I} X_i$ which are refined by a cover $\bigcup_{i\in I}\mathcal{A}_i$, $\mathcal{A}_i\in\mu_i$, together with the canonical injections is the coproduct resp. the sum of the spaces (X_i,μ_i). The forgetful functor $T : \mathit{Mero} \to \mathit{Set}$ is amnestic and topological. Unif is a bireflective subcategory of Mero, a merotopic space (X,μ) is uniform iff every $\mathcal{A}\in\mu$ admits a star-refinement, i.e. there is a cover $\mathcal{V}\in\mu$ such that $\{St(V,\mathcal{V}) \mid V\ \mathcal{V}\}$ is a refinement of $\mathcal{A}$ lying in μ with $St(V,\mathcal{V}) = \cup\{W\in\mathcal{V} \mid W\cap V \neq \emptyset\}$ (cp. [7]).

A merotopic space (X,μ) is called a nearness space iff the following condition holds: for every $\mathcal{A}\in\mu$ we have $\mathrm{int}_\mu\mathcal{A} := \{\mathrm{int}_\mu A \mid A\in\mathcal{A}\}\in\mu$ with $\mathrm{int}_\mu A = \{x \mid \exists\mathcal{V}\in\mu,\ St(x,\mathcal{V})\subset A\}$. Nearness spaces are a full bireflective subcategory of Mero, it is easy to check, that the reflector $R : \mathit{Mero} \to \mathit{Near}$ can be constructed by ordinal induction in the following way:

<u>2.6 Remark:</u> For (X,μ) ObMero define $\mu_o := \mu$. For an ordinal α take $\mu_{\alpha+1} := \{\mathcal{A}\in\mu_\alpha \mid \mathrm{int}_{\mu_\alpha}\mathcal{A}\in\mu_\alpha\}$ and for a limit ordinal α set $\mu_\alpha := \cap\{\mu_\beta \mid \beta<\alpha\}$. Then $R(X,\mu) := (X, \bigcap_\alpha \mu_\alpha)$ is a nearness space and the canonical morphism $\eta_X : (X,\mu) \to R(X,\mu)$ has the universal property of a reflection.

For every symmetric topological space X (R_o-space) the set of all open covers of X generates a nearness structure. It is easy to prove that R_o-spaces are a bicoreflective subcategory of Near, i.e. we have an

(2.6.1) "underlying topology" functor $V : \mathit{Near} \to \mathit{Top}_o$ being right adjoint to the inclusion $J : \mathit{Top}_o \to \mathit{Near}$.

An obvious way of getting monoidal closed structures in Near and Mero would be to define a semi-uniform product and a structure of uniform convergence like in Unif. But in absence of the star-refinement axiom it is difficult to obtain an exponential law. So we have to get along with the following

<u>2.7 Proposition:</u> For $(X,\mu),(Y,\nu)\in Ob\mathit{Mero}$ we define the semi-uniform product of X and Y by $X\circledast Y := (X\times Y,\mu\circledast\nu)$, where $\mu\circledast\nu$ is the set of all covers refined by some cover $A\circledast(\mathcal{B}_A) := \{A\times B_A | A\in\mathcal{A}, B_A\in\mathcal{B}_A\}$, $\mathcal{A}\in\mu$, $\mathcal{B}_A\in\mu$. $\circledast$ is a monoidal closed structure on Mero.

<u>Proof:</u> Obviously $\circledast$ is a tensorproduct in Mero, it remains to show, that $\circledast$ has a right adjoint. According to 0.1 it suffices to show, that for every $(Y,\nu)\in Ob\mathit{Mero}$ $\square\circledast Y$ preserves coproducts and final epimorphisms: Take $\coprod_{i\in I} X_i = (\bigcup_{i\in I} X_i, \coprod_{i\in I}\mu_i)$ to be the coproduct of $(X_i,\mu_i)\in Ob\mathit{Mero}$.
Then $(\coprod_{i\in I}\mu_i)\circledast\nu$ is generated by all covers
$\bigcup_{i\in I} \mathcal{A}_i\circledast(\mathcal{B}_{A_i}) = \{A_i\times B_{A_i} | A_i\in\mathcal{A}_i, B_{A_i}\in\mathcal{B}_{A_i}, i\in I\} = \bigcup_{i\in I}(\mathcal{A}_i\circledast(\mathcal{B}_{A_i}))$ with $\mathcal{A}_i\in\mu_i, \mathcal{B}_{A_i}\in\nu$, hence $(\coprod_{i\in I}\mu_i)\circledast\nu = \coprod_{i\in I}(\mu_i\circledast\nu)$ and $\square\circledast Y$ preserves coproducts.
Now take $e : (X,\mu) \to (X',\mu')$ to be a final epimorphism and assume $\mathcal{D}$ to be a cover of $X'\times Y$ with $(e\times Y)^{-1}(\mathcal{D})\in\mu\circledast\nu$. Then there is a refinement $\mathcal{A}\circledast(\mathcal{B}_A)$ of $(e\circledast Y)^{-1}(\mathcal{D})$ in $\mu\circledast\nu$ and $e(\mathcal{A})\circledast(\mathcal{B}_A)$ is a refinement of $\mathcal{D}$, i.e. $\mathcal{D}\in\mu'\circledast\nu$, so $e\circledast Y$ is final.

In order to state a result corresponding to 2.7 for nearness spaces we have to prove the following

<u>2.8 Lemma:</u> Assume $(X,\mu),(Y,\nu)\in Ob\mathit{Mero}$. Then the following holds:
(a) $\mathrm{int}_{\mu\circledast\nu}(\mathcal{A}\circledast(\mathcal{B}_A)) = \mathrm{int}_\mu\mathcal{A}\circledast(\mathrm{int}_\nu\mathcal{B}_A)$ for $\mathcal{A}\in\mu$, $\mathcal{B}_A\in\nu$.
(b) $X,Y\in Ob\mathit{Near}$ implies $X\circledast Y\in Ob\mathit{Near}$.
(c) For $Y\in Ob\mathit{Near}$ we get $R(X\circledast Y) = RX\circledast Y$.

<u>Proof:</u> (a) Take $(x,y)\in\mathrm{int}_{\mu\circledast\nu}(A\times B_A)$, then there are covers $\mathcal{U}\in\mu$, $\mathcal{V}_U\in\nu$, for each $U\in\mathcal{U}$ such that $St((x,y),\mathcal{U}\circledast(\mathcal{V}_U))\subset A\times B_A$. That means $\bigcup\{U\times V_U | (x,y)\in U\times V_U\}\subset A\times B_A$, hence $St(x,\mathcal{U})\times\bigcup_{U,x\in U} St(y,\mathcal{V}_U)\subset A\times B_A$ and we get $x\in\mathrm{int}_\mu A$, $y\in\mathrm{int}_\nu B_A$. Other way round given $(x,y)\in\mathrm{int}_\mu A\times\mathrm{int}_\nu B_A$ there are covers $\mathcal{U}\in\mu$, $\mathcal{V}\in\nu$ with $St(x,\mathcal{U})\subset A$, $St(y,\mathcal{V})\subset B_A$, so $St((x,y),\mathcal{U}\circledast\mathcal{V})\subset St(x,\mathcal{U})\times St(y,\mathcal{V})\subset A\times B_A$, i.e. $(x,y)\in\mathrm{int}_{\mu\circledast\nu}(A\times B_A)$.
($\mathcal{U}\circledast\mathcal{V} = \{U\times V | U\in\mathcal{U}, V\in\mathcal{V}\}$).
(b) is an immediate consequence of (a).
(c) We show $(\mu\circledast\nu)_\alpha = \mu_\alpha\circledast\nu$ for every ordinal α (cp. 2.6). For $\alpha = 0$ there is nothing to prove. For an arbitrary ordinal α we get $\mathcal{A}\circledast(\mathcal{B}_A)\in(\mu\circledast\nu)_{\alpha+1}$ iff $\mathcal{A}\circledast(\mathcal{B}_A)\in\mu_\alpha\circledast\nu$ and $\mathrm{int}_{\mu_\alpha\circledast\nu}(\mathcal{A}\circledast(\mathcal{B}_A))\in\mu_\alpha\circledast\nu$. Because of (a) this is equivalent to $\mathcal{A}\circledast(\mathcal{B}_A)\in\mu_\alpha\circledast\nu$ and $\mathrm{int}_{\mu_\alpha}\mathcal{A}\circledast(\mathrm{int}_\nu\mathcal{B}_A)\in\mu_\alpha\circledast\nu$, hence

$\mathrm{int}_{\mu_\alpha} A\in\mu_\alpha$, i.e. $A\in\mu_{\alpha+1}$ and $A\circledast(\mathcal{B}_A)\in\mu_{\alpha+1}\circledast\nu$. The other direction is obvious. Of course, for a limit ordinal α $(\mu\circledast\nu)_\alpha = \mu_\alpha\circledast\nu$ holds, hence we get $\bigcap_\alpha(\mu\circledast\nu)_\alpha = \bigcap_\alpha(\mu_\alpha\circledast\nu) = (\bigcap_\alpha\mu_\alpha)\circledast\nu$ and this proves our assumption.

The following proposition yields the monoidal closed structure of uniform convergence for nearness spaces:

<u>2.9 Proposition:</u> The semi-uniform product is a monoidal closed structure on *Near*.

<u>Proof:</u> *Near* is an amnestic topological category and obviously by 2.8 $\circledast$ is a tensorproduct in *Near*. It suffices to show that for each $X\in \mathrm{Ob}\,Near$ $\square\circledast X$ preserves final epi-sinks: Take $(f_i : Y_i \to Y)$ to be a final epi-sink in *Near*. Then there is a merotopic space Y' with the same underlying set as Y such that $(f_i : Y_i \to Y')$ is a final epi-sink in *Mero* and we get by 0.1 $RY' = Y$, hence
$(f_i\circledast X : Y_i\circledast X \to Y\circledast X) = (Rf_i\circledast X : Y_i\circledast X \to RY'\circledast X) = (R(f_i\circledast X) : Y_i\circledast X \to R(Y'\circledast X))$
by 2.8 R preserves final epi-sinks, hence the sink of all $f_i\circledast X$ is final.

<u>2.10 Theorem:</u> Take M to be a rigid class of non finite compact spaces. Then there are as many different monoidal closed structures in *Near* as there are non empty subclasses of M.

<u>Proof:</u> Take K to be a compact Hausdorff space, X an R_0-space. Then for an open cover A of X and open covers $\mathcal{B}_A$, $A\in A$, of K obviously $A\circledast(\mathcal{B}_A)$ is an open cover of the direct product $X\pi_0 K$ in the category of R_0-spaces, i.e. $X\pi_0 K = X\circledast K$. where $\circledast$ denotes the semi-uniform product in *Near*. Therefore assuming $C_n(\square,\square)$ to be right adjoint to $\circledast$ we can conclude $VC_n(K,Y) = C_{co}(K,VY)$ for $Y\in \mathrm{Ob}\,Near$ (cp. (2.6.1)) because $J(\square\pi_0 K)$ is left adjoint to $C_{co}(K,V)$. Furthermore we have $K\pi K = K\circledast K$, π being the product in *Near*, so analoguously to the proof of 2.5 we get for each $\emptyset \neq K\subset M$ a partial monoidal closed structure $(\circledast_p, H_p)$ which is extendable to a monoidal closed structure $\circledast^K$. For nearness spaces Y,Z the adjoint function space $H^K(Y,Z)$ is constructed by requiring the source of all $H^K(Y,Z)\xrightarrow{H^K(f,Z)} C_n(K,Z)$ to be initial. The functor $V : Near \to Top_0$ preserves initial sources, hence the argument in the proof 2.5 shows that $VH^K(X,\mathbb{R})$ coincides with the corresponding function space of 2.4 (X compact), thus different non-empty

subclasses of M yield different monoidal closed structures.

2.11 Theorem: Take M to be a rigid class of non finite compact spaces. Then there are as many different monoidal closed structures in *Mero* as there are non-empty subclasses of M.

Proof: The reflector R : *Mero* → *Near* preserves final epi-sinks. Hence for a class K of compact Hausdorff spaces we can construct a monoidal closed structure $\otimes^K$ in *Mero* such that $R\otimes^K$ yields the corresponding structure in *Near* (cp. proof of 2.10). So for different $\emptyset \neq P$, $N \subset M$ we get different monoidal closed structures $\otimes^P$ and $\otimes^N$ in *Mero*.

Finally a result of Trnkova [14], which constitutes the existence of a proper class of rigid spaces yields the following

2.12 Corollary: If there is no measurable cardinal, the number of monoidal closed structures in *Top*, *Unif*, *Near* and *Mero* is equipotent to the conglomerate of all subclasses of a proper class.

Let us remark, that the categories mentioned in this section give only a small choice of examples. A result corresponding to 2.4,2.5, 2.10 resp. 2.11, obviously holds for the category of general convergence spaces and probably can be obtained for proximity spaces.

In this paper we don't want to consider any more the symmetric case. It should be mentioned that 1.4 can be applied to MacNeille completions of classes of compact spaces or in a special form to Heyting algebras which are cartesian closed complete lattices, hence topological categories over the full subcategory $\{\emptyset\} < Set$.

REFERENCES

[1] Adamek, Herrlich, Strecker: Least and largest initial competions, Comment. Math. Univ. Carolin. 20,1, 43-58, 1979.

[2] Booth, Tillotson: Monoidal closed, cartesian closed and convenient categories of topological spaces, Pacific J. Math., 88, 35-53, 1980.

[3] Čincura: Tensorproducts in the category of topological spaces, Comment. Math. Univ. Carolin. 20, 431-446, 1979.

[4] Eilenberg, Kelly: Closed categories, Proc. Conf. on Categorical Algebra, La Jolla 1965, Springer Verlag, Berlin, 1969.

[5] Frölicher: Durch Monoide erzeugte kartesisch abgeschlossene Kategorien, Seminarberichte aus dem Fachbereich Mathematik der Fernuniversität, Nr. 5, 1979.

[6] Herrlich: Topologische Reflexionen und Coreflexionen, Lecture Notes in Math. 78, Springer, Berlin, 1968.

[7] Herrlich: Topological structures, Math. Centre Tracts 52, Amsterdam, 59-122, 1974.

[8] Isbell: Uniform spaces, Mathematical surveys No. 12, American Mathematical Society, Providence, XI, 1964.

[9] Kannan, Rajagopalan: Constructions and applications of rigid spaces, Advances in Math. 29, 89-130, 1978.

[10] Linton: Autonomuous equational categories, J. Math. Mech. 15, 637-642, 1966.

[11] MacLane: Categories for the working mathematician, Springer, Berlin, 1977.

[12] Pavelka: Tensorproducts in the category of convergence spaces, Comment. Math. Univ. Carolin. 13, 4, 693-709, 1972.

[13] Tholen: Semitopological functors I, J. Pure and Appl. Algebra 15, 53-73, 1979.

[14] Trnkova: Non-constant continuous maps of metric or compact Hausdorff spaces, Comment. Math. Univ. Carolin. 13,2, 283-295, 1972.

[15] Wischnewsky: On monoidal closed topological categories I, Proc. Conf. on Categorical Topology, Lecture Notes in Math. 540, Springer, Berlin, 676-686, 1976.

Georg Greve
Fachbereich Mathematik
Fernuniversität Hagen
Postfach 940
5800 Hagen 1
Fed. Rep. of Germany a.l.

THE FUNDAMENTAL GROUPOID AND THE HOMOTOPY CROSSED COMPLEX OF AN ORBIT SPACE

P.J. HIGGINS and J. TAYLOR

1. Introduction

Let G be a group acting on a topological space X and let X/G be the orbit space. Armstrong and Rhodes [1,2,10] have shown how to determine $\pi_1(X/G, *)$ in special cases from information on $\pi_1(X, *)$ and the action of G. We show here that the use of fundamental groupoids in place of fundamental groups provides a more functorial setting for these results and thus simplifies both their statements and their proofs. This in turn opens the way to analogous results in higher dimensions, the fundamental groupoids being replaced by the homotopy crossed complexes of suitable filtrations of the spaces, as in the higher dimensional Seifert-Van Kampen theorem proved in [6,7].

We first list the cases in which the fundamental group of X/G is known to have an easy algebraic description, and the reader will notice that in each case the action of G is assumed to be discontinuous in some sense. It is this assumption that isolates the algebraic aspects of the problem from the topological ones, and we will not consider more general types of action except for the purpose of setting up general machinery and formulating some natural questions.

We write $q: X \to X/G$ for the quotient map, and $x \cdot g$ for the image of $x \in X$ under the action of $g \in G$ (with $x \cdot gh = (x \cdot g) \cdot h$). The orbit $g(x) = x \cdot G$ will also be denoted by $\bar{x}$.

Case 1. Let X be connected and let the action of G be properly discontinuous (that is, every $x \in X$ has a neighbourhood N_x such that, for $g \neq 1$ in G, $N_x \cap N_x \cdot g = \emptyset$). Then G acts freely, the quotient map $q: X \to X/G$ is a regular covering projection, and $\pi_1(X/G, \bar{x}_o)$ is an extension of the group $\pi_1(X, x_o)$ by the group G. (See, for example, Spanier [11, p.87]).

Case 2. (Armstrong [1]). Let K be a connected simplicial complex and let $X = |K|$ be its polyhedron. Let G act on X by simplicial maps. If X is simply connected then $\pi_1(X/G, \bar{x}_o) \cong G/H$, where H is the normal subgroup of G generated by those elements which have fixed points.

Case 3. (Armstrong [2]). Let X be a path-connected locally compact metric space. Assume that G acts so that (i) the stabiliser of each point of X is finite, and (ii) each point $x \in X$ has a neighbourhood N_x such that $N_x \cap N_x \cdot g = \emptyset$ for all g not in the stabilizer of x. If X is simply connected, then $\pi_1(X/G, \bar{x}_o) \cong G/H$ as in Case 2.

Case 4. (Rhodes [10]). For any group G acting on a space X, define the fundamental group of the G-space X as follows. A path of type g in X at the base-point x_o is a pair (α, g), where $g \in G$ and α is a path in X from x_o to $x_o \cdot g^{-1}$. For fixed g, let $[\alpha, g]$ denote the set of paths of type g homotopic to α relative to end-points. These homotopy classes form a group $\sigma = \sigma(X, x_o, G)$ with respect to the composition defined by

$$[\alpha,g] + [\beta,h] = [\alpha + \beta\cdot g^{-1},gh].$$

Let σ' be the normal subgroup of σ generated by all elements of the form $[\alpha - \alpha\cdot g^{-1},g]$ where α is a path from x_o to x_1 and g fixes x_1. If $X = |K|$ is a connected polyhedron with G acting simplicially as in Case 2 (but without the assumption that X is simply connected) then $\pi_1(X/G,\bar{x}_o) = \sigma/\sigma'$.

These examples suggest that the fundamental group behaves rather erratically in relation to group actions, but we will show that order can be restored by the introduction of the fundamental groupoid. The indications that groupoids may be helpful are the following.

(i) The fundamental group is a functor on spaces with base-point, whereas we certainly want to consider group actions $G \to$ Top, where Top is the category of spaces without base-point. To abandon the base-point means to consider paths instead of loops and therefore to use fundamental groupoids.

(ii) Rhodes' construction seems unnatural in the context of groups; he is really using groupoids in disguise.

(iii) A simple example shows that, even in the case when G fixes the base-point the fundamental group does not behave well on passing to the orbit space, whereas a suitable fundamental groupoid does. Let X be a circle and let G be cyclic of order 2, its non-trivial element g acting by reflection in a diameter. If the fixed points are x_o, x_1 then the fundamental group $\pi_1(X,x_o)$ is infinite cyclic, with generator α, say, and G acts on this group with $\alpha\cdot g = -\alpha$. The result of killing this action is a group of order 2, whereas the fundamental group $\pi_1(X/G,\bar{x}_o)$ is trivial. However, the fundamental groupoid of X with respect to the pair of points x_o, x_1 (see §2 below for definitions) is a free groupoid with two vertices x_o, x_1 and two generators α, β from x_o to x_1. The group G acts on this groupoid with $\alpha\cdot g = \beta$, $\beta\cdot g = \alpha$ and the result of killing this action is a free groupoid with two vertices $\bar{x}_o$, $\bar{x}_1$ and one generator $\bar{\alpha}$ from $\bar{x}_o$ to $\bar{x}_1$. Since X/G is a closed interval joining $\bar{x}_o$ and $\bar{x}_1$, its fundamental groupoid relative to $\bar{x}_o$ and $\bar{x}_1$, is also free on one generator and we see that the algebra models the geometry more closely. The fundamental group of X/G is now obtained by retricting attention to a particular vertex after killing the action of G.

2. Fundamental groupoids

We recall that a groupoid is a small category in which all morphisms have inverses We refer to vertices and arrows rather than objects and morphisms, and we use additive notation. The fundamental groupoid $\pi_1(X,X_o)$ of a space X relative to a subspace X_o (to be thought of as a set of base-points) has X_o as its set of vertices; an arrow from x to y is a homotopy class, relative to end-points, of paths from x to y. Addition is induced by the usual addition of paths.

Given a G-space X, let X_o be a subspace stable under G. Then G acts on the fundamental groupoid $\pi_1(X,X_o)$. Also, G acts trivially on the groupoid $\pi_1(X/G,X_o/G_o)$, and the quotient map $q:X \to X/G$ induces a morphism of G-groupoids

$$q' : \pi_1(X,X_o) \to \pi_1(X/G,X_o/G).$$

Now, if Γ is any G-groupoid, we can "kill the action of G" to obtain a new groupoid $\Gamma /\!/ G$ on which G acts trivially. There is a canonical morphism of G-groupoids $\tau : \Gamma \to \Gamma /\!/ G$ and the construction is characterised by the universal property : every G-morphism $\rho : \Gamma \to \Delta$, where Δ is a G-groupoid with trivial action of G, factorises through τ. The existence of such a groupoid is ensured by the co-completeness of the category of groupoids. If Γ is presented as a groupoid by generators and relations, then $\Gamma /\!/ G$ is obtained by adding the relations $x = x\cdot g$ for $x \in \Gamma$, $g \in G$. However, the reader unfamiliar with the algebra of groupoids should be warned that, although the canonical morphism $\tau : \Gamma \to \Gamma /\!/ G$ is surjective, $\Gamma /\!/ G$ is not necessarily a quotient groupoid of Γ by a normal subgroupoid. (The morphism τ is in fact a fibration of groupoids [5]). We shall give an explicit description of $\Gamma /\!/ G$ later.

Using this purely algebraic construction we now have, in all cases, an induced morphism of groupoids

$$q_* : \pi_1(X,X_o) /\!/ G \to \pi_1(X/G,X_o/G).$$

<u>2.1 THEOREM.</u> Let X be a CW-complex and let X_o be its 0-skeleton. Let the group G act on X by cellular maps in such a way that if an element g of G stabilises a cell then it fixes that cell pointwise (in other words, the fixed-point set of every element of G is a subcomplex of X). Then

$$q_* : \pi_1(X,X_o) /\!/ G \to \pi_1(X/G,X_o/G)$$

is an isomorphism of groupoids.

<u>Proof.</u> The fundamental group of a CW-complex can be computed as its edge-path group; if the complex has more than one 0-cell, the resulting presentation involves the choice of a maximal tree in the 1-skeleton. For the fundamental groupoid the corresponding statement is more straightforward. First choose an orientation for each 1-cell and 2-cell, and a base-point in X_o for each 2-cell. Then the groupoid $\pi_1(X,X_o)$ has a presentation with one vertex for each 0-cell, one generator for each oriented 1-cell (with end-points as in the complex) and one relation for each oriented 2-cell, given by the attaching map on its boundary (starting at the base-point). This fact can be deduced from the standard result for the fundamental group, or it can be easily proved using R.Brown's groupoid version of the Seifert-Van Kampen theorem [4].

If G acts on X in the manner described in the theorem, we can choose orientations and base-points which are compatible with the group action. We can also re-choose the characteristic maps of the cells so that they are compatible with the action of G, without changing the topology of X (see Bredon [3], Ch.II; the details are given in [12]). With this renormalisation, two things are apparent. Firstly, the orbit space X/G is a CW-complex with one n-cell for each G-orbit of n-cells of X, and with characteristic maps induced by the normalised characteristic maps of X. Secondly, G acts, not only on $\pi_1(X,X_o)$, but on its standard presentation described above; it permutes the vertices, the generators and the relations. The result of killing this action is therefore a groupoid with one vertex for each G-orbit of 0-cells, one generator for each G-orbit of 1-cells and one defining relation for each G-orbit of 2-cells, given by the attaching map for any cell in the orbit. But this is just the standard presentation of $\pi_1(X/G, X_0/G)$. $\square$

Any simplicial action of G on a polyhedron $X = |K|$ satisfies the hypotheses of Theorem 2.1 if we take as its cell structure a suitable subdivision of the simplicial complex K (the first derived complex of K will suffice). To recover the theorems of Armstrong and Rhodes in Cases 2 and 4 above it is therefore enough to describe the groupoid $\Gamma /\!/ G$, where $\Gamma = \pi_1(X, X_0)$, in such a way that its vertex group at a particular point can be recognised as the group $\pi_1(X/G, \bar{x}_0)$ in the form described by them. To do this we use semi-direct products.

If a group G acts on a groupoid Γ, the semi-direct product $\Gamma \tilde{\times} G$ is a groupoid with the same vertices as Γ, whose arrows from x to y are pairs (α, g), where $g \in G$ and α is an arrow in Γ from x to $y \cdot g^{-1}$. Addition of arrows is defined by

$$(\alpha, g) + (\gamma, h) = (\alpha + \gamma \cdot g^{-1}, gh).$$

[A more symmetric form of this definition is obtained by taking as the arrows all triples (α, g, β) where $\alpha \cdot g = \beta$. The source of this arrow is the source of α and its target is the target of β. Addition is defined by

$$(\alpha, g, \beta) + (\gamma, h, \delta) = (\alpha + \gamma \cdot g^{-1}, gh, \beta \cdot h + \delta).$$

The asymmetric definition omits the superfluous last entry in each triple].

2.2 PROPOSITION. If a group G acts on a groupoid Γ, then

$$\Gamma /\!/ G = (\Gamma \tilde{\times} G)/N,$$

where N is the normal subgroupoid of $\Gamma \tilde{\times} G$ generated by all arrows $(0_x, g)$, 0_x being the zero element at a vertex x of Γ.

Proof. Write $\Sigma = \Gamma \tilde{\times} G$ and let $\mu : \Gamma \to \Sigma$ be the canonical injection $\alpha \longmapsto (\alpha, 1)$. The group G acts on Σ by the rule $(\alpha, g) \cdot h = (\alpha \cdot h, h^{-1} g h)$, and μ is a morphism of G-groupoids. The normal subgroupoid N of Σ generated by all arrows $(0_x, g)$ is clearly a G-subgroupoid, since G permutes these arrows. Hence Σ/N inherits an action of G, and the quotient map $\rho : \Sigma \to \Sigma/N$ is also a morphism of G-groupoids. We therefore have a canonical morphism of G-groupoids

$$\tau = \rho \circ \mu : \Gamma \to \Sigma/N$$

and we show that this has the universal property necessary to identify Σ/N as $\Gamma /\!/ G$. (This gives an independent proof that $\Gamma /\!/ G$ exists).

First, the action of G on Σ/N is trivial because, in Σ,

$$\begin{aligned}(\alpha, g) \cdot h &= (\alpha \cdot h, h^{-1} g h) \\ &= (0_x, h^{-1}) + (\alpha . g) + (0_y, h)\end{aligned}$$

for appropriate vertices x,y of Γ, and the first and last terms of this sum are in N.

If now $\theta : \Gamma \to \Delta$ is any morphism of G-groupoids, where G acts trivially on Δ,

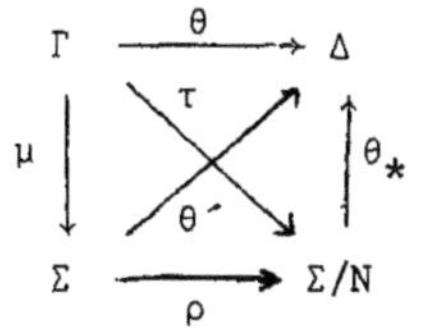

then the map $\theta' : \Sigma \to \Delta$ defined by

$$\theta'(\alpha,g) = \theta(\alpha)$$

is a morphism of G-groupoids. The kernel of θ' contains N, so there is a morphism of groupoids $\theta_* : \Sigma/N \to \Delta$ such that $\theta_* \circ \rho = \theta'$, whence $\theta_* \circ \tau = \theta$. To see that θ_* is unique subject to this last equation, we only need to show that $\tau : \Gamma \to \Sigma/N$ is surjective. But, in Σ,

$$(\alpha,g) = (\alpha,1) + (0_x,g)$$

for an appropriate vertex x, so

$$\rho(\alpha,g) = \rho(\alpha,1) = \tau(\alpha). \quad \Box$$

For any vertex x_o of Γ, the vertex group $\Sigma(x_o)$ of Σ at x_o consists of all arrows from x_o to x_o, that is, all pairs (α,g) where α is an arrow of Γ from x_o to $x_o \cdot g^{-1}$. The vertex group $N(x_o)$ of N is easily seen to be generated by all conjugates lying at x_o in Σ of loops $(0_x,g)$, where g fixes the vertex x. Such conjugates are of the form

$$u = (\alpha,h) + (0_x,g) + (-\alpha \cdot h,\ h^{-1}),$$

where h is an arbitrary element of G, and α is an arrow in Γ from x_o to $x\,h^{-1}$. Writing $x_1 = x \cdot h^{-1}$, $g_1 = h\,g\,h^{-1}$, we see that g_1 fixes x_1 and $u = (\alpha - \alpha \cdot g_1^{-1},\ g_1)$. It is now clear that, in the case $\Gamma = \pi_1(X,x_o)$, the groups $\Sigma(x_o)$ and $N(x_o)$ are precisely Rhodes' groups $\sigma(X,x_oG)$ and $\sigma'(X,x_o,G)$. We therefore recover his theorem that $\pi_1(X/G,\bar{x}_o) \cong \sigma/\sigma'$, but in the more general context of CW-complexes. Armstrong's theorem in Case 2 is an immediate corollary.

The results in Cases 1 and 3 also admit a reformulation in terms of groupoids (though the proofs are not thereby simplified), and the conclusion is the same in all cases, namely, that $q_* : \pi_1(X,X_o) /\!/ G \to \pi_1(X/G,X_o/G)$ is an isomorphism for sufficiently large G-stable subspaces X_o. In particular, writing $\pi_1(X)$ for the full fundamental groupoid $\pi_1(X,X)$, we have, in Cases 1-4 and under the conditions of Theorem 2.1,

(*) $\qquad q_* : \pi_1(X) /\!/ G \to \pi_1(X/G)$ is an isomorphism.

The assertion (*) is equivalent, by a purely algebraic argument, to the assertion that $q_* : \pi_1(X,X_o) /\!/ G \to \pi_1(X/G,X_o/G)$ is an isomorphism for all G-stable subspaces X_o which contain, for each $g \in G$, at least one point of each path-component of the fixed-point set of g (a condition which is satisfied by the X_o of Theorem 2.1). For computational purposes one would use this form of the assertion with X_o as small as possible.

The interesting question remains: if a group G acts on a space X, under what conditions is the assertion (*) valid? For example, as indicated in [2], the condition that X be simply connected in Case 3 can be omitted (and the conclusion (*) follows) if X satisfies local conditions ensuring that a simply-connected covering space exists. It is not clear whether these conditions are really necessary.

3. Crossed complexes

The groupoid version of the Seifert-Van Kampen theorem [4] states that the fundamental groupoid functor preserves suitable colimits. Theorem 2.1 uses this result and is of the same general type. It was proved in [7] that the homotopy crossed complex of a filtered space also preserves suitable colimits, giving a generalisation of the Seifert-Van Kampen theorem to higher dimensions. We now show that crossed complexes can be used similarly to prove a higher-dimensional analogue of Theorem 2.1.

If X is a filtered space $X_0 \subset X_1 \subset \ldots \subset X_n \subset \ldots \subset X$, the homotopy crossed complex $\pi\underset{\sim}{X}$ of $\underset{\sim}{X}$ is an algebraic structure with the following constituents. In dimension 1 lies the fundamental groupoid $\pi_1\underset{\sim}{X} = \pi_1(X_1, X_0)$. In dimension $n \geqslant 2$ lies the family of relative homotopy groups $\pi_n\underset{\sim}{X} = \{\pi_n(X_n, X_{n-1}, x); x \in X_0\}$ which we view as a (disconnected) groupoid with X_0 as its set of vertices. There are boundary morphisms $\delta : \pi_n\underset{\sim}{X} \to \pi_{n-1}\underset{\sim}{X}$ $(n \geqslant 2)$ and an action of $\pi_1\underset{\sim}{X}$ on $\pi_n\underset{\sim}{X}$ $(n \geqslant 2)$. The algebraic laws which hold in this example and are used as the defining laws of a crossed complex are written out fully in [8] in this volume. In summary, they say that $\pi_2\underset{\sim}{X}$ is a crossed module over $\pi_1\underset{\sim}{X}$, that $\delta\pi_2\underset{\sim}{X}$ acts trivially on $\pi_n\underset{\sim}{X}$ $(n \geqslant 3)$, that δ preserves the action of $\pi_1\underset{\sim}{X}$ and that $\delta\delta = 0$.

If a group G acts on $\underset{\sim}{X}$ by filtered maps, that is, each X_n is G-stable, then G also acts on $\pi\underset{\sim}{X}$. Since crossed complexes are equationally defined, they form a co-complete category, so we can kill the action of G, as in Section 2, to obtain a crossed complex $(\pi\underset{\sim}{X})/\!/G$ with the obvious universal property. Again, the quotient map $q : X \to X/G$ induces a morphism of crossed complexes $q_* : (\pi\underset{\sim}{X})/\!/G \to \pi(\underset{\sim}{X}/G)$.

3.1 THEOREM. Let G be a group acting on a CW-complex X and satisfying the conditions of Theorem 2.1. Let $\underset{\sim}{X}$ be the filtered space determined by the skeletons X_n $(n \geqslant 0)$ of X. Then

$$q_* : (\pi\underset{\sim}{X})/\!/G \to \pi(\underset{\sim}{X}/G)$$

is an isomorphism of crossed complexes.

Proof. As in the proof of Theorem 2.1, we may choose base-points, orientations and characteristic maps, for all cells of X, compatible with the action of G. Then X/G becomes a CW-complex whose cells are the orbits of cells of X with base-points orientations and characteristic maps induced by those of X.

Now the cell structure of X gives rise to a standard presentation of the crossed complex $\pi\underset{\sim}{X}$, which can be described as follows. $\pi_1\underset{\sim}{X}$ is the free groupoid with one vertex for each 0-cell and one generator for each (oriented) 1-cell. $\pi_2\underset{\sim}{X}$ is the free crossed module over $\pi_1\underset{\sim}{X}$ with one generator for each 2-cell, located at the vertex corresponding to its base-point, with $\delta : \pi_2\underset{\sim}{X} \to \pi_1\underset{\sim}{X}$ given by the attaching maps of the boundaries of the 2-cells. We write H for the groupoid $\pi_1\underset{\sim}{X}/\delta\pi_2\underset{\sim}{X}$ (which is the fundamental groupoid $\pi_1(X,X_o)$). For $n \geqslant 3$, $\pi_n\underset{\sim}{X}$ is the free H-module with one generator for each n-cell, located at the vertex corresponding to its base-point, with $\delta : \pi_n\underset{\sim}{X} \to \pi_{n-1}\underset{\sim}{X}$ given by the attaching maps of the boundaries of the n-cells, using the homotopy addition lemma. This description of $\pi\underset{\sim}{X}$ follows from classical results, or it can be deduced from Corollary 5.2 of [7]. It is clear that the action of G is compatible with all the constituents of this presentation so that $(\pi\underset{\sim}{X}) /\!/ G$ has a presentation of the same type with one generator in dimension n for each orbit of n-cells. This latter presentation is equivalent to the standard presentation of $\pi(\underset{\sim}{X}/G)$. □

In order to apply this result one needs to analyse the crossed complex $(\pi\underset{\sim}{X}) /\!/ G$. If

$$C : \ldots \longrightarrow C_n \xrightarrow{\delta} C_{n-1} \longrightarrow \ldots \longrightarrow C_2 \xrightarrow{\delta} C_1$$

is any crossed complex and G acts on C, we form the crossed complex $D = C \tilde{\times} G$ as follows. First, $D_1 = C_1 \tilde{\times} G$ as defined in Section 2. For $n \geqslant 2$, $D_n = C_n$, with boundary map $D_2 \to D_1$ given by $x \longmapsto (\delta x, 1)$ and the action of D_1 on D_n given by

$$x^{(\alpha,g)} = x^{\alpha}\cdot g = (x\cdot g)^{\alpha\cdot g} .$$

The following extension of Proposition 2.2 is easily verified (the details are in [12]).

<u>3.2 PROPOSITION.</u> If the group G acts on the crossed complex C, then the crossed complex $D = C /\!/ G$ is a quotient complex of $C \tilde{\times} G$ given by

(i) $D_1 = (C_1 \tilde{\times} G)/N_1$ where N_1 is the normal subgroupoid of $C_1 \tilde{\times} G$ generated by all elements $(0_x, g)$ $(x \in C_o,\ g \in G)$ as in Section 2;

(ii) $D_2 = C_2/N_2$, where N_2 is the normal C_1-subgroupoid of C_2 generated by all elements $c^u - c$ $(c \in C_2,\ u \in N_1)$;

(iii) for $m \geqslant 3$, $D_m = C_m/N_m$, where N_m is the C_1-submodule of C_m generated by all elements $c^u - c$ $(c \in C_m,\ u \in N_1)$;

(iv) the boundary maps and the action of D_1 are induced from those of C.

In particular, the canonical map $q_n : C_n \to D_n$ is a surjection for $n \geqslant 2$. □

Applying this proposition to the situation of Theorem 3.1, and translating the result into the language of groups we obtain the following recipe for the second relative homotopy group of the orbit complex.

<u>3.3 COROLLARY.</u> If G acts on the CW-complex X as in Theorems 2.1 and 3.1, then

$$\pi_2(X_2/G,\ X_1/G,\ \bar{x}) \cong \pi_2(X_2,\ X_1,\ x)/K,$$

where K is the normal $\pi_1(X_1,x)$ - subgroup of $\pi_2(X_2,X_1,x)$ generated by all elements $c\cdot g - c$ and $c^a - c$, where $c \in \pi_2(X_2,X_1,x)$, g is an element of G fixing x, and $a \in \mathrm{Ker}[\pi_1(X_1,x) \to \pi_1(X_1/G,\bar{x})]$. □

In principle, $\pi_2(X_2/G,\bar{x})$ can now be found as the kernel of the boundary map

$$\partial : \pi_2(X_2/G, X_1/G, \bar{x}) \to \pi_1(X_1/G, \bar{x}),$$

but this may be difficult because the relative group is usually rather large and the Reidemeister-Schreier method is needed to obtain a presentation of the absolute group.

We observe that Corollary 3.3 and similar results in higher dimensions will be deducible whenever the assertion

$$(**) \qquad q_* : (\pi \underset{\sim}{X}) /\!/ G \to \pi(\underset{\sim}{X}/G) \text{ is an isomorphism}$$

is true, and it is natural to ask under what more general conditions this is so. For example, it would be interesting to know whether conditions similar to Armstrong's in Case 3 together, perhaps, with a homotopy fullness condition on $\underset{\sim}{X}$, would be sufficient for (**) to hold.

Acknowledgement. The second author is grateful to the Science and Engineering Research Council for a Research Studentship covering the period when this work was done.

References

1. M.A.Armstrong, On the fundamental group of an orbit space, Proc. Cambridge Philos. Soc. 61 (1965) 639-646.

2. M.A.Armstrong, The fundamental group of the orbit space of a discontinuous group, Proc. Cambridge Philos. Soc. 64 (1968) 299-301.

3. G.E.Bredon, Equivariant Cohomology Theories, Springer Lecture Notes in Math. 34 (1967).

4. R.Brown, Groupoids and van Kampen's theorem, Proc.London Math. Soc. (3) 17 (1967) 385-401.

5. R.Brown, Fibrations of groupoids, J.Algebra 15 (1970) 103-132.

6. R.Brown and P.J.Higgins, On the algebra of cubes, J.Pure Appl. Algebra 21 (1981) 233-260.

7. R.Brown and P.J.Higgins, Colimit theorems for relative homotopy groups, J.Pure Appl. Algebra 22 (1981) 11-41.

8. R.Brown and P.J.Higgins, Crossed complexes and non-abelian extensions, this volume.

9. P.J.Higgins, Notes on Categories and Groupoids, Van Nostrand Mathematical Studies 32 (1971).

10. F.Rhodes, On the fundamental group of a transformation group, Proc.London Math. Soc. (3) 16 (1966) 635-650.

11. E.H.Spanier, Algebraic Topology, McGraw-Hill Series in Higher Mathematics (1966).

12. J.Taylor, Group actions on ω-groupoids and crossed complexes and the homotopy groups of orbit spaces, Ph.D. thesis, Univ. of Durham (1982).

Department of Mathematics,
University of Durham,
Science Laboratories,
South Road,
Durham DH1 3LE,
U.K.

Minimal Topological Completion of ${}_{\mathbb{K}}Ban_1 \to {}_{\mathbb{K}}Vec$*)

Rudolf-E.Hoffmann

The concept of a "norm" on a vector space V over $\mathbb{K}=\mathbb{R}$ (i.e. the real numbers) or $\mathbb{K}=\mathbb{C}$ (i.e. the complex numbers) is certainly a "topological" notion (in a non-technical sense). However, the obvious forgetful functor

$${}_{\mathbb{K}}Norm \to {}_{\mathbb{K}}Vec$$

from the category ${}_{\mathbb{K}}Norm$ (or $Norm$) of normed $\mathbb{K}$-vector spaces and $\mathbb{K}$-linear maps of norm at most one into the category ${}_{\mathbb{K}}Vec$ (or Vec) of $\mathbb{K}$-vector spaces and $\mathbb{K}$-linear maps fails to be a topological functor, since it does not preserve the product in $Norm$, the l_∞-join ([S] 11.3, p.196). Indeed, there is no topological functor $Norm \to \underline{X}$ into any balanced category $\underline{X}$ (proposition 1o).

Thus it seems natural to ask for a description of the "minimal topological completion" of the faithful functor $Norm \to Vec$.
The minimal topological completion of a faithful functor $V:\underline{A} \to \underline{C}$ consists of a topological functor $T:\underline{B} \to \underline{C}$ and a full and faithful functor $E:\underline{A} \to \underline{B}$ with $V \cong T\cdot E$ such that, whenever $V \cong T'\cdot E'$ for some topological functor $T':\underline{B}' \to \underline{C}$ and some full and faithful functor $E':\underline{A} \to \underline{B}'$, then there exists a full and faithful functor $F:\underline{B} \to \underline{B}'$ such that

$$T'\cdot F \cong T, \quad \text{and} \quad F\cdot E \cong E'.$$

Up to equivalences (of categories) and natural isomorphisms (of functors), the minimal topological completion is uniquely determined; it always exists, though transition to a higher universe $\underline{U}^+$ (containing the given universe $\underline{U}$ as an element) may be necessary, since $\underline{B}$ need not be $\underline{U}$-legitimate. The construction, given in [Ho_5] and [He], relies upon ideas of Ph.Antoine [A] (cf.also [W]).

The key idea of [Ho_5] and [He] is based upon an observation in [Ho_1]:
A functor $T:\underline{A} \to \underline{1}$ is a topological functor iff $\underline{A}$ is equivalent to a complete lattice (made into a category in the usual way, cf. [ML] I.2,p.11)
- where $\underline{1}$ denotes the category consisting of a unique morphism.

It is readily clear, then, that the above mentioned construction must extend the MacNeille completion of a poset P [MN] to the level of faithful functors $V:\underline{A} \to \underline{C}$ (note that the unique functor $\underline{A} \to \underline{1}$ is

---

*) The main result of this paper is briefly announced in [Ho_7] p.214.

faithful iff $\underline{A}$ is equivalent to a - not necessarily $\underline{U}$-small - partially ordered set).

Indeed, many of the results in [BB] characterizing the MacNeille completion carry over to the minimal topological completion. The following result has its root in a theorem of [B] (cf.also [BB]):

Suppose $T \cdot E \cong V$ for a faithful functor $V:\underline{A} \to \underline{C}$, a full and faithful functor $E:\underline{A} \to \underline{B}$ and a topological functor $T:\underline{B} \to \underline{C}$ such that

i) $E:\underline{A} \to \underline{B}$ is T-dense, i.e. for every $B \in Ob\underline{B}$ the co-cone

$$(\{f:E(A) \to B \text{ in } Mor\underline{B} \mid A \in Ob\underline{A}\}, B)$$

indexed over the (not necessarily $\underline{U}$-small) set

$$\{(A,f:E(A) \to B) \mid A \in Ob\underline{A}, f \in Mor\underline{B}\}$$

is T-identifying [1].

ii) $E:\underline{A} \to \underline{B}$ is T-co-dense (i.e. $E^{op}:\underline{A}^{op} \to \underline{B}^{op}$ is T^{op}-dense), then (E,T) is equivalent to the minimal topological completion of $V:\underline{A} \to \underline{C}$.

This result, obtained independently in [Ho_5] and [He], will be needed in the following.

The basic terminology used here ("topological functor", "V-(co-) identifying cone/lift", "V-(co-)discrete object") is that introduced in [Ho_1] (see e.g. [Ho_2] §o for an easily accessible reference). Other authors (following Bourbaki's terminology) prefer "...-initial" instead of "...-co-identifying" and "...-final" (or "...-co-initial") instead of "...-identifying".

1. A "quasi-norm" on a vector space V over $\mathbb{K}=\mathbb{R}$ or $\mathbb{K}=\mathbb{C}$ is a mapping

$$\| ? \| : V \to [o,\infty]$$

such that

i) $\|\lambda x\| \leq |\lambda| \cdot \|x\|$ [2]

ii) $\|x+y\| \leq \|x\| + \|y\|$

for all $x,y \in V, \lambda \in \mathbb{K}$ - where $o \cdot \infty = o$, $a \cdot \infty = b + \infty = \infty + b = \infty$ for $a,b \in [o,\infty]$ with $a \neq o$

---

1) In order to obtain an appropriate definition of a T-dense functor $E:\underline{A} \to \underline{B}$ with regard to a non-faithful functor $T:\underline{B} \to \underline{C}$, one has to use an index category $\underline{X}$ whose class of objects is

$$\{(A,f:E(A) \to B) \mid A \in Ob\underline{A}, f \in Mor\underline{B}\}$$

and whose morphisms $(A_1,f_1) \to (A_2,f_2)$ are induced by those $\underline{A}$-morphisms $g:A_1 \to A_2$ with $f_2 E(g) = f_1$.
For $\underline{C}=\mathit{1}$, this is the classical definition, due to J.R.Isbell [I] (for full embeddings $E:\underline{A} \to \underline{B}$, "left adequate subcategories"; cf. also [U] where the term "dense" is used).

The category of "quasi-normed" $\mathbb{K}$-vector spaces and non-expansive maps[3] is denoted by $_{\mathbb{K}}QNorm$ or, more simply, by

$$QNorm.$$

It contains the categories

$$_{\mathbb{K}}Norm \text{ and } _{\mathbb{K}}Ban_1$$

of normed vector spaces over $\mathbb{K}$ and Banach spaces over $\mathbb{K}$, respectively, as full subcategories - where the morphisms are the $\mathbb{K}$-linear maps of norm at most one [3]. In order to simplify notation we shall usually write

$$Norm \text{ and } Ban_1$$

instead of $_{\mathbb{K}}Norm$ and $_{\mathbb{K}}Ban_1$, respectively.

The obvious forgetful functor from $QNorm$ into the category Vec of vector spaces over $\mathbb{K}$ and $\mathbb{K}$-linear maps is denoted by

$$U: QNorm \to Vec.$$

2. Lemma:

For $\mathbb{K}$-linear spaces A and A_i $(i \in I)$, a family $f_i: A \to A_i$ $(i \in I)$ of $\mathbb{K}$-linear maps, and quasi-norms $\|.\|_i$ on A_i for every $i \in I$,

$$((A, \|.\|), \{f_i : (A, \|.\|) \to (A_i, \|.\|_i)\}_{i \in I})$$

is a U-co-identifying cone in $QNorm$ iff $\|.\|$ is given by

$$x = \sup\{\|f_i(x)\|_i \mid i \in I\}$$

for every $x \in A$.

(The proof - which uses standard techniques - may be left to the reader.)

It may be noted that, for $I=\emptyset$, we obtain the "U-co-discrete" quasi-norm on the $\mathbb{K}$-vector space A, defined by

$$\|x\| = 0$$

for every $x \in A$.

Now one easily deduces the following result ([Ma], [Ho_3]).

---

2) Equivalently, $\|\lambda x\| = |\lambda| \cdot \|x\|$.

3) or "maps of norm at most 1", i.e.

$$\|f(x)\|_2 \leq \|x\|_1$$

for every $x \in V$.

3. Proposition:

The functor $U: \mathcal{QNorm} \to \mathcal{Vec}$ is a topological functor.

4. Remark:

By the duality theorem for topological functors ([A],[R]), it is clear that all U-identifying lifts exist.

a) For vector spaces A and B, a linear map $g:A \to B$, and a quasi-norm $\|.\|_A$ on A,

$$g:(A,\|.\|_A) \to (B,\|.\|_B)$$

is a U-identifying morphism iff $\|.\|_B$ is defined by

$$\|y\|_B = \inf\{\|u\|_A \mid u \in A,\ g(u)=y\}$$

for every $y \in B$ - where $\inf\emptyset = \infty$.

b) The "U-discrete" quasi-norm $\|.\|$ on a vector space A is given by

$$\|x\| = \infty \text{ iff } x \neq o \text{ and } \|o\| = o.$$

(The proofs are routine and may be left to the reader.)

5. A quasi-norm $\|.\|$ on a vector space V is said to be separated iff

$$\|x\| = o \text{ implies } x = o$$

for every $x \in V$.

For a quasi-normed linear space V over $\mathbb{K}$, $V_o := \{x \in V \mid \|x\| = o\}$ is a linear subspace. The linear quotient space

$$V_{sep} := {}^{V}/_{V_o}$$

receives the U-identifying quasi-norm $\|.\|_{sep}$

$$\|[x]\|_{sep} = \|x\|.$$

Evidently, V_{sep} is separated with regard to $\|.\|_{sep}$. Moreover, the projection map

$$(V,\|.\|) \to (V_{sep},\|.\|_{sep})$$

is a U-co-identifying morphism.

6. Lemma:

Let $(V,\|.\|)$ be a quasi-normed space over $\mathbb{K}$. There is a countable family of norms $\|.\|_n$ $(n \in \mathbb{N})$ on V_{sep} such that the projection maps

$$p:(V,\|.\|) \to (V_{sep},\|.\|_n)$$

form a U-co-identifying cone indexed over $\mathbb{N}$.

Proof:

In view of 5., we may assume w.l.o.g. that $(V,\|.\|)$ is separated. We choose a basis

$$\{v_i \mid i \in I\} \cup \{w_k \mid k \in K\}$$

of V such that $\{v_i \mid i \in I\}$ is a basis of the linear subspace

$$V' = \{v \in V \mid \|v\| < \infty\} .$$

By

$$\|\sum_j \lambda_j w_{k_j}\|_n := n \cdot \sum_j |\lambda_j|$$

we define a norm $\|.\|_n$ on the linear subspace W generated by $\{w_k \mid k \in K\}$ for every $n \in \mathbb{N}$.

Obviously,

$$\|v+w\|_n := \|v\| + \|w\|_n$$

where $v \in V'$ and $w \in W$, defines a norm on V for every $n \in \mathbb{N}$, since, for every $x \in V$, there is a unique representation $x=v+w$ with $v \in V'$ and $w \in W$.

It is readily clear that $\|.\|$ is the supremum of the family of norms $\|.\|_n$ $(n \in \mathbb{N})$.

Since every normed space $(V,\|.\|)$ is a subspace of a Banach space, the Cauchy-completion of $(V,\|.\|)$, with the induced norm, we have thus established that the embedding

$$\mathcal{B}an_1 \to \mathcal{QN}orm$$

is U-co-dense relative to $U:\mathcal{QN}orm \to \mathcal{V}ec$. It remains to show that this embedding is also U-dense.

7. Lemma:

Let $(V,\|.\|)$ be a quasi-normed linear space over $\mathbb{K}$. Then there exists a U-identifying co-cone

$$(\{g_i:(\mathbb{K},|.|) \to (V,\|.\|)\}_{i \in I},(V,\|.\|))$$

where $|.|$ denotes the ordinary norm on $\mathbb{K}$.

Proof:

Let

$$I=\{x \in V \mid \|x\| \neq 0,\infty\} \cup \{(y,n) \in V \times \mathbb{N} \mid y \in V, \|y\|=0\}$$

For $x \in V$ with $\|x\| \neq 0,\infty$, we define $g_x:\mathbb{K} \to V$ by

$$g_x(\lambda) = \frac{\lambda}{\|x\|} \cdot x$$

for every $\lambda \in \mathbb{K}$.

For $y \in V$ with $\|y\| = 0$ and $n \in \mathbb{N}$, let

$$g_{(y,n)}(\lambda) = (n \cdot \lambda)y$$

for every $\lambda \in K$.

It is easily seen that every $g_i\,(i \in I)$ is $\mathbb{K}$-linear and non-expansive. In order to show that $(\{g_i\}_{i \in I}, (V, \|.\|))$ is a U-identifying co-cone, let

$$\begin{array}{ccc} & \xrightarrow{f_i} & (W, \|.\|_W) \\ \mathbb{K} & & \uparrow u \\ & \xrightarrow{g_i} & (V, \|.\|) \end{array}$$

commute for some quasi-normed linear space $(W, \|.\|_W)$, some linear map $u: V \to W$ and a family of linear non-expansive maps $f_i: \mathbb{K} \to W$. For $x \in V$ with $\|x\| \neq o, \infty$, we have

$$u(x) = ug_x(\|x\|) = f_x(\|x\|),$$

hence

$$\|u(x)\|_W = \|f_x(\|x\|)\|_W \leq \|x\|,$$

since f_x is non-expansive. If $\|x\| = o$, then

$$u(x) = ug_{(x,n)}\left(\frac{1}{n}\right) = f_{(x,n)}\left(\frac{1}{n}\right)$$

for every $n \in \mathbb{N}$, hence

$$\|u(x)\|_W = \|f_{(x,n)}\left(\frac{1}{n}\right)\|_W \leq \frac{1}{n}$$

for every $n \in \mathbb{N}$, hence

$$\|u(x)\|_W = o\ .$$

In all this says that $u: (V, \|.\|) \to (W, \|.\|_W)$ is non-expansive. This completes the proof.

The characterization of the minimal topological completion mentioned above together with 6. and 7. now yields

8. Theorem:

The minimal topological completion of

$${}_{\mathbb{K}}\mathcal{B}an_1 \to {}_{\mathbb{K}}\mathcal{V}ec$$

is

$${}_{\mathbb{K}}\mathcal{QN}orm \to {}_{\mathbb{K}}\mathcal{V}ec$$

where ${}_{\mathbb{K}}\mathcal{B}an_1$ is canonically embedded into ${}_{\mathbb{K}}\mathcal{QN}orm$.

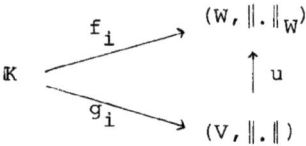

9. Remark:

i) It may be noted that a quasi-normed vector space $(V,\|.\|)$ is "separated" iff every U-co-identifying cone

$$((V,\|.\|),\{f_i:(V,\|.\|) \to (X_i,\|.\|_i)\}_{i\in I})$$

in *QNorm* with domain $(V,\|.\|)$ separates the points of $(V,\|.\|)$ or, equivalently, iff every U-co-identifying morphism in *QNorm* with domain $(V,\|.\|)$ is one-to-one [4].

ii) A quasi-normed vector space $(V,\|.\|)$ enjoys the property that

$$\|x\| < \infty$$

for every $x \in V$ iff $(V,\|.\|)$ is "co-separated", i.e. iff every U-identifying co-cone

$$(\{g_i:(X_i,\|.\|_i) \to (V,\|.\|)\}_{i\in I},(V,\|.\|))$$

in *QNorm* is (i.e. induces) an epimorphic co-cone in *Vec* , or, equivalently, iff every U-identifying morphism with co-domain $(V,\|.\|)$ is onto [4].

iii) A quasi-normed vector space $(V,\|.\|)$ is complete in the sense that every Cauchy sequence converges to a unique point iff $(V,\|.\|)$ is separated and every U-co-identifying epimorphism with domain $(V,\|.\|)$ in the category

sepQNorm

of separated quasi-normed vector spaces over $\mathbb{K}$ and non-expansive linear maps is an isomorphism.

It results from (i)-(iii) and theorem 8 that $Ban_1 \to Vec$, though it fails to be right adjoint, is externally reconstructible from its MacNeille completion, an exceptional phenomenon for which no strict parallel is known in any other situation.
(The present example plays a great role in the theory of "separated" objects with regard to a topological functor, developed in $[Ho_3]$ [5] as well as in the theory of "complete" objects developed in $[Ho_4]$).

It may be also worth pointing out that the separated quasi-normed vector spaces and non-expansive linear maps form the (E,M)-topological hull of Ban_1 over *Vec* (with regard to the obvious functors and E,M = {epimorphisms} and {monomorphisms} in *Vec*, respectively) and that the complete (+ separated) quasi-normed vector spaces (in which Ban_1 is coreflective) form the semi-topological hull of

---

4) This latter concept of a "(co-)separated" object with regard to a functor $U:\underline{X} \to Ens$ is essentially due to [Br] (cf.also $[Ho_3]$).

5) A brief indication of the results in $[Ho_3]$ may be found in $[Ho_4]$ §2.

Ban_1 over Vec - cf. [Ho_6] p.73.

(iv) The question of the "size" of the minimal topological completion does not concern us in theorem 8, since the completion has a $\underline{U}$-legitimate domain, indeed $\underline{U}$-small fibers by the very construction.

The failure of topologicity of $Norm \to Vec$ and $Ban_1 \to Vec$ cannot be remedied by any reasonable change of these functors or their co-domain Vec, leaving the domains unchanged.

Recall that a category $\underline{C}$ is <u>balanced</u> iff every bimorphism (= a morphism which is both an epimorphism and a monomorphism) is an isomorphism. It is easy to see that Vec is balanced (indeed every abelian category is balanced).

<u>10.</u> <u>Proposition</u>:

There is no topological functor T

$${}_{\mathbb{K}}Norm \to \underline{X} \text{ or } {}_{\mathbb{K}}Ban_1 \to \underline{X}$$

into a balanced category $\underline{X}$.

<u>Proof</u>:

Suppose, on the contrary, that such a topological functor $T:\underline{A} \to \underline{X}$ is given - where $\underline{A}$ is either $Norm$ or Ban_1.

By [Ho_2] 1.3 and 2.1, a normed vector space (resp., a Banach space) $(V, \|.\|)$ over $\mathbb{K}$ is T-discrete iff every bimorphism with co-domain $(V,\|.\|)$ is an isomorphism. Clearly, the natural map

$$(V, 2\cdot\|.\|) \to (V,\|.\|)$$

(mapping every point identically) is a bimorhphism in $\underline{A}$. This is an isomorphism iff $V \cong \{o\}$, as one easily checks. Thus $(V,\|.\|)$ is a T-discrete object if and only if $V \cong \{o\}$ (the validity of "if" is straightforward).

Since there is an isomorphism between the full subcategory of T-discrete objects and $\underline{X}$, we obtain

$$\underline{X} \simeq \underline{1} .$$

This contradicts the fact that T is a topological functor, since $T:\underline{A} \to \underline{1}$ is a topological functor iff $\underline{A}$ is equivalent to a complete lattice ([Ho_1]).

References

A Antoine,Ph.: Etude élémentaire des catégories d'ensembles structurés. Bull.Soc.Math.Belgique 18, 142-164 and 387-414(1966).

B Banaschewski,B.: Hüllensysteme und Erweiterung von Quasi-Ordnungen.Z.Math.Logik Grundlagen Math.2, 117-13o (1956).

BB Banaschewski,B. and G.Bruns: Categorical characterization of the MacNeille completion. Archiv d.Math.18, 369-377(1967).

Br Brümmer,G.C.L.: A categorial study of initiality in uniform topology. Ph.D.thesis, University of Cape Town,1971.

He Herrlich,H.: Initial completions. Math.Z.15o, 1o1-11o(1976) (also in: "Kategorienseminar" (Hagen,1975), pp.3-25(1976)).

Ho_1 Hoffmann,R.-E.: Die kategorielle Auffassung der Initial- und Finaltopologie. Dissertation, Universität Bochum 1972.

Ho_2 --: Topological functors and factorizations. Archiv d.Math.26, 1-7(1975).

Ho_3 --: (E,M)-universally topological functors. Habilitationsschrift, Universität Düsseldorf 1974.

Ho_4 --: Topological functors admitting generalized Cauchy-completions. In: Categorical Topology. Proceedings of a Conference (Mannheim, 1975), pp.286-344. Lecture Notes in Math.54o. Berlin-Heidelberg-New York: Springer 1976.

Ho_5 --: Topological completion of faithful functors (1975; unpublished). Summary (=§o) in: "Kategorienseminar" (Hagen, 1975), pp.26-37(1976).

Ho_6 --: Note on semi-topological functors. Math.Z.16o, 69-74(1978).

Ho_7 --: Note on universal topological completion. Cahiers Topologie Géom.Différentielle 2o, 199-216(1979).

I Isbell,J.R.: Adequate subcategories. Illinois J.Math.4, 541-552(196o).

Ma Manes,E.G.: A pullback theorem for triples in a lattice fibering with applications to algebra and analysis. Algebra Universalis 2, 7-17(1972).

ML MacLane,S.: Categories for the working mathematician. Berlin-Heidelberg-New York: Springer 1971.

MN MacNeille,H.: Partially ordered sets. Trans.Amer.Math.Soc.42, 416-46o(1937).

R Roberts,J.E.: A characterization of initial functors. J.Algebra 8,181-193(1968).

S Semadeni,Z.: Banach Spaces of Continuous Functions. Warszawa: PWN-Polish Scientific Publishers 1971.

U Ulmer,F.: Properties of dense and relative adjoint functors J.Algebra 8, 77-95(1968).

W Wyler,O.: Are there topoi in topology? in: Categorical Topology. Proceedings of a Conference (Mannheim, 1975), pp.699-719. Lecture Notes in Math.54o. Berlin-Heidelberg-New York: Springer 1976.

Rudolf-E.Hoffmann
Fachbereich Mathematik
Universität Bremen
D-28 Bremen
Federal Republic of Germany

ON THE FREENESS OF WHITEHEAD-DIAGRAMS

Michael Höppner

1. Introduction

Diagrams over a partially ordered set I with values in a category R-Mod of modules over some ring R may also be considered as modules over a *ring with several objects* (i.e. a small additive category). The study of functor categories in the spirit of module theory was initiated by Mitchell [7] and in the following there has been some interest in homological properties of posets and diagrams. As was shown in [5] any projective diagram is free, i.e. a direct sum of representables, and when analizing the proof it became apparent that freeness of a diagram is related to what is known as *Whitehead's property* in abelian group theory.

It is well-known that any Whitehead-group of countable rank is free [4], but in general the freeness of a Whitehead-group is undecidable in the usual ZFC-set theory by a result of Shelah [10] (see also [3]). For diagrams the situation is as follows: In the hereditary case Whitehead-diagrams are exactly the flat diagrams (Proposition 2.3.) and therefore there are some examples of non-free Whitehead-diagrams easily to be obtained within ZFC. However, we derive at the following analogue of abelian group theory.

Theorem W. *Let* D: I $\rightarrow$ F-Mod *be a Whitehead-diagram of countable rank without divisible elements, then* D *is free.*

Whitehead-diagrams over the ordered set of integers $\mathbb{Z}$ have already been studied in [6].

2. Whitehead-diagrams over hereditary posets

To keep in track with abelian group theory we have to restrict to categories of diagrams (I, R-Mod) that are *hereditary*, i.e. of global dimension one. Therefore, by a result of Brune [2] the ring R has to be semi-simple and the poset I shall neither contain $(\mathbb{N}+1)^{op}$ nor $\overline{2}\times\overline{2}$ where $\mathbb{N}$ and $\overline{2}$ denote the ordered sets of natural numbers and $\{1<2\}$, respectively. So I has to be a *tree with well-ordered intervalls* and without loss of generality we will further assume R=F a *field.*

For every subset J of I and any vectorspace X we define the *constant diagram* $\Delta_J X$ with value X on J and value 0 elsewhere. As special cases we note the *quasi-simple diagrams* $E_i X$ with J={i} and the *representable diagrams* $S_i X$ with $J=\{j\in I \mid i\leq j\}$. Because S_i: F-Mod → (I, F-Mod) is left adjoint to evaluation at i, the diagrams $S_i X$ are *projective* and every projective diagram P is of the form $P=\bigoplus_{i\in I} S_i Q_i$ for some spaces Q_i, i.e. P is *free* [5].

The following definition is in obvious analogy with abelian group theory.

<u>2.1. Definition.</u> A diagram D: I → F-Mod is called a *Whitehead-diagram* (W-diagram) if Ext $(D, S_i F)=0$ for every $i\in I$.

By heredity the class of W-diagrams is closed by taking subdiagrams. Of course, any projective diagram is a W-diagram but so is every flat diagram, because every $S_i F$ is injective with respect to pure-exact sequences. The last fact follows easily from the splitting of the canonical embedding $S_i F \to (S_i F)^{**}$, where $^*:(I, \text{F-Mod}) \to (I^{op}, \text{Mod-F})$ denotes the obvious duality with $D^*_i=\text{Hom}_F(D_i, F)$ for every diagram D and $i\in I$. We will see in a moment that there are no other W-diagrams.

<u>2.2. Lemma.</u> *Let* D: I → F-Mod *be a W-diagram, then* D *is a diagram of monomorphisms.*

<u>Proof.</u> Suppose there is a non-trivial $x\in\text{Ker}(D_i\to D_j)$, then consider the subdiagram U of D generated by x and an exact sequence $0 \to V \to S_i F \to U \to 0$. Because $V=\bigoplus_{k\in K} S_k F$ for some subset K of I bounded by i, we have $F=\text{Hom}(V, S_k F)=\text{Ext}(U, S_k F)=0$ for every $k\in K$, a contradiction. □

<u>2.3. Proposition.</u> *Let* D: I → F-Mod *be a diagram of monomorphisms. Then the following are equivalent*

1) D *is a W-diagram*
2) Ext $(D, E_i F)=0$ *for every* $i\in I$
3) $\varinjlim_{j\in J} D_j \to D_i$ *is a monomorphism for every* $i\in I$ *and* $J=\{j\in I \mid j<i\}$
4) D *is flat*
5) Ext $(D, S_i X)=0$ *for every* $i\in I$ *and space* X.

<u>Proof.</u> 1)⟹ 2) by heredity.
2)⟹ 3) Set $\bar{J}=\{j\in I \mid j\leq i\}$ and consider the exact sequences

$$0 \to E_i F \to \Delta_{\bar{J}} F \to \Delta_J F \to 0 \quad \text{and}$$

$$\to \mathrm{Hom}\,(D,\ \Delta_{\bar J}F) \xrightarrow{f} \mathrm{Hom}\,(D,\ \Delta_J F) \to \mathrm{Ext}\,(D,\ E_i F) \to$$

Note that f may be identified with the dual of the canonical morphism $\varinjlim_{j\in J} D_j \to D_i$ which therefore must be monomorphic.

3)=> 4) $\varinjlim_{j\in J} D_j = \bigoplus_Z (\bigcup_{j\in Z} D_j)$ with sum taken over the (right filtered) connected components of J. Now use the flatness criterion in [6] and induction on the number of generators of a left-open subset K of I bounded by i.

4)=> 5) by the splitting criterion for pure-injectivity above. □

The proposition has a lot of interesting consequences.

2.4. Corollary. *Let* D: I → F-Mod *be a finitely presented W-diagram, then* D *is free.*

For any W-diagram to be free (projective) the category of diagrams (I, F-Mod) has to be *perfect*. By a well-known theorem of Bass [1, 8] this is equivalent to the condition that all descending chains $(S_{i_n}F)_{n\in\mathbb{N}}$ in (I^{op}, Mod-F) become stationary.

2.5. Corollary. *Any W-diagram in* (I, F-Mod) *is free iff* I *is artinean.*

By Proposition 2.3. and the flatness criterion in [6] a diagram D is a W-diagram iff for every $i\in I$ the restriction of D to $J=\{j\in I \mid j\leq i\}$ is a W-diagram, so being a W-diagram is a *local property*. Since there is an analogous result for projective diagrams by Mitchell [9], we may restrict ourselves to hereditary posets I with a *maximal element* o when studying freeness of W-diagrams. Note that by this restriction the category (I, F-Mod) becomes *noetherian*.

3. Whitehead-diagrams without divisible elements

The following are all non-free W-diagrams.

3.1. Counterexamples. a) $D^{(1)} = \prod_{n\in\mathbb{N}} S_n F\colon\ \mathbb{N}^{op} \to$ F-Mod.

b) $D^{(2)} = \Delta_{\mathbb{N}^{op}} F\colon\ \mathbb{N}^{op} \to$ F-Mod.

c) Let $I = \mathbb{N}\times\bar{2}$ be ordered by $(n,i) < (m,j)$ iff $m<n$ and $j=1$. Then consider the subdiagram $D^{(3)}$ of $D = \prod_{n\in\mathbb{N}} S_{(n,2)}F\colon I \to$ F-Mod generated by the elements $(1,1,\ldots)\in D_{(n,1)}$ and $1\in D_{(n,2)}$.

Since any non-artinean poset I must contain $\mathbb{N}^{op}$ the first two examples may be transfered to any such set by Kan-extensions.

The second example is not a counterexample really. Remember that any abelian W-group G is *separable*, i.e. every element of G is contained in a free direct summand of G. On the other hand, for any complete discrete valuation ring the quotient field (which is analogous to the diagram $D^{(2)}$, see [6]) is a W-module. So in general we have to exclude divisible elements to obtain free modules. While the second example has a module-theoretic counterpart, the third seems to be a speciality of diagrams but should be seen in the same light. Note that an element of a free diagram cannot be branched as wild as $(1,1,\dots)\in D^{(3)}_{(o,1)}$.

3.2. Definition. Let D: I → F-Mod be a diagram (of monomorphisms). A non-trivial element $x\in D_i$ is called *divisible*, if either

a) there is a countable descending chain J in I such that $x\in \bigcap_{j\in J} D_j$ or

b) for countable many $n\in\mathbb{N}$ there are pairwise non-comparable elements $j_1,\dots,j_n<i$ and non-trivial $x_k\in D_{j_k}$ such that $x=x_1+\dots+x_n$.

3.3. Proposition. *Let* D: I → F-Mod *be a W-diagram, then the following are equivalent*

1) D *does not have divisible elements.*

2) *For every* $x\in D_i$ *there are pairwise non-comparable elements* $j_1,\dots,j_n\leq i$ *and* $x_l\in D_{j_l}\setminus \sum_{k<j_l} D_k$ *such that* $x=x_1+\dots+x_n$.

3) D *is separable.*

Proof. 3)⟹ 1)⟹ 2) are obvious.

2)⟹ 3) Choose direct decompositions $D_{j_l} = \hat{D}_{j_l} \oplus \sum_{k<j_l} D_k$ such that $x_l\in\hat{D}_{j_l}$.

Because D is a W-diagram there is a homomorphism $f\colon D \to \bigoplus_{l=1}^{n} S_{j_l}\hat{D}_{j_l}$ which lifts the obvious morphism $D\to \bigoplus_{l=1}^{n} E_{j_l}\hat{D}_{j_l}$ along the projective cover $\bigoplus_{l=1}^{n} S_{j_l}\hat{D}_{j_l} \to \bigoplus_{l=1}^{n} E_{j_l}\hat{D}_{j_l}$. Moreover $D=\bigoplus_{l=1}^{n} S_{j_l}\hat{D}_{j_l}\oplus \operatorname{Ker} f$ and $x\in\bigoplus_{l=1}^{n} S_{j_l}\hat{D}_{j_l}$. □

In the following sense the remaining first example is of too great cardinality.

3.4. Definition. Let D: I → F-Mod be a diagram. Then $\dim_F D_o$ is called the *rank* of D.

Now we are prepared for the

Proof of Theorem W. Let $\{x_n \mid n \in \mathbb{N}\}$ be a generating set for D_o. Then by induction there are free direct summands $P^{(n)}$ of D such that $D = P^{(1)} \oplus \ldots P^{(n)} \oplus D^{(n)}$ and $x_1, \ldots, x_n \in P^{(1)} \oplus \ldots \oplus P^{(n)}$. Because $P = \bigoplus_{n \in \mathbb{N}} P^{(n)}$ is a pure subdiagram of D the quotient diagram is flat. Since $(D/P)_o = 0$ we must have $D = P$. □

In abelian group theory Theorem W is usually proved by establishing *Pontryagin's criterion* for freeness [4]. For diagrams this seems to work well for downward filtered posets or for posets with finitely many branching points only.

REFERENCES

[1] H. Bass: Finitistic dimension and a generalization of semi-primary rings. Trans. Amer. Math. Soc. 95 (1960) 466-488

[2] H. Brune: On projective representations of ordered sets. to appear

[3] P.C. Eklof: Whitehead's problem is undecidable. Amer. Math. Monthly 83 (1976) 755-788

[4] L. Fuchs: Infinite Abelian Groups I, II. Academic Press, New York 1970, 1973

[5] M. Höppner and H. Lenzing: Projective diagrams over partially ordered sets are free. J. pure appl. Alg. 20 (1981) 7-12

[6] -,- : Diagrams over ordered sets: a simple model for abelian group theory. Abelian Group Theory (Proc. Oberwolfach Conf. 1981), Springer Lect. Notes Math. 874 (1981) 417-430

[7] B. Mitchell: Rings with several objects. Advances Math. 8 (1972) 1-161

[8] - : Some applications of module theory to functor categories. Bull. Amer. Math. Soc. 84 (1978) 867-885

[9] - : A remark on projectives in functor categories. J. Alg. 69 (1981) 24-31

[10] S. Shelah: Infinite abelian groups, Whitehead's problem, and some constructions. Israel J. Math. 18 (1974) 243-256

Michael Höppner
Fachbereich Mathematik-Informatik
Universität-Gesamthochschule-Paderborn
D-4790 Paderborn
Germany

APPLICATIONS OF CATEGORY THEORY TO UNIFORM STRUCTURES

M.Hušek, Praha

Applications of category theory to some special structures have usually two components: a categorical background and its convenient application in the structures,which need not be straightforward and one must often use special properties of the structures.I want to show here one such procedure in uniform spaces (also in topological spaces) concerning reflections and coreflections: any ordered set may be embedded into the conglomerate of all coreflections or bireflections on the category *Unif* of all uniform spaces (or *Top* of all topological spaces) with some additional properties (e.g.that the greatest lower bound of coreflections need not be a coreflection).At the end I will add some new results concerning cartesian closed subcategories of *Unif* and ordered functors into *Unif* and suggest two possibilities of investigation in category theory coming from *Unif* or *Top*.

Six years ago in the Conference on Categorical Topology held in Mannheim I have introduced a "lattice"of epi or monotransformations, see [Hu₂].Let me repeat the main definitions and properties arranged for *Unif* and *Top*.Let K be a topological category over *Set*.By M we denote the conglomerate of all natural monotransformations (i.e.all the components η_X are monomorphisms) $\eta:F\longrightarrow 1_K$ for some F,together with the order

$$\eta_1<\eta_2 \text{ if } \eta_1=\eta_2\circ\varepsilon \text{ for some } \varepsilon.$$

Clearly,M *is a complete lattice* (i.e.,every $N\subset M$ has a least upper bound and a greatest lower bound in the usual sense) and its *subconglomerate* C *of all coreflections is join-stable in* M (i.e.,C is also complete and the least upper bounds of $D\subset C$ in C and M coincide). The dual situation: *The conglomerate* E *of all epitransformations* $\eta:1_K\longrightarrow F$ *is complete* in the order

$$\eta_1<\eta_2 \text{ if } \eta_2=\varepsilon\circ\eta_1 \text{ for some } \varepsilon;$$

its subconglomerate R *of all epireflections is complete and meet-stable in* E.We shall show now a method how to find examples that C is not meet-stable in M and R is not join-stable in E even for nice subfamilies in K=*Unif* or K=*Top*.

At first we repeat briefly the construction of special functors from [Hu$_1$], adapted for our case. If G' is a functor from a full subcategory K_1 of K into K, then the functor $G: K \longrightarrow K$ *projectively generated* by G' is defined as follows: if $X \in \mathrm{ob}K$, then GX is that object of K projectively generated by all $f \in K(X,Y)$ into $G'Y$, where $Y \in \mathrm{ob}K_1$. Dually one may define *inductively generated* functors. The functor G extends G'.

By a *strongly rigid class* of objects of a concrete category we mean a class such that the full subcategory generated by the class consists of identities and constant maps only. A *concrete map* is the map preserving underlying sets. By an *order in concrete categories* we always mean the relation $X<Y$ which means that X is finer than Y (i.e. the underlying sets $|X|, |Y|$ coincide and the map $1_{|X|}$ is a morphism from X into Y).

PROPOSITION. *Let A, B be topological categories, $F: A \longrightarrow B$ a concrete functor, $S \subset \mathrm{ob}B$ be a strongly rigid class. Then any concrete map G' from a part of $F^{-1}(S)$ into $\mathrm{ob}A$ preserving the order can be extended to a concrete functor $G: A \longrightarrow A$. If G' is idempotent and bigger or smaller than 1_A, then G can be found to be a bireflection or a coreflection.*

Proof. Since S is strongly rigid and G' preserves the order, G' generates in a clear way the functor (denoted also by G') from the full subcategory of A generated by the domain of G' into A. The extension G of G' can be defined as any functor in between the functor G_1 inductively generated by G' and G_2 projectively generated by G'. Suppose now that G' is idempotent and bigger than 1_A. We shall prove that G_2 is a bireflection. The fact that G_2 is bigger than 1_A follows directly from the fact that $T<G'T$ for T from the domain of G' (realize that G_2X is the coarsest object such that every $f \in A(X,T)$ belongs to $A(G_2X, G'T)$). If G' is idempotent, then the map f belongs to $A(G_2G_2X, G'T)$ for any $f \in A(X,T)$, hence $G_2 \circ G_2 < G_2$, which proves $G_2 \circ G_2 = G_2$. Similarly, if G' is idempotent and smaller than 1_A, then G_1 is a coreflection extending G'.

It follows directly from the construction that if we have two maps G', H' in our Proposition with the same domain, and $G'<H'$, then the corresponding extremal functors satisfy the same relation (i.e., $G_i < H_i$ for $i=1,2$).

The constructed functor G_2 is the biggest bireflection extending G' (provided G' is idempotent and bigger than 1_A); it follows from the fact mentioned above that in that case there is also the smallest bireflection extending G'. Dually for coreflections.

The following construction prepares the situation for our application of the Proposition into *Unif* and *Top*.

CONSTRUCTION. Let A,B be topological categories, $F:A\longrightarrow B$ a concrete functor, P be a strongly rigid object in B with $|P|>1$, $T\subset F^{-1}(P)$ be such that $F(\sup T)=F(\inf T)=P$ and $\inf T\notin T$. For $S\in T$, define $G'_S:T\cup\{\sup T\}\longrightarrow A$ as follows

$$G'_S(T)=G'_S(\sup T)=S \quad \text{if } T>S,$$
$$G'_S(T)=T \quad \text{if } T\not>S.$$

Then the inductive extension G_S of G'_S is a coreflection, the family $\{G_S|S\in T\}$ is order-isomorphic to T, and $G=\inf\{G_S|S\in T\}$ is not a coreflection because

$$G(\sup T)=\inf\{G_S(\sup T)|S\in T\}=\inf T$$
$$G(\inf T)=\inf\{G_S(\inf T)|S\in T\}\neq\inf T$$

(in fact, the last object is the A-discrete object on $|P|$).
One can proceed similarly to obtain a family $\{H_S|S\in T\}$ of bireflections in A order-isomorphic to T such that $\sup\{H_S|S\in T\}$ is not a bireflection (instead of $\inf T\notin T$ we must assume $\sup T\notin T$).

In the following we identify the category *Prox* with the full bireflective subcategory of *Unif* composed of all totally bounded uniform spaces.

THEOREM. *Let $(Z,\leq)$ be an ordered set without the least (or biggest) element. If K=Top or Prox or Unif, then there is a family $\{G_z|z\in Z\}$ of coreflections (or of bireflections) in K such that $\{z\longrightarrow G_z\}$ is an order-isomorphism and $\inf\{G_z|z\in Z\}$ in M is not a coreflection (or, $\sup\{G_z|z\in Z\}$ in E is not a bireflection, resp.).*

Proof. Suppose first that K=*Unif* and F:*Unif*$\longrightarrow$*Top* is the standard functor. By the preceding construction it suffices to find a strongly rigid space P in *Top* such that $(Z,\leq)$ can be embedded into $(F^{-1}(P),<)$. At first we shall add a point 0 to Z as the smallest element of $Z\cup(0)=Z_0$. Then we take a metrizable strongly rigid topological space P containing an infinite closed discrete subspace C of cardinality at least $|Z|$ (by [T], there are metrizable strongly rigid topological spaces of arbitrarily large cardinality). Now, we shall identify the set Z_0 with a closed discrete subset in $\beta C-C=\overline{C}^{\beta C}-C$. For $S\subset Z$ let $\beta_S P$ be the compactification of P obtained from βP by shrinking S to a point, and u_S be the proximity on P corresponding to $\beta_S P$. Then, if $S(z)=[0,z]$ for $z\in Z$, the family $\{u_{S(z)}|z\in Z\}$ is order-isomorphic to $(Z,\leq)$. Indeed, if $z\leq x$ in Z, then $S(z)\subset S(x)$ and $u_{S(z)}$ is finer than $u_{S(x)}$. If $a\in S(z)-S(x)$, then there is $A\subset C$ with $\overline{A}\ni a$, $\overline{A}\cap S(x)=\emptyset$; take a $B\subset C$ with $\overline{B}\ni 0$ $\overline{B}\cap Z=\emptyset$, $B\cap A=\emptyset$, then A is proximal to B in $u_{S(z)}$ (since $\overline{A}\cap\overline{B}\cap S(z)\neq\emptyset$) but A is far from B in $u_{S(x)}$ (since $\overline{A}\cap\overline{B}=\emptyset$ in βP, $\overline{A}\cap\overline{B}\cap S(x)=\emptyset$), hence $u_{S(z)}$ is not finer than $u_{S(x)}$. This completes the proof for the case when K=*Prox* or K=*Unif*.

If $K=Top$, we shall use for the functor F the sequential modification $Top \longrightarrow Top$. Take again the previous situation with P,C,Z, choose a point $y \in P-C$ with the neighborhood system U_y, and for $x \in Z$ denote by U_x the trace of βP-neighborhood system at x on P. For $S \subset Z$ let t_S be the topology on $|P|$ coinciding with that of P on $|P|-(y)$ and such that

$$U_y \cap \bigcap\{U_s \mid s \in S\} \text{ is the neighborhood system at } y.$$

Then $F(|P|,t_S)=P$, t_S is Hausdorff and again $(Z,\leq)$ is order-isomorphic to $\{(|P|,t_{S(z)}) \mid z \in Z\}$: if $z \leq x$ in Z, then $S(z) \subset S(x)$ and $t_{S(z)}$ is finer than $t_{S(x)}$, and if there is some $a \in S(z)-S(x)$, then some neighborhood of y in $t_{S(x)}$ contains no member of U_a, so that $t_{S(z)}$ is not finer than $t_{S(x)}$.

COROLLARY 1. *There are infinite well-ordered arbitrarily long families of coreflections (or bireflections) on Unif, Prox or Top the meet in M (or join in E) of which is not a coreflection (or a bireflection, resp).*

By the same way one obtains for instance an example of two bireflections F,G such that $\sup(F,G)$ is not a bireflection (hence $F \circ G \neq G \circ F$). Such functors may be described directly, e.g. in Top we put FX (or GX) to be the space projectively generated by $C(X,R)$ (or $C(X,T)$ where T is a countable coarsest T_1-space); then for the set X of reals with $([x-\varepsilon,x+\varepsilon] \cap \{\text{rationals}\}) \cup \{x\}$, $\varepsilon>0$, as a neighborhood base at $x \in X$, one has $GX=X$, $FX=R$ the standard space of reals, and GR is the indiscrete space.

Suppose now that $F \in E$. Then we may define by induction the functors $F^{\xi} \in E$, ξ an arbitrary ordinal number, in the following way:

$$F^0=1_K,\quad F^{\xi+1}=F \circ F^{\xi}, \text{ and } F^{\xi}=\sup\{F^{\alpha} \mid \alpha \in \xi\} \text{ for limit } \xi .$$

Clearly, $F^{\alpha}<F^{\beta}$ for $\alpha<\beta$. Dually we define F^{ξ} for $F \in M$ (then $F^{\alpha}>F^{\beta}$ for $\alpha<\beta$).

COROLLARY 2. *For any ordinal ξ there is a concrete functor G in Unif, Prox or Top such that $1 \lneq G \lneq G^2 \lneq \ldots \lneq G^{\xi}=G^{\xi+1}$; and similarly for decreasing families.*

Proof. Take for K either $Unif$ or $Prox$ or Top and for $F: K \longrightarrow Top$ the functor from the proof of our Theorem. For some strongly rigid topological space P there is a subset $\{T_{\alpha} \mid \alpha \in \xi+1\} \subset F^{-1}(P)$ which is order-isomorphic to the well-ordered set $\xi+1$ of all ordinals less or equal to ξ. It suffices to put now $G'T_{\alpha}=T_{\alpha+1}$ for $\alpha \in \xi$, $G'T_{\xi}=T_{\xi}$, and to take the functor $G: K \longrightarrow K$ projectively generated by G'.

Using strongly rigid proper classes in Top, one can construct e.g. a functor which "never stops", or a *proper class of mutually incomparable coreflections or bireflections*. We shall prove here only the first case:

There are concrete functors G,H *in* $\mathcal{Unif}$, $\mathcal{Prox}$ *or* $\mathcal{Top}$ *such that* $G^{\alpha}\lneq G^{\beta}$, $H^{\alpha}\gneq H^{\beta}$ *whenever* $\alpha<\beta$.
The proof is similar to the preceding one, only one must take a strongly rigid class $\{P_{\kappa}|\kappa$ infinite cardinal$\}$ of paracompact spaces constructed in [K] such that P_{κ} contains a closed discrete subspace of cardinality κ. It follows from the proof of our Theorem that in $F^{-1}(P_{\kappa})$ there is a strictly increasing family $\{X_{\kappa,\alpha}|\alpha\in\kappa\}$, and define $G'X_{\kappa,\alpha}=X_{\kappa,\alpha+1}$. Then the functor G projectively generated by G' has the required property. Using decreasing families and the inductive generation, one gets the functor H.

Some other lattice properties of M,C,E,R were investigated in [Hu2]. I would like to add here the following two notes.

The above examples of families F of bireflections such that $\sup F$ is not a bireflection, give also various examples when, for a functor H, the functor $\sup\{F\in R|HF=H\}$ is not a bireflection - it suffices to put $H=\sup F$ (see [Hu2] for the motivation and properties of the functors of the type $\sup\{F\in R|HF=H\}$).

The above lattices contain many gaps. Take a topological space X such that for any two points $a,b\in X$, any open set U and $x\in U$ there exists a continuous map $f:X\longrightarrow X$ with $fx=a$, $f(X-U)=\{b\}$ (e.g. the space of reals or the coarsest T_1-space or the connected two-point T_0-space). If we define $J_1=\inf\{F\in E|F>1_{Top}, FX$ is indiscrete$\}$ in E, $J_2=\sup\{F\in E|FX=X\}$ in E, then there are no bireflections in the intervals $]\bar{J}_1\wedge J_2,\bar{J}_1[$, $]J_2,\overline{J_1\vee J_2}[$ where the bar means the bireflective modification (realize that $J_2=\bar{J}_2$, but e.g. for $X=R$ and Y the long line together with the last point one has J_1^2Y indiscrete and J_1Y not indiscrete) - the assertion follows from the fact that if $FX>X$ then either $FX=X$ or FX is indiscrete. Similarly, the intervals $]J_1\wedge J_2,J_1[$, $]J_2,J_1\vee J_2[$ in E are empty.

Now, I would like to mention two new "categorical" results concerning uniform spaces. The first one answers the question posed by F.Schwarz in the Ottawa Categorical Conference in the last year: Is there a nontrivial epireflective subcategory of $\mathcal{Unif}$ which is cartesian closed? The answer is no and the details are contained in the joint paper with H.Brandenburg who proved the same assertion for Top (published in this Volume). In fact, we have proved more:

<u>THEOREM</u>. *There is no nontrivial epireflective subcategory of an epireflective subcategory of* $\mathcal{Unif}$ *or* $\mathcal{Top}$ *which is cartesian closed.*

<u>COROLLARY</u>. *There is no cartesian closed nontrivial epireflective subcategory of uniform or topological Hausdorff spaces.*

The other result concerns orderable uniform spaces (i.e. those uniform spaces having a base of uniform covers composed of convex sets with respect to some order). Unlike in *Top*, where there is exactly one concrete functor $\mathcal{O}rd \longrightarrow \mathcal{T}op$ that assigns to $(X,<)$ the topological space orderable with respect to the order $<$ (for morphisms in *Ord* we take all monotone maps preserving greatest lower bounds and least upper bounds), the situation in *Unif* is quite different (see [Hu₃] for details):

THEOREM. *There is a proper class of concrete functors* $\mathcal{O}rd \longrightarrow \mathcal{U}nif$ *that assign to* $(X,<)$ *a uniform space orderable with respect to* $<$.

We add that all the functors from the last Theorem are situated in between the functor which assigns to $(X,<)$ the topological fine uniformity of the orderable topology on $(X,<)$, and the functor assigning to $(X,<)$ the proximity of the biggest Dedekind completion (i.e., the finest orderable compactification) of $(X,<)$.

At the end I would like to mention two categorical problems connected with uniform and topological spaces.

We can say that a topological category K over *Set* is of *bounded type* if there is a class C of objects in K and a cardinal κ such that

(i) K is projectively generated by C;

(ii) for each $\{X_i \mid i\in I\} \subset \mathrm{ob}K$, $C\in C$, $f\in K(\Pi\{X_i \mid i\in I\}, C)$ there is a $J\subset I$, $g\in K(\Pi\{X_i \mid i\in J\}, C)$ such that $|J|\le\kappa$, $f=g\circ \mathrm{pr}_J$ (i.e., f depends on at most κ coordinates).

We say then that the category K is of the type κ.

It is known [Hu_4] that *Unif* and *Prox* are of countable type (C is the class of metrizable uniform spaces or the unit interval, resp.), but *Top* is of unbounded type (perhaps the simplest proof is the following: if κ is a given infinite cardinal, C projectively generates *Top*, we take a $C\in C$ with $|C|>1$, we choose two points a,b in C and take the "characteristic function" f of the set $G=\{g\in(\kappa^+)^{\kappa^+} \mid g(g0)=g0\}$, that means $f(g)=a$ if $g\in G$, $f(g)=b$ if $g\notin G$; then $f:(\kappa^+)^{\kappa^+}\longrightarrow C$ does not depend on κ coordinates).

QUESTION 1. What are special features of categories of bounded type? Can these categories be characterized more categorically? What categorical properties of topological categories depend on the fact that the category is or is not of bounded type?

Another possible investigation in topological categories deals with sequential structures. In *Top*, sequential spaces are inductively generated by $\omega+1$, the convergent sequence, and the connected two-point

T_0-space which on the other hand projectively generate the whole *Top*. In *Unif*, sequential spaces are inductively generated by metrizable uniform spaces which again projectively generate the whole *Unif*, and also every $f \in U(\Pi X_i, M)$, M metrizable, depends on countably many coordinates. In *Top*, one can use also the factorization property: every $f \in C((\omega+1)^\kappa, \omega+1)$ depends on countably many coordinates.

<u>QUESTION 2</u>. Can the above and perhaps some additional properties be used for a general description of sequential structures in any topological category?

REFERENCES

[BH] H.Brandenburg and M.Hušek, Note on cartesian closed subcategories of Unif and Top, this Volume.

[Hu_1] M.Hušek, Construction of special functors and its applications, Comment.Math.Univ.Carolinae 8 (1967) 555-566.

[Hu_2] M.Hušek, Lattices of reflections and coreflections in continuous structures, Categorical Topology (Proc.Int.Conf.Mannheim 1975), Springer Lecture Notes in Math.540 (1976) 404-424.

[Hu_3] M.Hušek, Categories of orderable spaces, to appear in Proc. Int.Categ.Conf.Ottawa 1980.

[Hu_4] M. Husek, Factorizations of mappings (products of proximally fine spaces), Seminar Uniform Spaces 1973/74, Academia Prague (1975), 173-190.

[K] V.Koubek, Each concrete category has a representation by T_2 paracompact topological spaces, Comment.Math.Univ.Carolinae 15 (1974) 655-664.

[T] V.Trnková, Non-constant continuous mappings of metric or compact Hausdorff spaces, Comment.Math.Univ.Carolinae 13 (1972) 283-295.

Mathematical Institute of Charles University
Sokolovská 83
186 00 Praha
Czechoslovakia

A Categorical Framework for Interpolation Theory

Sten Kaijser and Joan Wick Pelletier*

0. Introduction

Interpolation theory, a branch of functional analysis which has applications to partial differential equations, numerical analysis, and approximation theory, has at its core a very categorical theorem (the Aronszajn-Gagliardo theorem "A-G") and various duality theorems, making it irresistible material for category theorists.

Briefly, the traditional formulation considers pairs of Banach spaces (A_0,A_1) contained in a third vector space, made into a category (the category of compatible couples) by taking pairs of bounded linear maps $(T_i: A_i \to B_i)$ which yield bounded linear sum maps $A_0 + A_1 \to B_0 + B_1$. An intermediate Banach space A, $A_0 \cap A_1 \subset A \subset A_0 + A_1$, where $\subset$ denotes a monomorphism only, is called an interpolation space with respect to (A_0,A_1) if each map $T: (A_0,A_1) \to (A_0,A_1)$ in the category restricts to a bounded linear map $T: A \to A$. The results of Aronszajn and Gagliardo (see [1]) show that given an interpolation space A with respect to (A_0,A_1) there exist interpolation functors F and H on the category of compatible couples to the category of Banach spaces (i.e. $F(X_0,X_1)$, $H(X_0,X_1)$ are interpolation spaces with respect to (X_0,X_1)) such that $F(A_0,A_1) = H(A_0,A_1) = A$ and F and H are, respectively, minimal and maximal among interpolation functors having this property.

We generalize the above setting in a natural way by considering the so-called doolittle diagrams of Banach spaces. Since the "dual" of a doolittle diagram is again a doolittle diagram, this eliminates the classical difficulty of (A_0',A_0') not necessarily being a "dual couple" of the Banach couple (A_0,A_1). Moreover, we shall show that the A-G result in this context yields functors related by the theory of dual functors, introduced by Mityagin and Svarc [6].

---

*Both authors acknowledge partial support from the Natural Science and Engineering Research Council of Canada.

1. A new setting for interpolation theory

Let $\mathcal{B}$ denote the category of Banach spaces over I (= real or complex field) and norm-decreasing linear maps. By $\overline{\mathcal{B}}$ we denote the category of doolittle diagrams in $\mathcal{B}$, i.e. objects are pushout-pullback diagrams

$$\overline{X} = \quad \begin{array}{ccc} & \Delta_0 & \\ X_\Delta & \rightarrow & X_0 \\ \Delta_1 \downarrow & & \downarrow \Sigma_0 \\ X_1 & \rightarrow & X_\Sigma \\ & \Sigma_1 & \end{array}$$

and morphisms $T: \overline{X} \rightarrow \overline{Y}$ are pairs $(T_i : X_i \rightarrow Y_i)_{i=0,1}$ satisfying $\Sigma_0 \circ T_0 \circ \Delta_0 = \Sigma_1 \circ T_1 \circ \Delta_1$. (We neglect to index the maps Δ_i, Σ_i by $\overline{X}$ or $\overline{Y}$, hoping that the context will clarify the meaning.)

1.1. Example. Let (X_0, X_1) be a pair of Banach spaces each continuously embedded into some Hausdorff topological vector space $\mathcal{V}$. Let Δ be the intersection $X_0 \cap X_1 \subset \mathcal{V}$ and let Σ be the sum $X_0 + X_1 \subset \mathcal{V}$. Then

$$\begin{array}{ccc} \Delta & \rightarrow & X_0 \\ \downarrow & & \downarrow \\ X_1 & \rightarrow & \Sigma \end{array}$$

is an element of $\overline{\mathcal{B}}$. Thus, we see that our setting for interpolation theory contains the usual setting. ❐

Given an element $\overline{X} \in \overline{\mathcal{B}}$, X_Δ and X_Σ will often be denoted by $\Delta\overline{X}, \Sigma\overline{X}$, respectively, Δ and Σ being functors $\overline{\mathcal{B}} \rightarrow \mathcal{B}$. In fact, letting $\delta : \mathcal{B} \rightarrow \overline{\mathcal{B}}$ be the embedding functor which sends X to the identity doolittle diagram, we have the following easily verified proposition.

1.2. Proposition. δ has left adjoint Σ and right adjoint Δ. ❐

It is well known that $\mathcal{B}$ is a closed monoidal category with Hom functor $L: \mathcal{B}^{op} \times \mathcal{B} \rightarrow \mathcal{B}$ (the unit ball of $L(X,Y)$ is $\mathcal{B}(X,Y)$). Happily, we see that $\overline{\mathcal{B}}$ inherits this feature.

1.3. Proposition. $\overline{\mathcal{B}}$ is a $\mathcal{B}$-category and, moreover, $\overline{\mathcal{B}}$ is a closed monoidal category.

Proof. We define the $\mathcal{B}$-valued Hom functor $L: \overline{\mathcal{B}}^{op} \times \overline{\mathcal{B}} \rightarrow \mathcal{B}$ to be the pullback

$L(\bar{X},\bar{Y})$ of the diagram

$$\begin{array}{ccc} & & L(X_0,Y_0) \\ & & \downarrow \\ L(X_1,Y_1) & \rightarrow & L(X_\Delta,Y_\Sigma) \end{array} .$$

The $\bar{\mathfrak{B}}$-valued lifting $\bar{L}(\bar{X},\bar{Y})$ is then obtained by completing the diagram below by forming its pushout:

$$\begin{array}{ccc} L(\bar{X},\bar{Y}) & \rightarrow & L(X_0,Y_0) \\ \downarrow & & \\ L(X_1,Y_1) & & \end{array}$$

Similarly, $\bar{X} \otimes \bar{Y} \in \mathfrak{B}$ is defined to be the pushout of

$$\begin{array}{ccc} X_\Delta \otimes Y_\Delta & \rightarrow & X_0 \otimes Y_0 \\ \downarrow & & \\ X_1 \otimes Y_1 & & \end{array}$$

where $\otimes$ denotes the projective tensor product in $\mathfrak{B}$ and $\bar{X} \mathbin{\bar{\otimes}} \bar{Y} \in \mathfrak{B}$ is defined by taking the pullback of the following diagram to obtain a doolittle diagram:

$$\begin{array}{ccc} & & X_0 \otimes Y_0 \\ & & \downarrow \\ X_1 \otimes Y_1 & \rightarrow & \bar{X} \otimes \bar{Y} \end{array} .$$

We shall use the notation $\bar{X} \otimes \bar{Y} = \Sigma(\bar{X},\bar{Y}) = \Sigma_{\bar{X}}\bar{Y}$ and $\bar{X} \mathbin{\bar{\otimes}} \bar{Y} = \bar{\Sigma}(\bar{X},\bar{Y})$, pointing out that the Σ of 1.2 is in accord with this notation since $X_\Sigma = \Sigma\bar{X} = \Sigma_{\bar{I}}\bar{X}$, where $\bar{I} = \delta I$.

The adjointness relation $\bar{\mathfrak{B}}(\bar{X} \mathbin{\bar{\otimes}} \bar{Y} , \bar{Z}) \cong \bar{\mathfrak{B}}(\bar{X},\bar{L}(\bar{Y},\bar{Z}))$ is a pushout-pullback exercise left to the interested reader. Finally, the adjointness $\mathfrak{B}(\bar{X} \otimes \bar{Y},Z) \cong \bar{\mathfrak{B}}(\bar{X},\bar{L}(\bar{Y},\delta Z))$ follows from the above result by using the composition of adjoints:

$$\bar{\mathfrak{B}} \;\underset{\bar{L}(\bar{Y},-)}{\overset{\bar{\Sigma}_{\bar{Y}}}{\rightleftarrows}}\; \bar{\mathfrak{B}} \;\underset{\delta}{\overset{\Sigma_{\bar{I}}}{\rightleftarrows}}\; \mathfrak{B} . \qquad \square$$

It now follows according to Linton [5] that there is available a theory of dual functors for functors $\bar{\mathcal{B}} \to \mathcal{B}$. Functors will always be assumed to be strong in the sense that the assignment by F from $L(\bar{X},\bar{Y})$ to $L(F\bar{X},F\bar{Y})$ is continuous and norm-decreasing. The dual functor DF of F is defined as follows: for $\bar{X} \in \bar{\mathcal{B}}$, $DF\bar{X} = NAT(F,\Sigma_{\bar{X}})$, the set of natural transformations made into a Banach space in a natural way. We shall see that the dual functor turns out to be useful in describing the minimal and maximal A-G functors.

2. The Aronszajn-Gagliardo functors

Let us for a moment consider the most general formulation of the A-G theorem. Let $\bar{A} = (A_0,A_1) \in \bar{\mathcal{B}}$ and let $\mathcal{K} = \{\bar{A}\} \subset \bar{\mathcal{B}}$. A functor $F: \mathcal{K} \to \mathcal{B}$ picks out a Banach space $A = F\bar{A}$ which is an $L(\bar{A})$-module, i.e. if $T: \bar{A} \to \bar{A}$, then $T: A \to A$. Conversely, every $L(\bar{A})$-module A gives us a functor $\mathcal{K} \to \mathcal{B}$. Now the restriction functor

$$U : \mathcal{B}^{\bar{\mathcal{B}}} \to \mathcal{B}^{\mathcal{K}}$$

has both left and right adjoints, the left and right Kan extension along the inclusion $\mathcal{K} \subset \bar{\mathcal{B}}$, $Lan_{\mathcal{K}}$, $Ran_{\mathcal{K}}$, as they are usually denoted. In fact we may calculate these functors directly.

2.1. Theorem. Let $F \in \mathcal{B}^{\mathcal{K}}$, $\bar{X} \in \bar{\mathcal{B}}$, $A = F\bar{A}$. Then $Lan_{\mathcal{K}}F\bar{X} = L(\bar{A},\bar{X}) \otimes_{L\bar{A}} A$ and $Ran_{\mathcal{K}}F\bar{X} = L_{L\bar{A}}(L(\bar{X},\bar{A}),A)$, where $\otimes_{L\bar{A}}$ and $L_{L\bar{A}}$ denote, respectively, the Banach module tensor product over $L\bar{A}$ and the $L\bar{A}$-module linear maps (see [4]).

Proof. The first statement follows from the two isomorphisms below, the first of which is proved in [4], the second of which follows from the Yoneda lemma available in this setting [5]:

$$NAT_{\mathcal{B}^{\bar{\mathcal{B}}}}(L(\bar{A},-) \otimes_{L\bar{A}} A,G) \cong L_{L\bar{A}}(A,NAT(L(\bar{A},-),G)) \cong L_{L\bar{A}}(A,G\bar{A}) = L_{L\bar{A}}(F\bar{A},G\bar{A}) =$$

$$NAT_{\mathcal{B}^{\mathcal{K}}}(F, UG) .$$

To prove the second statement we note as above that $NAT_{\mathcal{B}^{\mathcal{K}}}(UG,F) = L_{L\bar{A}}(G\bar{A},F\bar{A}) = L_{L\bar{A}}(G\bar{A},A)$, and we proceed to define an isomorphism

$$\Phi : NAT_{\mathcal{B}^{\bar{\mathcal{B}}}}(G,L_{L\bar{A}}(L(-,\bar{A}),A)) \to L_{L\bar{A}}(G\bar{A},A)$$

and its inverse Ψ . We define $\Phi(t)(a) = t_{\bar{A}}(a)(1_{\bar{A}})$ and $\Psi(S)_{\bar{X}}(x)(g) = \langle S, Gg(x)\rangle$, where $t\colon G \to L_{L\bar{A}}(L(-,\bar{A}),A)$ $a \in G\bar{A}$, $S \in L_{L\bar{A}}(G\bar{A},A)$, $x \in G\bar{X}$, $g \in L(\bar{X},\bar{A})$. We leave to the reader the tedious verification that Φ and Ψ are inverses. □

We remark that the Kan extensions give us a situation similar to the A-G theorem since (1) $\mathrm{Lan}_{\mathcal{K}} F\bar{A} = A = \mathrm{Ran}_{\mathcal{K}} F\bar{A}$ (we omit this proof) and (2) given any other functor $G\colon \bar{\mathcal{B}} \to \mathcal{B}$ such that $G\bar{A} = A$, there are natural transformations (not necessarily monos)

$$\mathrm{Lan}_{\mathcal{K}} F \to G \to \mathrm{Ran}_{\mathcal{K}} F$$

corresponding to the identity of $L_{L\bar{A}}(A,A)$ via adjointness. However, $\mathrm{Lan}_{\mathcal{K}} F$ and $\mathrm{Ran}_{\mathcal{K}} F$ are not exactly what we are seeking for our generalization of the A-G theorem, as we shall see below.

We now specify that a generalized interpolation space A with respect to $\bar{A}$ will be an $L\bar{A}$-module. A generalized interpolation functor F must satisfy the property $F\circ\delta = 1_{\mathcal{B}}$. We note that for all $\bar{X} \in \bar{\mathcal{B}}$ there are maps $X_\Delta \xrightarrow{\tau} FX \xrightarrow{\sigma} X_\Sigma$, deriving from the above property, such that the following diagram commutes:

$$\begin{array}{ccc} X_\Delta & \to & X_0 \\ \downarrow & \searrow \; F\bar{X} \; \searrow & \downarrow \\ X_1 & \to & X_\Sigma \end{array} .$$

Given a generalized interpolation space A with respect to $\bar{A}$, we define $F_A\bar{X}$ to be the coimage of the map

$$d\colon L(\bar{A},\bar{X}) \otimes_{L\bar{A}} A \to \Sigma\bar{X}$$

defined by $d(S\otimes a) = S(\sigma a)$, where $\sigma\colon A \to \Sigma\bar{A}$ is the above map. (This definition makes sense since $S\colon \bar{A} \to \bar{X}$ gives rise to $S\colon \Sigma\bar{A} \to \Sigma\bar{X}$.) $H_A\bar{X}$ is defined to be the following pullback:

$$\begin{array}{ccc} H_A\bar{X} & \dashrightarrow & L_{L\bar{A}}(L(\bar{X},\bar{A}),A) \\ \downarrow & & \downarrow L_{L\bar{A}}(L(\bar{X},\bar{A}),\sigma) \\ \Sigma\bar{X} & \xrightarrow[e]{} & L_{L\bar{A}}(L(\bar{X},\bar{A}),\Sigma\bar{A}) \end{array} ,$$

where $e(x)(S) = S(x)$. It can be verified that $F_A \circ \delta = 1_{\mathcal{B}}$ and $H_A \circ \delta = 1_{\mathcal{B}}$.

We now give our version of the A-G theorem.

2.2. Theorem. If A is a generalized interpolation space with respect to $\bar{A}$, then the functors F_A and H_A , defined above, are generalized interpolation functors satisfying $F_A\bar{A} = A = H_A\bar{A}$. Furthermore, F_A and H_A are, respectively, the minimal and maximal such generalized interpolation functors.

Proof. Only the second assertion needs discussion. We note that for any generalized interpolation functor G such that $G\bar{A} = A$, we have maps

$$L(\bar{A},\bar{X}) \otimes_{L\bar{A}} A \underset{f}{\rightarrow} G\bar{X} \underset{h}{\rightarrow} L_{L\bar{A}}(L(\bar{X},\bar{A}),A) \quad ,$$

where $f(S\otimes a) = G(S)(a)$, $h(x)(T) = G(T)(x)$. Moreover, f clearly lifts to a map from $F_A\bar{X}$ to $G\bar{X}$, and h together with $\sigma\colon G\bar{X} \rightarrow \Sigma\bar{X}$ equalize e and $L_{L\bar{A}}(L(\bar{X},\bar{A}),\sigma)$ yielding a map, by definition of $H_A\bar{X}$, from $G\bar{X}$ to $H_A\bar{X}$. Hence, we have

$$F_A\bar{X} \rightarrow G\bar{X} \rightarrow H_A\bar{X} \quad . \qquad \square$$

3. Duality results

Duality results in interpolation theory usually require density conditions on the elements of $\bar{\mathcal{B}}$, e.g. $\Delta\bar{A}$ is dense in $\bar{A}$, meaning in A_0 and A_1 , or on the intermediate space A , e.g. $\Delta\bar{A}$ is dense in A . One reason for this is that the following result, true in our context without hypotheses, is true in the traditional context only when $\Delta\bar{A}$ is dense in A .

3.1. Proposition. $\bar{\mathcal{B}}$ is closed under formation of dual diagrams, i.e. if $\bar{A} \in \bar{\mathcal{B}}$, $\Delta(\bar{A})' = \Sigma(\bar{A}')$ and $\Sigma(\bar{A})' = \Delta(\bar{A}')$.

Proof. The second assertion follows from the fact that duals of pushouts are pullbacks. Although duals of pullbacks need not be pushouts in general, examination of the doolittle diagram shows that duality does preserve them, hence, proving the first assertion. $\square$

Our main result of this section is to show that the maximal and minimal A-G functors are connected by the notion of dual functors.

3.2. <u>Theorem</u>. Let A be a generalized interpolation space with respect to $\overline{A}$ such that $\Delta\overline{A}$ is dense in A and $A \subset \Sigma\overline{A}$. Then $DF_A\overline{X} = H_{A'}\overline{X}$ for all $\overline{X} \in \overline{\mathcal{B}}$ such that $\Delta\overline{X}$ is dense in $\overline{X}$.

<u>Proof</u>. We must establish an isomorphism

$$\Phi : \mathrm{NAT}(F_A, \Sigma_{\overline{X}}) \to H_{A'}(\overline{X}) .$$

We first observe that there are maps $\eta: \mathrm{NAT}(L(\overline{A},-) \otimes_{L\overline{A}} A, \Sigma_{\overline{X}}) \to L_{\overline{LA}'}((\overline{X}\otimes\overline{A})', A')$, $\xi: \mathrm{NAT}(L(\overline{A},-) \otimes_{L\overline{A}}, A, \Sigma_{\overline{X}}) \to \Sigma\overline{X}$, given by $\eta(T)(g)(a) = g(T_{\overline{A}}(a))$ and $\xi(T) = T_{\overline{I}}(1)$, recalling that $L(\overline{A},\overline{A}) \otimes_{L\overline{A}} A = A$ and $L(\overline{A},\overline{I}) \otimes_{L\overline{A}} A = I$. Moreover,
$(e\circ\xi)(T)(g)(a) = e(T_{\overline{I}}(1))(g)(a) = g(T_{\overline{I}}(1) \otimes a)$ and
$(L_{\overline{LA}'}((\overline{X}\otimes\overline{A})', \sigma)\circ\eta)(T)(g)(a) = g(T_{\overline{A}}(a))$. However, by the naturality of T, we have (thinking of a both as an element of $\Delta\overline{A}$ and as a map $\overline{I} \to \overline{A}$)

$$((\overline{X}\otimes a)\circ T_{\overline{I}})(1) = T_{\overline{I}}(1) \otimes a = (T_{\overline{A}}\circ L(A,a) \otimes A)(1)$$
$$= T_{\overline{A}}(a) :$$

$$\begin{array}{ccc} L(\overline{A},\overline{I}) \otimes A & \xrightarrow{T_{\overline{I}}} & \overline{X}\otimes\overline{I} \\ {\scriptstyle L(\overline{A},a)\otimes A}\downarrow & & \downarrow{\scriptstyle \overline{X}\otimes a} \\ L(\overline{A},\overline{A}) \otimes A & \xrightarrow{T_{\overline{A}}} & \overline{X}\otimes\overline{A} \end{array} .$$

Hence, $e\circ\xi = L_{\overline{LA}'}((\overline{X}\otimes\overline{A})', \sigma)\circ\eta$, and, thus, there is a map $\varphi: \mathrm{NAT}(L(\overline{A},-) \otimes_{L\overline{A}} A, \Sigma_{\overline{X}}) \to H_{A'}(\overline{X})$. Let $\Phi = \varphi \circ Dd: \mathrm{NAT}(F_A, \Sigma_{\overline{X}}) \to H_{A'}(\overline{X})$.

A map

$$\psi: H_{A'}(\overline{X}) \to \mathrm{NAT}(L(\overline{A},-) \otimes_{L\overline{A}} A, \Sigma_{\overline{X}})$$

is given for $(T,x) \in H_{A'}(\overline{X})$, $S \in L(\overline{A},\overline{Y})$, $a \in \Delta\overline{A}$, by

$$\psi(T,x)_{\overline{Y}}(S\otimes a) = x_i \otimes S(a)_i , \quad i = 0,1 .$$

We note that if $d(S\otimes a) = 0$, then $\psi(T,x)_{\overline{Y}}(S\otimes a) = 0$, so we get $\Psi: H_{A'}(\overline{X}) \to \mathrm{NAT}(F_A, \Sigma_{\overline{X}})$. Φ and Ψ may be readily checked to be inverses. □

As an important corollary of 3.2, we obtain by means of a duality theorem for dual functors on $\overline{\mathcal{B}}$ a result similar to Theorem 1 of Janson [3]. The hypotheses (and, of course, the generality) differ somewhat, however, but we shall not explore here the exact differences. Briefly, we define a functor $F\colon \overline{\mathfrak{B}} \to \mathfrak{B}$ to be computable (à la [2]) if

$$F\overline{X} = \lim \{F\overline{Y} \mid \overline{Y} \text{ finite dimensional subspace of } \overline{X}\} .$$

A finite dimensional element $\overline{Y}$ of $\overline{\mathfrak{B}}$ is one in which Y_0 and Y_1 (and, hence, $\Delta\overline{Y}$ and $\Sigma\overline{Y}$) are finite dimensional. Our result is similar to Theorem 1.9 of Herz-Pelletier [2].

3.3. <u>Theorem</u>. If $F\colon \overline{\mathfrak{B}} \to \mathfrak{B}$ is computable, then $DF(\overline{X}') = (F\overline{X})'$ for all $\overline{X}$ such that $\Delta\overline{X}$ is dense in $\overline{X}$. □

3.4. <u>Corollary</u>. If F_A is computable, then $(F_A\overline{X})' = H_{A'}(\overline{X}')$.

<u>Proof</u>. This follows from 3.2 since $DF_A\overline{X}' = H_{A'}\overline{X}'$. □

We remark in closing that 3.4 is a slightly stronger result than Theorem 1 of [3]. Hence, our setting promises to be advantageous to the study of interpolation theory.

<u>References</u>

[1] J. Bergh and J. Löfström, Interpolation Spaces, Grundlehren der mathematischen Wissenschaften 223, Springer-Verlag Berlin, Heidelberg, New York (1976).

[2] C. Herz and J. Wick Pelletier, Dual Functors and Integral Operators in the Category of Banach spaces, J. Pure Appl. Alg. 8 (1976), 5-22.

[3] S. Janson, Minimal and Maximal Methods of Interpolation, Report No. 6, Institut Mittag-Laffler (1980).

[4] S. Kaijser, On Banach Modules I, Math. Proc. Camb. Phil. Soc., to appear.

[5] F.E.J. Linton, Autonomous Categories and Duality of Functors, J. Alg. 2 (1965), 315-349.

[6] B.S. Mityagin and A.S. Švarc, Functors in categories of Banach Spaces, Russ. Math. Surveys 19 No. 2 (1964), 65-127.

Uppsala University
Uppsala, Sweden

York University
Toronto, Ontario, Canada

Toposes are monadic over categories.

by

J. Lambek[0)]

McGill University and University of Oxford.

In [L2, LS1, LS2] a study has been made of the category Top whose objects are toposes with canonical subobjects and whose morphisms are (strict) logical functors which preserve these canonical subobjects on the nose. An auxiliary notion, the concept of a dogma, was introduced in [L2] to capture the algebraic aspects of toposes. In the present context it will be convenient to assume that dogmas are Cartesian closed.[1)] Dogmas are the objects of a category Dog whose morphisms are also called logical functors in [LS2] (orthodox functors in [L2]). Actually, Top is already equational in a wider sense, as we shall see, over the category Cat of categories.

Cartesian closed categories, closed monoidal categories and (Cartesian closed) dogmas are all monadic (tripleable) over Grph, the category of graphs, or Cat.[2)] They are all examples of the "graphical algebras" of Albert Burroni [B] in a very specific way: they are graphs (or categories) with operations and equations. The notion of graphical algebra as envisaged by Burroni is much more general: it includes toposes, provided these are "equipped with a choice of finite limits and colimits and even 'parts' - a representative choice of monomorphisms". It would therefore be of interest to give an independent verification that the category Top discussed above is monadic (tripleable) over Grph. Instead, we shall prove that it is monadic over Cat, in the sense that the forgetful functor U: Top → Cat has a left adjoint F with adjunctions η and ε such that the comparison functor $K\colon \mathcal{A} \to \mathcal{A}^T$ is an isomorphism, where $T = (UF, \eta, U\varepsilon F)$.

In what follows, we shall suppress explicit mention of the functor U. According to Beck's Theorem [M], we must then verify the following statement: Given two arrows $F,G\colon \mathcal{A} \rightrightarrows \mathcal{B}$ in Top having a split coequalizer in Cat, each (split) coequalizer $\mathcal{A} \rightrightarrows \mathcal{B} \xrightarrow{H} \mathcal{C}$ in Cat is a coequalizer in Top. Since Dog is monadic over Cat, we may assume that $\mathcal{A} \rightrightarrows \mathcal{B} \to \mathcal{C}$ is a coequalizer in Dog. It remains to show:

(I) $\mathcal{C}$ is a topos with canonical subobjects,

(II) H is a coequalizer of (F,G) in Top.

(Ia) On the way to proving (I), we shall first show that $\mathcal{C}$ has canonical equalizers and, incidentally, that H preserves them. We know of course that $\mathcal{A}$ and $\mathcal{B}$ have canonical equalizers [L2]. To be precise, let E be the category $\cdot \rightrightarrows \cdot$ and $Q_{\mathcal{A}}\colon \mathcal{A}^E \to \mathcal{A}$ the functor which sends any object of $\mathcal{A}^E$, essentially a pair $f,f'\colon A_0 \rightrightarrows A_1$ of arrows in $\mathcal{A}$, onto its canonical equalizing object:

$$Q_{\mathcal{A}}(f,f') = \mathrm{Ker}(\delta_{A_1} < f,f' >) .$$

Now consider the following diagram in Cat:

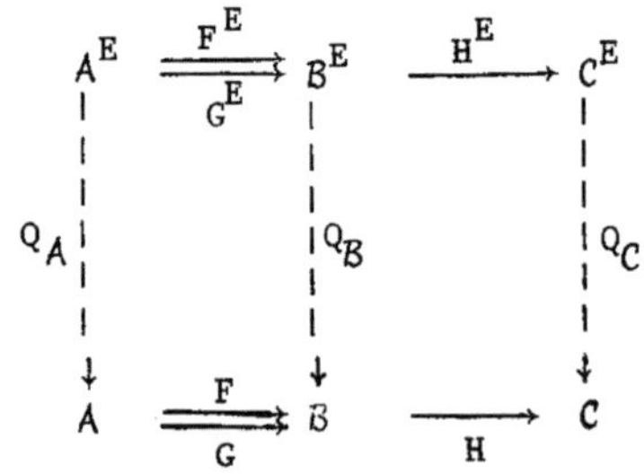

Since F and G preserve canonical equalizers [L2], the two squares on the left commute. Since H is an absolute coequalizer of (F,G), H^E is a coequalizer of (F^E,G^E), hence there exists a unique functor $Q_{\mathcal{C}}: \mathcal{C}^E \to \mathcal{C}$ such that the square on the right commutes. It remains to show that $Q_{\mathcal{C}}$ sends any object of $\mathcal{C}^E$, namely a pair of arrows $h,h': C_0 \rightrightarrows C_1$, onto its equalizing object $Q_{\mathcal{C}}(h,h')$.

The defining property of $Q_{\mathcal{A}}$ may be expressed by saying that $Q_{\mathcal{A}}$ is right adjoint to the constant functor $K_{\mathcal{A}}: \mathcal{A} \to \mathcal{A}^E$. This fact amounts to a pair of equations involving the natural transformations $\eta_{\mathcal{A}}: 1_{\mathcal{A}} \to Q_{\mathcal{A}} K_{\mathcal{A}}$ and $\varepsilon_{\mathcal{A}}: K_{\mathcal{A}} Q_{\mathcal{A}} \to 1_{\mathcal{A}^E}$. The same is true for $Q_{\mathcal{B}}$. We define $\eta_{\mathcal{C}}$ and $\varepsilon_{\mathcal{C}}$ as follows, where S is the given splitting of H in Cat:

$$\eta_{\mathcal{C}} = H\eta_{\mathcal{B}}S, \quad \varepsilon_{\mathcal{C}} = H^E\varepsilon_{\mathcal{B}}S^E.$$

It is then routine to derive the equations which render $Q_{\mathcal{C}}$ right adjoint to $K_{\mathcal{C}}$ with adjunctions $\eta_{\mathcal{C}}$ and $\varepsilon_{\mathcal{C}}$, so that $\mathcal{C}$ has equalizers also, which we may call "canonical". Moreover H preserves canonical equalizers in the sense that

$$HQ_{\mathcal{B}} = Q_{\mathcal{C}}H^E, \quad H\eta_{\mathcal{B}} = \eta_{\mathcal{C}}H, \quad H^E\varepsilon_{\mathcal{B}} = \varepsilon_{\mathcal{C}}H^E.$$

The calculations, which are omitted here, also make use of the arrow $T: \mathcal{B} \to \mathcal{A}$ of the given split coequalizer diagram in Cat.

Incidentally, the above argument establishes that categories with canonical equalizers are monadic over Cat. Similar arguments apply to categories with canonical coequalizers, limits, colimits and any combination of these, even if additional equations are prescribed.

(Ib) Next we shall prove that $\mathcal{C}$ is a "quasi-topos" in the following sense, probably equivalent to the notion of Penon [P]: it is a dogma with canonical kernels such that char(ker h). = . h for all $h: C \to \Omega$ in $\mathcal{C}$. We recall [L2] that in any dogma the characteristic morphism of a monomorphism $m: C_0 \to C$ is defined by

$$\ulcorner \text{char } m \urcorner \;.=.\; \{z \in C \mid \exists_{z_0 \in C_0}\; z = m\, z_0\} .$$

For any $h: C \to \Omega$ we put

$$\text{Ker } h \;=\; Q_{\mathcal{C}}(h, TO_C) .$$

This is just the equalizing object; it comes equipped with an arrow $\ker h: \text{Ker } C \to \Omega$ which is related to $\varepsilon_{\mathcal{C}}(h, TO_C)$.

For any $g: B \to \Omega$ in the topos $\mathcal{B}$, we have char(ker g) . = . g . Take $B = S(C)$, $g = S(h)$ and recall that $HS = 1_{\mathcal{C}}$. Then

$$H(\text{char}(\ker S(h))) \;.=.\; h .$$

Now the logical functor H preserves characteristic morphisms and, by (Ia), also kernels. Therefore

$$\text{char}(\ker h) \;.=.\; h ,$$

as required.

(Ic) To show that $\mathcal{C}$ is a topos, it remains to verify that every monomorphism is, up to isomorphism, of the form ker h.

Let $\iota_C: C \to PC$ be the singleton morphism. Putting $B = S(C)$ as above, we have $C = H(B)$. Now in the topos $\mathcal{B}$

$$\iota_B \;.\!\cong\!.\; \ker(\mathrm{char}\ \iota_B)\ ,$$

hence, applying the logical functor H, we obtain

$$\iota_C \;.\!\cong\!.\; \ker(\mathrm{char}\ \iota_C).$$

This shows that there is an equalizer diagram $C \to PC \rightrightarrows \Omega$, and we know from [L2] that the canonical logical functor $H_{\mathcal{C}}\colon \mathcal{C} \to T(\mathcal{C})$, where $T(\mathcal{C})$ is the topos generated by the dogma $\mathcal{C}$, is then full and faithful, or, equivalently, that description holds in $\mathcal{C}$, in particular, internal equality implies external equality.

Now let $m\colon C_0 \to C$ be any mono in $\mathcal{C}$ and put $h \;.\!=\!.\; \mathrm{char}\ m$, that is,

$$\ulcorner h \urcorner \equiv \{z \in C \mid \exists_{z_0 \in C_0}\ m z_0 = z\}\ .$$

We shall prove that m is an equalizer of h and TO_C, that is, the following square is a pullback.

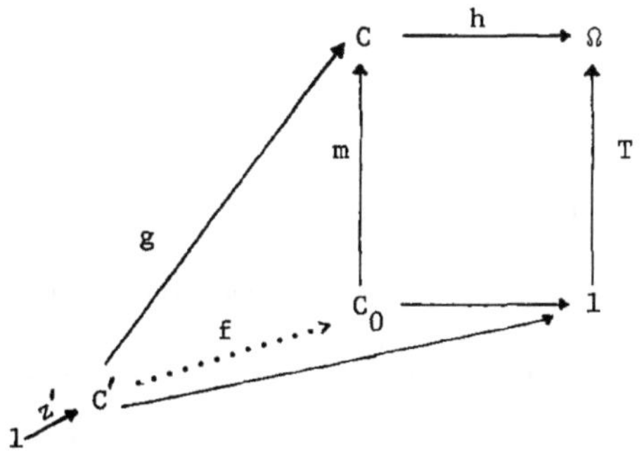

Indeed, suppose $g: C' \rightarrow C$ is such that $hg \;.=.\; TO_{C'}$. Let z' be an indeterminate arrow $1 \rightarrow C'$, then $h\,g\,z' \;.=.\; T$. Therefore $\vdash_{z'} g\,z' \in \ulcorner h \urcorner$, that is, $\vdash \forall_{z' \in C'} \exists_{z_0 \in C_0}\, mz_0 = gz'$. Suppose for the moment that we can replace $\exists$ by $\exists!$ in this. Then, by description, there exists a unique f such that $\vdash \forall_{z' \in C'}\, mf\,z' = g\,z'$. Since internal equality implies external equality, we actually have $mf \;.=.\; g$.

Why can we replace $\exists$ by $\exists!$? It suffices to show that $\vdash \forall_{x,y \in C_0} (mx = my \Rightarrow x = y)$. Look at the proof of [L3, Lemma 12.4], where this is shown in a topos. However, the proof holds in any dogma in which equalizers exist (Ia), internal equality implies external equality (see above) and characteristic morphisms are unique (an easy consequence of (Ib)).

(Id) To show that $\mathcal{C}$ has canonical subobjects, we call a monomorphism $m: C_0 \rightarrow C$ canonical if $m \;.=.\; \ker h$ for some $h: C \rightarrow \Omega$. It remains to check the conditions (i) to (v) of [L2,§9].

For example, let us verify that the product of two canonical subobjects of C in $\mathcal{C}$ is a canonical subobject. Now a canonical subobject of C in $\mathcal{C}$ has the form

$$\mathrm{Ker}\; h = Q_C(h, TO_C) = HQ_{\mathcal{B}}(S(h), S(TO_C)) = H(\mathrm{Ker}\; k),$$

where $k \;.=.\; \delta_{S(\Omega)} < S(h), S(TO_C) >$. The result now follows from the fact that the product of two canonical subobjects of $S(C)$ in $\mathcal{B}$ is one.

(II) We claim that $H: \mathcal{B} \rightarrow \mathcal{C}$ is a coequalizer of $(F,G): \mathcal{A} \rightrightarrows \mathcal{B}$ in Top. We know it is the coequalizer in Dog, so suppose $H': \mathcal{B} \rightarrow \mathcal{C}'$ is a morphism in Top such that $H'F = H'G$. Then there exists a unique logical functor $M: \mathcal{C} \longrightarrow \mathcal{C}'$ such that $MH = H'$. It remains to show that M is in Top , that is, that it preserves canonical equalizers. Now H does preserve canonical equalizers, that is, $HQ_{\mathcal{B}} = Q_{\mathcal{C}}H^E$ etc. Recalling that $HS = 1_{\mathcal{C}}$, we therefore have

$$\begin{aligned} MQ_{\mathcal{C}} &= MHQ_{\mathcal{B}}S^E \\ &= H'Q_{\mathcal{B}}S^E \\ &= Q_{\mathcal{C}'}H'^E S^E \\ &= Q_{\mathcal{C}'}M^E H^E S^E \\ &= Q_{\mathcal{C}'}M^E . \end{aligned}$$

Similarly one shows that $M\eta_{\mathcal{C}} = \eta_{\mathcal{C}'}M$ and $M^E\varepsilon_{\mathcal{C}} = \varepsilon_{\mathcal{C}'}M^E$, and so M preserves canonical equalizers

POSTSCRIPT.

The theorem stated in the title is one of those results which are easily proved once one is convinced of their truth. In particular, it is not necessary to base its proof on the notion of a dogma, with which the author happens to be familiar. A direct proof along some of the lines indicated in [B] is also possible.[3)]

In the published version [B] of his talk at Amiens, Burroni actually asserts that toposes are tripleable over graphs. It is instructive to see why the present argument fails to yield this apparently stronger result, even if dogmas are eliminated.

To simplify matters and to vary the approach a little, let us confine attention to the related assertion that categories with canonical equalizers are monadic over graphs. Let $\alpha(f,g)$ denote the "canonical" equalizer of $f,g\colon A \rightrightarrows B$. Then, for each $h\colon C \to A$ such that $fh \cdot{=}\cdot gh$, there is a unique arrow $\beta(f,g,h)\colon C \to \text{source } \alpha(f,g)$ such that $\alpha(f,g)\,\beta(f,g,h) \cdot{=}\cdot h$.

To equip a category $\mathcal{A}$ with <u>canonical equalizers</u> is to provide a right adjoint U to the functor $F\colon \mathcal{A} \longrightarrow \mathcal{A}^{\cdot\rightrightarrows\cdot}$ given by

$$F(A) = A \overset{1_A}{\underset{1_A}{\rightrightarrows}} A \;,\quad F(a) = a \begin{array}{ccc} A & \rightrightarrows & A \\ \downarrow & & \downarrow \\ A' & \rightrightarrows & A' \end{array} a \;,$$

where $a\colon A \longrightarrow A'$. Now it is easy to express U and the adjunctions $\eta\colon \mathrm{id} \to UF$ and $\varepsilon\colon FU \to \mathrm{id}$ in terms of α and β. The functoriality of U and the adjunction equations relating η and ε may then be written as equations involving α and β.

It readily follows from these considerations, as in the main text of the present article, that the category Equ of small categories with canonical equalizers is monadic over Cat. What goes wrong in trying to show monadicity over Grph is that the condition $fh \cdot{=}\cdot gh$ presupposed in the definition of $\beta(f,g,h)$ involves the fact that $\mathcal{A}$ is a category, not just a graph.

Indeed, in attempting to verify Beck's condition for monadicity over Grph, we assume that $F,G\colon \mathcal{A} \rightrightarrows \mathcal{B}$ in Equ have a split coequalizer in Grph and want to show that each (split) coequalizer $\mathcal{A} \rightrightarrows \mathcal{B} \xrightarrow{H} \mathcal{C}$ in Grph is a coequalizer in Equ. (Since Cat is monadic over Grph, we may assume that $\mathcal{A} \rightrightarrows \mathcal{B} \longrightarrow \mathcal{C}$ is a coequalizer in Cat, but not a split coequalizer.)

A crucial part of the argument is the construction of $\beta_{\mathcal{C}}$. Let $S: \mathcal{C} \longrightarrow \mathcal{B}$ be the arrow splitting H in Grph, then we try to define

$$\beta_{\mathcal{C}}(f,g,h) = H(\beta_{\mathcal{B}}(S(f),S(g),S(h)))$$

for any $f,g: A \rightrightarrows B$ in $\mathcal{C}$ such that $fh \cdot = \cdot gh$. Indeed, this definition is forced upon us if H is to preserve canonical equalizers, provided the right hand side makes sense. For this to be the case we must check that $S(f)S(h) \cdot = \cdot S(g)S(h)$. Unfortunately we cannot deduce this from $fh \cdot = \cdot gh$, since S is an arrow only in Grph, not in Cat.

Burroni has made the surprising discovery that only two special cases of β are needed, whose defining conditions are automatically satisfied, namely

$$\gamma(u,v,p): \text{source } \alpha(up,vp) \longrightarrow \text{source } \alpha(u,v)$$

and
$$\delta(u): \text{source } u \longrightarrow \text{source } \alpha(u,u).$$

One then defines $\beta(f,g,h) = \gamma(f,g,h)\delta(fh)$ in case $fh = gh$.

At first it seems that this ingenious observation won't resolve our problem with Beck's condition, as the description of the source of γ still involves composition, an operation in Cat not in Grph. However, following a suggestion by Burroni, we treat composition like the other operations on $\mathcal{C}$ and write $\mu_{\mathcal{C}}(u,p)$ for up in $\mathcal{C}$. Referring to the split coequalizer diagram

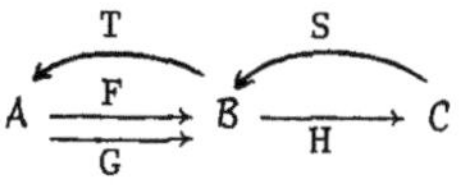

in Grph and using the equations $HF = HG$, $HS = 1$, $FT = 1$ and $SH = GT$, one easily calculates that[4)]

$$\text{source } \gamma_{\mathcal{C}}(u,v,p) = \text{source } \alpha_{\mathcal{C}}(\mu_{\mathcal{C}}(u,p),\ \mu_{\mathcal{C}}(v,p))$$

from the given fact that

$$\text{source } \gamma_{\mathcal{B}}(S(u),S(v),S(p)) = \text{source } \alpha_{\mathcal{B}}(\mu_{\mathcal{B}}(S(u),S(p)),\mu_{\mathcal{B}}(S(v),S(p))).$$

Thus, borrowing some of Burroni's ideas, we may recapture his result by our method.

For a quick exposition of the difference between the methods of [B] and [L1], note that any functor $A\times\mathcal{B} \to C$ may be used to obtain a structure-semantics adjointness between concrete categories over A and theories under $\mathcal{B}$.[5)] In particular, if $\theta\colon \mathcal{B} \to T$ is bijective on objects and one forms the pullback

$$\begin{array}{ccc} A_\theta & \longrightarrow & C^{T} \\ {\scriptstyle U_\theta}\downarrow & & \downarrow{\scriptstyle C^\theta} \\ A & \longrightarrow & C^{\mathcal{B}} \end{array},$$

U_θ will satisfy Beck's condition, although it won't in general have a left adjoint. In [L1] and [B] the functors $\mathrm{Cat}\times\mathrm{Cat}^{\mathrm{Cat}} \longrightarrow \mathrm{Cat}$ and $\mathrm{Grph}\times\mathrm{Grph}^{\mathrm{op}}_{\mathrm{fin}} \longrightarrow \mathrm{Sets}$ are exploited respectively. (The methods of [L1] should be amended in accordance with Footnote 2.)

REFERENCES

[B] A. Burroni, Algèbres graphiques, 3éme colloque sur les catégories, Cahiers de Topologie et Géométrie Différentielle 23 (1981), 249-265.[6)]

[L1] J. Lambek, Deductive systems and categories II, Lecture Notes in Mathematics 86(1969), 76-122.

[L2] J. Lambek, From types to sets, Advances in Math. 36(1980), 113-164.

[LS1] J. Lambek and P. Scott, Intuitionist type theory and the free topos, J. Pure and Applied Algebra 19(1980), 576-619.

[LS2] J. Lambek and P. Scott, Algebraic aspects of topos theory, 3[éme] colloque sur les catégories, Cahiers de Topologic et Géométrie Différentielle 22(1981), 129-140.

[Li] F.E.J. Linton, An outline of functorial semantics, Lecture Notes in Mathematics 80 (1969), 7-52.

[M] S. MacLane, Categories for the working mathematician (Springer, New York, 1971).

FOOTNOTES

0) The author wishes to acknowledge support from the National Science and Engineering Research Council of Canada and from the Social Sciences and Humanities Research Council of Canada. He is endebted to Albert Burroni for an exchange of ideas and to Max Kelly for insisting on the footnotes and for uncovering an error in [L1]. This paper was first presented in Sussex in November, 1980.

1) We briefly recall the definition of a dogma in [L2]. While dogmas there were not required to be Cartesian closed (only partially so), we shall insist on this property here. A (Cartesian closed) dogma is a Cartesian closed category with a specified object Ω and with specified arrows $T, \bot: 1 \to \Omega$, $\wedge, \vee, \Rightarrow: \Omega\times\Omega \to \Omega$ and $\forall_A, \exists_A: PA \to \Omega$, where $PA = \Omega^A$. These arrows satisfy a number of conditions to assure that, when interpreting higher order logic, all intuitionistic theorems are equal to the arrow T. It was shown in [L2, p. 124] that these conditions can be expressed as identities, in fact, as equations between constant arrows, e.g.,

$$\wedge\langle 1_\Omega, 1_\Omega\rangle = 1_\Omega, \quad \wedge = \langle \pi'_{\Omega,\Omega}, \pi_{\Omega,\Omega}\rangle$$

where $\pi_{A,B}: A\times B \to A$ and $\pi'_{A,B}: A\times B \to B$ are the usual projections. There are 17 such equations or families of equations.

2) For example, a Cartesian closed category $\mathcal{A}$ is a category with canonical finite products and a functor $\exp_{\mathcal{A}}: \mathcal{A}\times\mathcal{A}^{op} \to \mathcal{A}$, usually written $\exp_{\mathcal{A}}(A,A') = A^{A'}$ such that $\exp_{\mathcal{A}}(-,A')$ is right adjoint to $-\times A'$ by virtue of natural transformations $\eta_{\mathcal{A}}(-,A'): id_{\mathcal{A}} \to \exp(-\times A',A')$ and $\varepsilon_{\mathcal{A}}(-,A'): \exp_{\mathcal{A}}(-,A')\times A' \to id_{\mathcal{A}}$ satisfying the usual adjunction equations. We shall verify monadicity over Cat.

Given a pair of arrows $F,G: \mathcal{A} \rightrightarrows \mathcal{B}$ in Cart, the category of Cartesian closed categories, let $\mathcal{A} \rightrightarrows \mathcal{B} \xrightarrow{H} \mathcal{C}$ be a split, hence absolute coequalizer in Cat. One wishes to show that H is a coequalizer in Cart, in particular, that $\mathcal{C}$ has the structure of a Cartesian closed category and that H preserves it. As in [L1, p.102], we define $\exp_{\mathcal{C}}: \mathcal{C}\times\mathcal{C}^{op} \to \mathcal{C}$ as the unique functor such that $\exp_{\mathcal{C}}(H\times H^{op}) = H \exp_{\mathcal{B}}$

What was said about natural transformations in [L1, p.102] is incorrect, instead one should define

$$\varepsilon_{\mathcal{C}}(C,C') = H\varepsilon_{\mathcal{B}}(S(C),S(C'))$$

and similarly for $\eta_{\mathcal{C}}$. From this it easily follows that $H\varepsilon_{\mathcal{B}}(B,B') = \varepsilon_{\mathcal{C}}(H(B),H(B'))$, and similarly for η, and that $\varepsilon_{\mathcal{C}}(-,C')$ and $\eta_{\mathcal{C}}(-,C')$ are natural transformations satisfying the adjunction equations.

To show that Dog is monadic over Cat, we assume that $F,G\colon \mathcal{A} \rightrightarrows \mathcal{B}$ in Dog and have to show how to define the additional structure Ω, T, $\bot$, $\wedge,\vee$, $\Rightarrow$, $\forall_C$, $\exists_C$ in $\mathcal{C}$. One defines, for example,

$$\Omega_{\mathcal{C}} = H(\Omega_{\mathcal{B}}),\quad T_{\mathcal{C}} = H(T_{\mathcal{B}}),\quad \wedge_{\mathcal{C}} = H(\wedge_{\mathcal{B}})$$

and

$$\forall_{\mathcal{C}}(C) = H(\forall_{\mathcal{B}}(S(C)),$$

where we have written $\forall(C)$ in place of $\forall_C$ and S is the functor which splits H. One easily verifies that H preserves the additional structure and that the 17 equations hold in $\mathcal{C}$.

3) According to the revised version of [B], a topos is a bicartesian closed category with equalizers and coequalizers such that for every arrow $f\colon A \to B$ the canonically associated arrow $\mathrm{Coim}(f) \to \mathrm{Im}(f)$ is invertible and also with a specified object Ω and arrow $T\colon 1 \to \Omega$ such that for every arrow $f\colon A \to B$ there is associated a unique arrow $B \to \Omega$ such that the following square is a pullback:

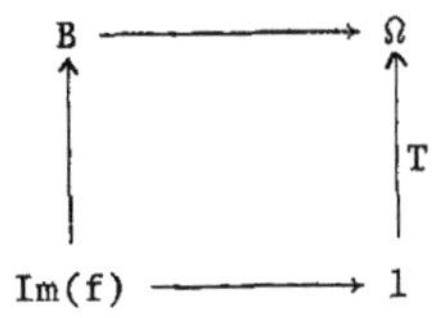

4) Indeed, we have

$$\begin{aligned}
\text{source } \gamma_{\mathcal{C}}(u,v,p) &= \text{source } H\gamma_{\mathcal{B}}(Su,Sv,Sp) \\
&= H \text{ source } \gamma_{\mathcal{B}}(Su,Sv,Sp) \\
&= H \text{ source } \alpha_{\mathcal{B}}(\mu_{\mathcal{B}}(Su,Sp),\mu_{\mathcal{B}}(Sv,Sp)) \\
&= \text{source } H\alpha_{\mathcal{B}}(FT\mu_{\mathcal{B}}(Su,Sp),\ldots\ldots) \\
&= \text{source } HF\alpha_{\mathcal{A}}(T\mu_{\mathcal{B}}(Su,Sp),\ldots\ldots) \\
&= \text{source } HG\alpha_{\mathcal{A}}(T\mu_{\mathcal{B}}(Su,Sp),\ldots\ldots) \\
&= \text{source } H\alpha_{\mathcal{B}}(GT\mu_{\mathcal{B}}(Su,Sp),\ldots\ldots) \\
&= \text{source } H\alpha_{\mathcal{B}}(SH\mu_{\mathcal{B}}(Su,Sp),\ldots\ldots) \\
&= \text{source } \alpha_{\mathcal{C}}(\mu_{\mathcal{C}}(u,p),\ldots\ldots) \quad .
\end{aligned}$$

5) This theme, implicit in the work of Fred Linton, was expounded by me at the Midwest Category Seminar in Waterloo and in a graduate seminar at McGill in 1968.

6) There exists also a more recent manuscript with the same title concerning Burroni's talk at Cambridge in July 1981.

ESSENTIALLY MONADIC ADJUNCTIONS

by John MacDonald and Arthur Stone

In an earlier paper [9] the authors show how the canonical regular epic decomposition of a morphism is paralleled by an analogous decomposition for adjunctions - the canonical monadic decomposition (cf. [2] and [7]).

The first section of this paper shows that for a faithful right adjoint U the essential length of the associated adjunction is the same as the regular length of the counit, when both are defined.

The "higher" Beck theorem, characterizing essentially monadic adjunctions is proved in the second section. An essentially monadic adjunction is one which can be written as a (canonical) composite of monadic adjunctions. Such adjunctions have faithful right adjoints.

1. Length of a faithful adjoint

This section shows that the essential length λ of the adjunction $\underline{N}$ associated to a faithful right adjoint U is equal to the regular length of the counit ε. For notational reasons we recall that $\underline{N} : \underline{\underline{X}} \to \underline{\underline{A}}$ consists of functors F and U, with $F : \underline{\underline{X}} \to \underline{\underline{A}}$ left adjoint to U, together with a choice of unit η and counit ε. Given adjunctions $\underline{N}^{\alpha} : \underline{\underline{X}} \to \underline{\underline{Y}}$ and $\underline{N}^{\beta} : \underline{\underline{Y}} \to \underline{\underline{Z}}$ the composite $\underline{N} = \underline{N}^{\beta}\Delta\underline{N}^{\alpha}$ is defined as in [9]. Using this composition a category $\underline{\underline{Adj}}$ of adjunctions is determined as in [10], page 102.

Let the ordinal λ be identified with the ordered set (considered as a category) or ordinals less than λ and write $\alpha\beta$ for the unique morphism $\alpha \to \beta$ when $\alpha < \beta < \lambda$. A <u>chain of object length</u> λ in the category $\underline{\underline{C}}$ is a functor $F : \lambda \to \underline{\underline{C}}$. A morphism $\underline{N}$ of $\underline{\underline{C}}$ is a λ <u>chain composite</u> of morphisms $\underline{N}^{\alpha\beta}$ $(\alpha < \lambda; \beta=\alpha+1)$ if there is a colimit preserving chain F of object length $\lambda + 1$ with $\underline{N}^{\alpha\beta} = F(\alpha\beta)$ and $\underline{N} = F(0\lambda)$.

An adjunction $\underline{N} = \underline{\underline{X}} \to \underline{\underline{A}}$ has a <u>canonical monadic decomposition</u> if it is the δ chain composite of morphisms $\underline{N}^{\alpha\beta}(\alpha < \delta; \beta=\alpha+1)$ in the comma category $(\underline{\underline{Adj}}, \underline{\underline{A}})$ pictured as in

(1.1)

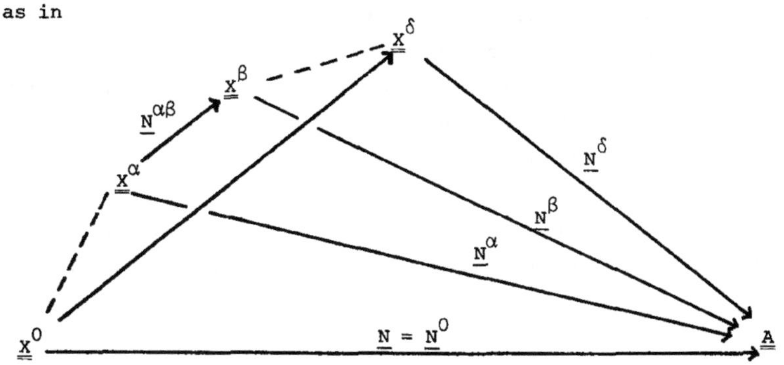

and subject to the following conditions. First of all, when $\beta = \alpha+1$ the adjunction $\underline{N}^{\alpha\beta}$ is a first monadic component of $\underline{N}^{\alpha}$, that is, $\underline{N}^{\alpha} = \underline{N}^{\beta}\Delta\underline{N}^{\alpha\beta}$, $\eta^{\alpha} = \eta^{\alpha\beta}$ and $F^{\alpha\beta} = U^{\beta}F^{\alpha}$. This leads to the well known universal property and identification of $\underline{\underline{X}}^{\beta}$ as the category of algebras for $\underline{N}^{\alpha}$ (cf. Beck's results cited in [9] and [10]). Secondly when κ is a limit ordinal $\underline{\underline{X}}^{\kappa}$ is the limit in $\underline{\underline{Cat}}$ of the diagram of right adjoints $U^{\alpha\beta}$ $(\alpha < \beta < \kappa)$, with $F^{\alpha\kappa} = U^{\kappa}F^{\alpha}$ $(\alpha < \kappa)$ and $\eta^{\alpha\kappa} = \eta^{\alpha}$ and, finally, η^{δ} has an inverse but η^{α} does not, for $\alpha < \delta$. The ordinal δ is called the <u>monadic length</u>.

An adjunction $\underline{N} : \underline{\underline{X}} \to \underline{\underline{A}}$ is <u>essentially monadic</u> (and $\underline{\underline{A}}$ is <u>essentially monadic</u> over $\underline{\underline{X}}$) if N has a canonical monadic decomposition 1.1 in which the last comparison adjunction $\underline{N}^{\delta}$ determines an equivalence of categories.

<u>Lemma 1.2</u>: Let $\ldots\underline{\underline{X}}^{\alpha} \xleftarrow{U^{\alpha\beta}} \underline{\underline{X}}^{\beta}\ldots$ $(\alpha < \beta < \kappa)$ represent a functor $\kappa^{op} \to \underline{\underline{Cat}}$ with limit $\underline{\underline{X}}^{\kappa}$ and projections $\underline{U}^{\alpha\kappa}$ in $\underline{\underline{Cat}}$.

(1) If the $U^{\alpha\beta}$ are faithful, then so are the $U^{\alpha\kappa}$.

(2) If the $U^{\alpha\beta}$ reflect isomorphisms, then so do the $U^{\alpha\kappa}$. □

The <u>essential length</u> of $\underline{N}$ is the smallest ordinal λ such that $\underline{N}^{\lambda}$ generates an idempotent monad, that is, λ is smallest so that $U^{\lambda}\varepsilon^{\lambda}F^{\lambda}$ is an isomorphism. But this is equivalent to $\eta^{\lambda}U^{\lambda}$ being an isomorphism (see 2.9 in [9]). Thus the essential length λ is less than or equal to the monadic length δ and in fact $\lambda \le \delta \le \lambda+1$ with $\delta = \lambda$ occurring in some examples and $\delta = \lambda+1$ in others, as shown in [9].

From [9], based on work of Isbell [6] and Kelly [7], we recall that a regular epimorphism f^{01} is a <u>regular epic component</u> of morphism f if $f = f^{1} . f^{01}$ for some f^{1} and whenever $f = h.g$ with g a regular epic, there is a (necessarily unique) j with $j.g = f^{01}$. A morphism e is <u>regular</u> if for each h with $hx = hy$ for all pairs (x,y) with $ex = ey$ there is a unique d with $h = de$.

<u>Lemma 1.3</u>: Let $\underline{N} : \underline{\underline{X}} \longrightarrow \underline{\underline{A}}$ be an adjunction with monadic decomposition, with essential length λ, and assume U is faithful. Then in 1.1 for every object A of $\underline{\underline{A}}$

(1) for $\alpha < \lambda$ and $\beta = \alpha+1$, the morphism $F^{\beta}\varepsilon^{\alpha\beta}U^{\beta}A$ is a regular epic component of $\varepsilon^{\alpha}A$, and

(2) the morphism $\varepsilon^{\lambda}A$ is monic.

<u>Proof</u>: It is well known (cf. [9] and [10]) that the comparison functor U^{1} (with $U^{1}A = \langle UA, U\varepsilon A\rangle$) has left adjoint F^{1} defined by the coequalizer diagram

$$FT^{01}U^{01} = FTU^{01} \underset{\varepsilon FU^{01}}{\overset{FU^{01}\varepsilon^{01}}{\rightrightarrows}} FU^{01} \xrightarrow{\quad T \quad} F^1$$

It turns out that $T = F^1\varepsilon^{01}$ and hence that $F^1\varepsilon^{01}U^1A$ coequalizes the pair $(G\varepsilon A, \varepsilon GA) = (FU^{01}\varepsilon^{01}U^1A, \varepsilon FU^{01}U^1A)$ where $G = FU$. Thus $F^1\varepsilon^{01}U^1A$ is a regular epimorphism. Clearly $\varepsilon.G\varepsilon = \varepsilon.\varepsilon G$. We let ε^1A be the unique morphism for which $\varepsilon^1A.F^1\varepsilon^{01}U^1A = \varepsilon A$. Of course ε^1 is the counit of $\underline{N}^1$ in 1.1. In the same way we see that $F^\beta\varepsilon^{\alpha\beta}U^\beta A$ is regular since it coequalizes $(G^\alpha\varepsilon^\alpha A, \varepsilon^\alpha G^\alpha A)$ when $\beta = \alpha+1$. Let $\varepsilon^\alpha A = h.g$ for g a regular epic. Then suppose $\varepsilon^\alpha A.x = \varepsilon^\alpha A.y$ where g is a coequalizer for x and y. The remainder of the proof is essentially the same if we replace (x,y) by a family of pairs $(x_\theta, y_\theta)_{\theta\in I}$. We must find a j so that $jg = F^\beta\varepsilon^{\alpha\beta}U^\alpha A$.

(1.4)

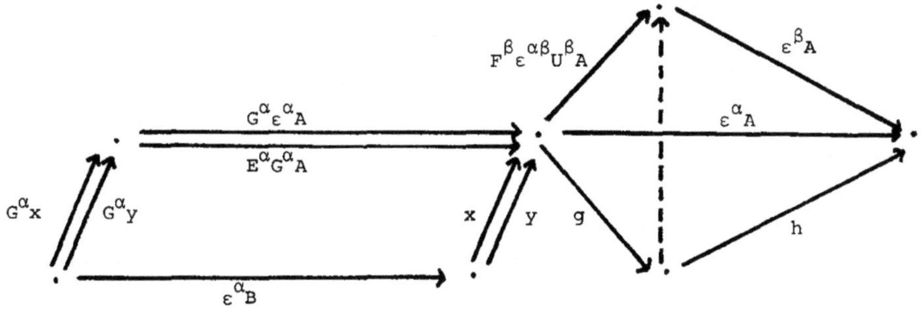

Since g is a coequalizer for x,y we have only to show that $F^\beta\varepsilon^{\alpha\beta}U^\beta A.x = F^\beta\varepsilon^{\alpha\beta}U^\beta A.y$. Let $B = \mathrm{Sce}\ x = \mathrm{Sce}\ y$. If U is faithful, then εB is epic (cf. [10], page 88), and $\varepsilon^\alpha B$ is epic (since $\varepsilon B = \varepsilon^\alpha B \,.\, F^\alpha\varepsilon^{\sigma\alpha}U^\alpha B$). So it suffices to show that the composites $F^\beta\varepsilon^{\alpha\beta}U^\beta A \,.\, x \,.\, \varepsilon^\alpha B$ and $F^\beta\varepsilon^{\alpha\beta}U^\beta A \,.\, y \,.\, \varepsilon^\alpha B$ are equal. By the naturality of ε^α, these are equal to the composites $F^\beta\varepsilon^{\alpha\beta}U^\beta A \,.\, \varepsilon^\alpha G^\alpha A \,.\, G^\alpha x$ and $F^\beta\varepsilon^{\alpha\beta}U^\beta A \,.\, \varepsilon^\alpha G^\alpha A \,.\, G^\alpha y$. Now use $F^\beta\varepsilon^{\alpha\beta}U^\beta A \,.\, \varepsilon^\alpha G^\alpha A = F^\beta\varepsilon^{\alpha\beta}U^\beta A \,.\, G^\alpha\varepsilon^\alpha A$ and $\varepsilon^\alpha A \,.\, x = \varepsilon^\alpha A \,.\, y$.

(2) We show that if $\varepsilon^\alpha A$ is not monic, then $\alpha < \lambda$. By the construction of 1.1 the ordinal λ is the first ordinal α such that $\varepsilon^\alpha = \varepsilon^\beta (\alpha < \beta)$ or, equivalently, $\varepsilon^\alpha = \varepsilon^\beta (\beta = \alpha+1)$. So λ is the first ordinal for which $F^\beta\varepsilon^{\alpha\beta}U^\beta A$ $(\beta=\alpha+1)$ is an isomorphism for all objects A.

If $\varepsilon^\alpha A$ is non-monic, then for some pair x,y with $x \neq y$ we have $\varepsilon^\alpha A \,.\, x = \varepsilon^\alpha A \,.\, y$. Let $B = \mathrm{Sce}\ x = \mathrm{Sce}\ y$. If U is faithful, then $\varepsilon^\alpha B$ is epic and $x \,.\, \varepsilon^\alpha B \neq y \,.\, \varepsilon^\alpha B$. Then by the naturality of ε^α we have $\varepsilon^\alpha G^\alpha A \,.\, G^\alpha x \neq \varepsilon^\alpha G^\alpha A \,.\, G^\alpha y$. Since $G^\alpha\varepsilon^\alpha A \,.\, G^\alpha x = G^\alpha\varepsilon^\alpha A \,.\, G^\alpha y$, this implies $\varepsilon^\alpha G^\alpha A \neq G^\alpha\varepsilon^\alpha A$. So the coequalizer $F^\beta\varepsilon^{\alpha\beta}U^\alpha A$ is not monic -- not an isomorphism. Hence $\alpha < \lambda$. □

Theorem 1.5: If $\underline{N} : \underline{\underline{X}} \longrightarrow \underline{\underline{A}}$ is an adjunction with monadic decomposition 1.1 and U is faithful, then the diagram

(1.6)

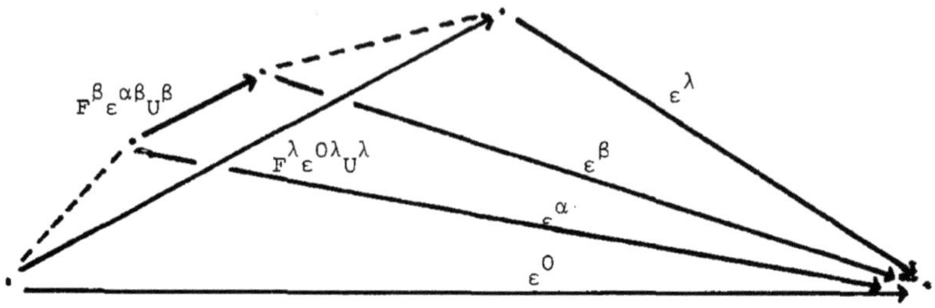

is a regular decomposition of ε in $(\underline{\underline{A}},\underline{\underline{A}})$ for λ the essential length of $\underline{N}$.

Proof: By 1.3 (1) , $F^{\beta}\varepsilon^{\alpha\beta}U^{\beta}$ is a regular epic component of ε^{α} when $\beta = \alpha+1$. By the construction of 1.1 at limit ordinals κ we have G^{κ} a colimit of the diagram of morphisms $F^{\beta}\varepsilon^{\alpha\beta}U^{\beta}$ $(\alpha < \beta < \kappa)$. (cf. [9]) . Also by the construction of 1.1, $G^{\alpha}\varepsilon^{\alpha} \neq \varepsilon^{\alpha}G^{\alpha}$ for $\alpha < \lambda$ and $\varepsilon^{\alpha} . G^{\alpha}\varepsilon^{\alpha} = \varepsilon^{\alpha} . \varepsilon^{\alpha}G^{\alpha}$; so ε^{α} is not monic for $\alpha < \lambda$. By 1.3 (2), ε^{λ} is monic. □

Corollary 1.7: If $\underline{N} : \underline{\underline{X}} \longrightarrow \underline{\underline{A}}$ is essentially monadic, then the essential length of $\underline{N}$ $\leq$ regular length of $\underline{\underline{A}}$.

Proof: Since monadic right adjoints U are faithful, by 1.2 essentially monadic right adjoints U are faithful and 1.5 applies. The regular length of ε in $(\underline{\underline{A}},\underline{\underline{A}})$ is the supremum of the regular lengths of the component morphisms εA in $\underline{\underline{A}}$. □

2. Characterization of essentially monadic adjunctions

This section presents the analogue for essentially monadic adjunctions of the Beck characterization theorem for monadic adjunctions (cf. MacLane [10], page 147) . To avoid obscuring the view of the forest with underbrush, we postpone to 2.2 and 2.4 a careful rephrasing of several familiar definitions. They will facilitate the proof of the following theorem and its corollaries.

Theorem 2.1: The adjunction $\underline{N} : \underline{\underline{X}} \longrightarrow \underline{\underline{A}}$ is essentially monadic if and only if $\underline{N}$ has monadic decomposition 1.1 and the following (then) equivalent conditions are satisfied.

(A) U reflects extremal epimorphisms

(B) U reflects strong epimorphisms, and

(C) U reflects isomorphisms.

Definition 2.2: (cf. Herrlich-Strecker [5]) . A set $\underline{\underline{L}}$ (not necessarily small) of objects of $\underline{\underline{A}}$ is separating at an object A if

(0) for every parallel pair f,g of morphisms with source A there is at least one object L of $\mathbb{L}$ and a morphism $x : L \longrightarrow A$ with $fx \neq gx$,

extremal separating at A if (0) and

(a) for every proper (= non-isomorphic) monomorphism m with target A there is at least one object L of $\mathbb{L}$ and a morphism $x : L \longrightarrow A$ that does not factor through m ,

and *strong* separating at A if (0) and

(b) if for a morphism b and a monomorphism m as in 2.3 there are, for every L in $\mathbb{L}$ and every $x : L \longrightarrow A$, morphisms a_x that make the square commute, then there is a morphism d that makes the diagram 2.3 commute for the given b and m , for every x .

(2.3)

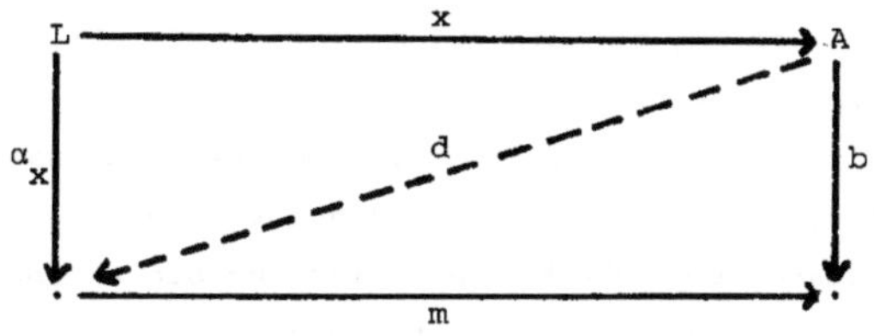

The set $\mathbb{L}$ is ... separating *in* $\mathbb{A}$ if it is ... separating *at* every object of $\mathbb{A}$. A functor $F : \mathbb{X} \longrightarrow \mathbb{A}$ is ... *separating* if $\operatorname{Im} F$ is.

When $\mathbb{A}$ has equalizers (a) implies (0) . To see that (b) implies (a) consider the case $b = 1_A$.

Definition 2.4: The functor $U : \mathbb{A} \longrightarrow \mathbb{X}$ *reflects* epimorphisms (extremal epimorphisms, strong epimorphisms) *at* an object A if Uf an epimorphism (extremal epimorphism, strong epimorphism) and $\text{Tgt}.f = A$ imply f an epimorphism (extremal epimorphism, strong epimorphism); and U is *faithful at* A if $g \neq h$, $\text{Sce}.g = \text{Sce}.h = A$, and $\text{Tgt}.g = \text{Tgt}.h$ imply $Ug \neq Uh$.

The functor U is, of course, faithful and only if it is faithful *at* every object of $\mathbb{A}$.

Underbrush Propositions: For an adjunction $\underline{N} : \mathbb{X} \longrightarrow \mathbb{A}$ and an object A of $\mathbb{A}$

2.5: The following are equivalent:

(0) U is faithful at A ,

(1) U reflects epimorphisms at A ,

(2) εA is epic

(3) F is separating at A , and

(4) (when $\underline{N}$ is model induced by $\mathbb{L}$) $\mathbb{L}$ is separating at A ;

2.6A: The following are equivalent (under the conditions specified):

(1) (when U is faithful) U reflects extremal epimorphisms at A,

(2) εA is an extremal epic

(3) F is extremal separating at A, and

(4) (when $\underline{N}$ is model induced by $\underline{\underline{L}}$) $\underline{\underline{L}}$ is extremal separating at A.

2.6B: The statements of 2.6A are equivalent if "extremal" is everywhere replaced by "strong".

The proofs are straightforward or familiar (e.g. Schubert [11], page 176). To emphasize why we assume U faithful in 2.6A and B we show the equivalence of 2.6B (1) and (2).

Since $U\varepsilon A$ is always a regular epic (indeed, a split coequalizer, split by ηUA) and regular implies strong, if U reflects strong epics, then εA is one. For the converse, let f be any morphism with $\mathrm{Tgt} A$, Uf strong epic, and $b.f = m.a$ (for some morphisms a,b and monomorphism m). We wish to show the existence of a diagonal d for the square $b.f = m.a$, i.e. a morphism d satisfying $d.f = a$ and $m.d = b$. Since Uf is strong epic and right adjoints preserve monics, there is a diagonal w for the square $Ub . Uf = Um . Ua$. Let $B = \mathrm{Sce}\ a$ and $C = \mathrm{Tgt}\ a$. If εA is strong epic, there is a diagonal d for the square $b \cdot \varepsilon A = m \cdot (\varepsilon C . Fw)$. To show that d is also a diagonal for the original square $b.f = m.a$, we use the naturality of ε and we must assume that εB is epic (which will be true if U is faithful). □

Proof: (of 2.1) Let $\underline{N} : \underline{\underline{X}} \longrightarrow \underline{\underline{A}}$ be essentially monadic, with essential and monadic lengths λ and δ. Since monadic functors are faithful, by 1.2 the right adjoint U is faithful and ε has a decomposition as in 1.5. Since $\underline{N}^{\delta}$ is an equivalence, ε^{δ} is an isomorphism, and hence so is ε^{λ}. This makes ε in 1.5 a generalized composite of strong epimorphisms, hence strong (and extremal) epic. Therefore, by 2.6B (and A) the functor U reflects strong epics (and extremal epics).

Since monadic right adjoints reflect isomorphisms, by 1.2, so does U.

Conversely, if U reflects extremal epics or strong epics, then by the Underbrush propositions U is faithful, and ε has the regular decomposition of 1.5. Since $U\varepsilon$ is a coequalizer of $U\varepsilon FU$ and $UFU\varepsilon$, it is regular epic, hence extremal and strong. So if U reflects extremal or strong epics, then ε is one. Then, since $F^{\lambda}\varepsilon^{0\lambda}U^{\lambda}$ (a generalized composite of strong epics) is epic, and $\varepsilon^{\lambda} . F^{\lambda}\varepsilon^{0\lambda}U^{\lambda} = \varepsilon$, the natural transformation ε^{λ} is extremal or strong. And monic extremal (or strong) epimorphisms are isomorphisms.

Since $\underline{N}^{\lambda}$ is idempotent, η^{δ} is an isomorphism. So ε^{λ} an isomorphism (hence ε^{δ} an isomorphism) makes $\underline{N}^{\delta}$ an equivalence of categories.

Now consider the case U reflects isomorphisms. Since $\underline{N}^\lambda$ is idempotent, $U^\lambda F^\lambda \eta^\lambda . U^\lambda \varepsilon^\lambda F^\lambda = 1$ and

1. $U^\lambda \varepsilon^\lambda F^\lambda U^\lambda = U^\lambda F^\lambda U^\lambda \varepsilon^\lambda . U^\lambda F^\lambda \eta^\lambda U^\lambda . U^\lambda \varepsilon^\lambda F^\lambda U^\lambda = U^\lambda F^\lambda U^\lambda \varepsilon^\lambda . 1$. Since $U^\lambda \varepsilon^\lambda$ is a co-equalizer of $U^\lambda \varepsilon^\lambda F^\lambda U^\lambda$ and $U^\lambda F^\lambda U^\lambda \varepsilon^\lambda$, this makes $U^\lambda \varepsilon^\lambda$, and hence $U^{0\lambda} U^\lambda \varepsilon^\lambda$, i.e. $U\varepsilon^\lambda$, an isomorphism. So if U reflects isomorphisms, then ε^λ is one. Again, this makes $\underline{N}^\delta$ an equivalence of categories. □

Corollary 2.1a: The category $\underline{\underline{A}}$ is essentially monadic over $\underline{\underline{Ens}}^m$ if and only if $\underline{\underline{A}}$ contains an extremely separating set $\underline{\underline{L}}$ of cardinality m and the decomposition 1.1 exists for $\underline{N}$ model induced by $\underline{\underline{L}}$.

Remark: M. Barr in [3] shows that when $\underline{\underline{A}} = \underline{\underline{Ens}}^m$, then $\underline{\underline{A}}$ does not contain any extremely separating set of cardinality smaller than m .

Definition 2.7: The category $\underline{\underline{A}}$ is Isbell cocomplete provided that it has colimits of long chains of strong epics as well as colimits of small diagrams.

Corollary 2.1b: Let $\underline{\underline{A}}$ be Isbell cocomplete with an extremely separating set $\underline{\underline{L}}$, then $\underline{\underline{A}}$ is Isbell complete.

This is an improvement on Herrlich-Strecker ([5], page 163) only when $\underline{\underline{A}}$ is not co-well-powered.

Definition 2.8: Let $\underline{\underline{L}}$ be a discrete subcategory of $\underline{\underline{A}}$. The full colimit closure of $\underline{\underline{L}}$ is the intersection of all full subcategories of $\underline{\underline{A}}$ containing $\underline{\underline{L}}$.

Corollary 2.1c: Suppose that $\underline{\underline{A}}$ is sufficiently cocomplete for $\underline{\underline{L}}$ and that the counit ε has an epic monic factorization. Then an object A is in the full colimit closure of $\underline{\underline{L}}$ if and only if $\underline{\underline{L}}$ is extremely separating at A .

Corollary 2.1d: Suppose that $\underline{N}$ is model induced by $\underline{\underline{L}}$ and 1.1 exists, that the counit ε has an epic monic factorization and that $\underline{\underline{L}}$ is extremely separating at A and B . Then A is isomorphic to B if and only if $U^\alpha A$ is isomorphic to $U^\alpha B$ for all ordinals $\alpha \leq \mu$.

Corollary 2.1e: Suppose that $\underline{N}$ is model induced by $\underline{\underline{L}}$ in $\underline{\underline{A}}$ and 1.1 exists and that ε has an epic monic factorization. Then $\underline{\underline{X}}^\mu$ is equivalent to the full colimit closure of $\underline{\underline{\text{Image}}}$ F in $\underline{\underline{A}}$ which is the same as the full colimit closure of $\underline{\underline{L}}$ in $\underline{\underline{A}}$.

Comment: This means that the operations in the tower suffice. We explore this in a subsequent paper.

Corollary 2.1f: The Δ-composite of essentially monadic adjunctions with Isbell cocomplete targets is again essentially monadic. (This is true for the generalized Δ-composite as well).

References

1. H. Applegate and M. Tierney, Categories with models, in: Lecture notes in Mathematics 80, (Springer-Verlag, 1969), 156-244.
2. H. Applegate and M. Tierney, Iterated cotriples, in: Lecture Notes in Mathematics 137, (Springer-Verlag, 1970), 56-99.
3. M. Barr, The point of the empty set, Cahier de top. et geom. diff. 13 (1972), 356-368.
4. B. Day, On adjoint-functor factorization, in: Lecture Notes in Mathematics 420, (Springer-Verlag, 1974), 1-19.
5. H. Herrlich and G. Strecker, Category theory, (Heldermann-Verlag, 1979).
6. J. Isbell, Structure of categories, B.A.M.S. 72 (1966), 619-655.
7. M. Kelly, Monomorphisms, epimorphisms and pullbacks. J. Austral. Math Soc. 9 (1969), 124-142.
8. J. MacDonald, Cohomology operations in a category, J. Pure Appl. Alg., 19 (1980), 275-297.
9. J. MacDonald and A. Stone, The tower and regular decomposition, Cahier de top. et. geom. diff., to appear.
10. S. MacLane, Categories for the working mathematician, (Springer-Verlag, 1971).
11. H. Schubert, Categories, (Springer-Verlag, 1972).
12. A. Stone, The heights of adjoint towers, Am. Math. Soc. Notices 21 (1974), A-81.

John MacDonald
Mathematics Department
University of British Columbia
Vancouver, B.C., Canada
V6T 1Y4

Arthur Stone
Mathematics Department
UC Davis
Davis, California

Decomposition of morphisms into infinitely many factors

John MacDonald and Walter Tholen

Considering iterated dominion factorizations Isbell has used first decompositions of morphisms into infinitely many factors (cf. [6]). In the dual situation Kelly [7] described the iterated regular factorization of a morphism. In the recent paper [8] by the first author and Stone the analogous behaviour of the regular decomposition of a morphism and the adjoint tower construction (factorizing an adjoint functor as often as possible over the induced Eilenberg-Moore category) was pointed out (see also [9]).

The present paper gives the precise reason for this analogy: both constructions (up to duality) arise from *locally orthogonal* $\mathcal{E}$*-factorizations* (with two factors) which were first considered more generally by the second author (cf. [13]). Locally orthogonal $\mathcal{E}$-factorizations of morphisms generalize the usually considered factorization systems with the unique diagonal-fill-in property so far as $\mathcal{E}$ need not be closed under composition; as soon as it is, the notion does not give anything new.

We can consider only a few examples. Besides the mentioned applications there is a wide range of examples in topology which are briefly subsumed here as *monotone-light factorizations* (see 2.2 below). They will be more intensively treated in a forthcoming paper which will include the investigation of locally orthogonal factorizations of cones (instead of just morphisms; see [13]).

After finishing the manuscript the authors got acquainted with an unpublished paper by H. Ehrbar and O. Wyler "On subobjects and images in categories" of 1968. In this paper the notion of a "strong $\mathcal{J}$-image of f" is introduced, which is precisely a locally coorthogonal $\mathcal{J}$-factorization in our terminology. It also contains Proposition 1.3 (in the dual form).

AMS classification: 18 A 32, 18 C 20, 54 C 10, 54 D 05

1. Locally orthogonal $\mathcal{E}$-factorizations of morphisms

Factorizations of morphisms and of cones or sources have been investigated by many authors, most successfully by use of the so called *diagonalization property*. This property necessarily makes the used factorization classes closed under composition. In what follows we will give a "non-compositive" weakening of the diagonalization property and will show its naturality and usefulness by some theorems and examples. It has been introduced before for factorization of cones in order to describe semitopological functors (cf.[13]). But for simplicity we restrict ourselves to morphisms in this paper.

Throughout, for a category $\mathcal{K}$, let $\mathcal{E}$ be a subclass of $\mathrm{Mor}\mathcal{K}$ containing all isomorphisms and being closed under composition with isomorphisms.

We recall some phrases. A *factorization* (of a morphism f) is a pair (e,m) of morphisms such that the composite me exists (and $f=me$); it is called an *$\mathcal{E}$-factorization* in case $e\in\mathcal{E}$. A morphism p *is orthogonal to a factorization* (e,m), if, for all commutative (outer) diagrams

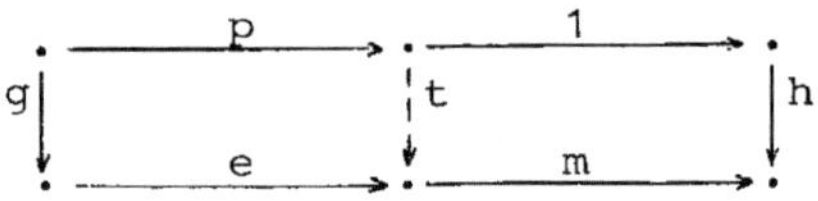

one has a unique t with $tp=eg$ and $mt=h$; we write $p\perp(e,m)$ in this case. An $\mathcal{E}$-factorization (e,m) is called *locally orthogonal*, if $p\perp(e,m)$ for all $p\in\mathcal{E}$; it is called *orthogonal*, if $p\perp(1,m)$ for all $p\in\mathcal{E}$.

The following properties can be easily proved: every orthogonal $\mathcal{E}$-factorization is locally orthogonal, and every locally orthogonal $\mathcal{E}$-factorization is *rigid*, that is, for an endomorphism t one has $te=e$ and $mt=m$ only if $t=1$. This property yields uniqueness (up to canonical isomorphisms) of locally orthogonal $\mathcal{E}$-factorizations.

We examine the precise relation between locally orthogonal and orthogonal $\mathcal{E}$-factorizations in Proposition 1.2 below. For that we need

Lemma 1.1: *Let every morphism in* $\mathcal{K}$ *have a locally orthogonal* $\mathcal{E}$*-factorization. Then,* $fe=e'$ *with* $e,e'\in\mathcal{E}$ *can hold only if* $f\in\mathcal{E}$.

Proof: Let (e'',m) be a locally orthogonal $\mathcal{E}$-factorization of f. Since $e'\in\mathcal{E}$ and $me''e=e'$ one has a unique t such that

$$\begin{array}{ccccc} \cdot & \xrightarrow{e'} & \cdot & \xrightarrow{1} & \cdot \\ {\scriptstyle e}\downarrow & & \downarrow{\scriptstyle t} & & \downarrow{\scriptstyle 1} \\ \cdot & \xrightarrow[e'']{} & \cdot & \xrightarrow[m]{} & \cdot \end{array}$$

commutes. The diagram

$$\begin{array}{ccccc} \cdot & \xrightarrow{e} & \cdot & \xrightarrow{1} & \cdot \\ {\scriptstyle e}\downarrow & & \downarrow{\scriptstyle d} & & \downarrow{\scriptstyle f} \\ \cdot & \xrightarrow[e'']{} & \cdot & \xrightarrow[m]{} & \cdot \end{array}$$

trivially commutes for $d = e''$, but also for $d = tf$. Therefore $tf = e''$. As (e'',m) is rigid, from

$$(tm)e'' = tf = e'' \quad \text{and} \quad m(tm) = m$$

it follows $tm = 1$. Thus m is an isomorphism and $f = me'' \in E$. □

Proposition 1.2: *Under the assumption of Lemma* 1.1 *the following assertions are equivalent:*

(i) *If* $M = \{m \in \operatorname{Mor}K \mid \exists e : (e,m)$ is a locally orthogonal E-factorization$\}$, *then* $E \cap M$ *consists of isomorphisms only.*

(ii) E *is closed under composition.*

(iii) *Locally orthogonal* E*-factorizations are orthogonal.*

Proof: (i) => (ii) For $e,e' \in E$ such that the composite $e'e$ exists, let (e'',m) be a locally orthogonal E-factorization of $e'e$. There is a unique t with $te = e''$ and $mt = e'$. By Lemma 1.1 one has $t \in E$ and therefore, again by Lemma 1.1, $m \in E$. By (i), m is an isomorphism, hence $e'e = me'' \in E$.

(ii) => (iii) In order to show that the locally orthogonal E-factorization (e,m) of f is orthogonal it suffices to show that, for a locally orthogonal E-factorization (e',m') of m, necessarily e' is an isomorphism. Since, by (ii), $e'e \in E$ one has a unique t such that

$$\begin{array}{ccccc} \cdot & \xrightarrow{e'e} & \cdot & \xrightarrow{1} & \cdot \\ {\scriptstyle 1}\downarrow & & \downarrow{\scriptstyle t} & & \downarrow{\scriptstyle m'} \\ \cdot & \xrightarrow[e]{} & \cdot & \xrightarrow[m]{} & \cdot \end{array}$$

commutes. From the rigidness of the two locally orthogonal E-factorizations one now easily gets the equations $te' = 1$ and $e't = 1$.

(iii) => (i) is well known (and trivial). □

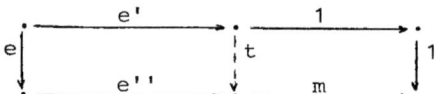

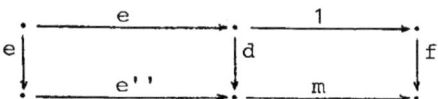

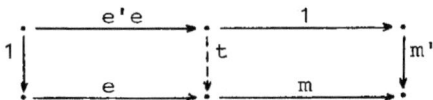

Remark: The equivalence of (ii), (iii) has been proved in [13] Lemma 7.3, for factorizations of cones. But in that case the proof is easier since then one has to consider only classes E of epimorphisms whereas, in the present context, E is arbitrary.

We mention without proof that E is closed under pushouts and multiple pushouts, if K admits locally orthogonal E-factorizations for its morphisms. More precisely, if in the pushout diagrams

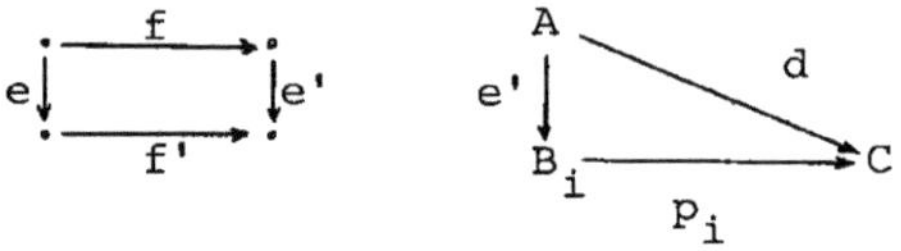

we have $e \in E$ and $e_i \in E$ for all $i \in I$ (I any set), we also have $e' \in E$ and $d \in E$. Vice versa, the existence of those pushouts yields the existence of locally orthogonal E-factorizations of all morphisms. This was shown more generally for factorization of cones in [13], Corollary 6.5(1).

We conclude this section by showing that the existence of locally orthogonal E-factorizations can be equivalently described by the existence of right adjoints to very natural functors.

Consider the <u>category</u> Mor$K$ of morphisms of K: objects are the morphisms of K, and a morphism $[u,v] : f \to g$ in Mor$K$ is given by a commutative K-diagram

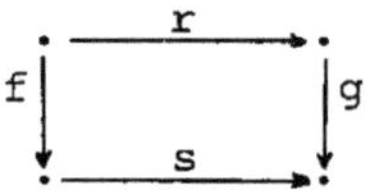

Composition is taken from K. Mor$K$ contains the full subcategory $\mathrm{Mor}_E K$ whose objects are all morphisms in E.

<u>Proposition 1.3</u>: *The following assertions are equivalent:*

(i) K *admits locally orthogonal* E*-factorizations for all morphisms.*

(ii) $\mathrm{Mor}_E K$ *is coreflective in* Mor K.

(iii) $\mathrm{Mor}_E K$ *is coreflective in* Mor K *with coreflections of type* $[1,m]$.

Proof: One easily proves that (e,m) is a locally orthogonal E-factorization of f, iff $[1,m] : e \to f$ is a coreflection. This gives (i) <=> (iii). In order to show (ii) => (iii), let $[k,m] : e \to f$ be any coreflection. Since $1 \in E$ the morphism $[1,f]: 1 \to f$ factorizes as $[k,m][r,s]=[1,f]$ with $[r,s]:1\to e$; in particular $kr=1$. Now, the two $\mathrm{Mor}_{\mathcal{E}}\mathcal{K}$-morphisms

$$[1,e],[rk,sk]:1\to e$$

are equalized by the coreflection $[1,m]$, so they must be equal; in particular $rk=1$. Therefore, k is an isomorphism and may be even chosen as an identity. □

Consider the category $\mathrm{Fact}\,\mathcal{K}$ of factorizations of $\mathcal{K}$: objects are all pairs (e,m) such that the domain of m is the codomain of e , and a morphism $[r,t,s]:(e,m)\to(p,n)$ in $\mathrm{Fact}\,\mathcal{K}$ is given by a commutative $\mathcal{K}$-diagram

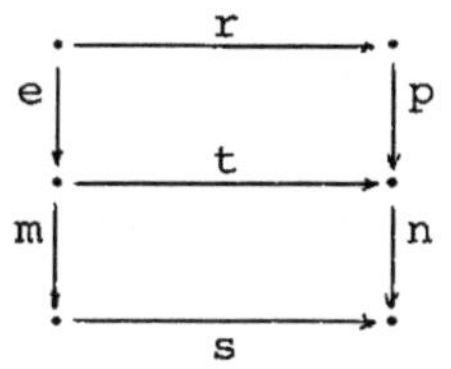

Composition is taken from $\mathcal{K}$. Let $\mathrm{Fact}_{\mathcal{E}}\mathcal{K}$ be the full subcategory of all $\mathcal{E}$-factorizations and let

$V_{\mathcal{E}}:\mathrm{Fact}_{\mathcal{E}}\mathcal{K}\to\mathrm{Mor}\mathcal{K}$, $(e,m)\mapsto me$,

be the *composition functor.*

<u>Proposition 1.4</u>: *The following assertions are equivalent:*

(i) $\mathcal{K}$ *admits locally orthogonal* $\mathcal{E}$*-factorizations for all morphisms.*

(ii) $V_{\mathcal{E}}$ *has a right adjoint.*

(iii) $V_{\mathcal{E}}$ *has a full and faithful right adjoint.*

Proof: The easy proof that (e,m) is a locally orthogonal $\mathcal{E}$-factorization of f iff $[1,1]:V_{\mathcal{E}}(e,m)\to f$ is $V_{\mathcal{E}}$-couniversal (i.e. a counit of $V_{\mathcal{E}}$ at f) is left to the reader. From this one obtains immediately the equivalence of (i) and (iii). So we still have to prove (ii) ⇒ (iii) and to consider an arbitrary $V_{\mathcal{E}}$-couniversal morphism $[k,h]:V_{\mathcal{E}}(e,m)\to f$. There is a unique $\mathrm{Fact}_{\mathcal{E}}\mathcal{K}$-morphism $[r,t,s]$: $(1,f)\to(e,m)$ with $[k,h]V_{\mathcal{E}}[r,t,s]=[1,1]$, in particular $kr=1$ and $hs=1$. The $\mathrm{Fact}_{\mathcal{E}}\mathcal{K}$-morphisms $[1,e,1]$, $[rk,tk,sh]:(1,me)\to(e,m)$ satisfy the equation $[k,h]V_{\mathcal{E}}[1,e,1]=[k,h]V_{\mathcal{E}}[rk,tk,sh]$, so they are equal. We have $rk=1$ and $sh=1$, so the $V_{\mathcal{E}}$-couniversal morphism is an isomorphsm. □

The third characterization of the existence of locally orthogonal $\mathcal{E}$-factorizations by adjoint functors requires the existence of pushouts in K. Let $\mathrm{Rel}_{\mathcal{E}}K$ be the following category: objects are pairs (e,f) with common domain and $e \in \mathcal{E}$, and a morphism $\{u,r,t\}:(e,f)\to(p,g)$ is given by a commutative K-diagram

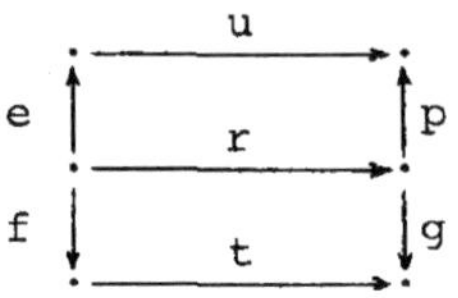

Composition is taken from K. There is a canonical functor

$$P_{\mathcal{E}} : \mathrm{Rel}_{\mathcal{E}}K \to \mathrm{Fact}\ K,\ (e,f)\mapsto(f,f(e));$$

here $f(e)$ denotes the (fixed) pushout of e along f.

<u>Proposition 1.5</u>: *The following assertions are equivalent:*

(i) K *admits locally orthogonal* $\mathcal{E}$*-factorizations for all morphisms.*

(ii) $P_{\mathcal{E}}$ *has a right adjoint.*

(iii) $P_{\mathcal{E}}$ *has a right adjoint with counits of type* $\{1,1,h\}$.

Proof: (i) $\Rightarrow$ (iii) Let (f,u) be an object in Fact K and let (e,m) be a locally orthogonal $\mathcal{E}$-factorization of uf. By the pushout property there is a unique h such that

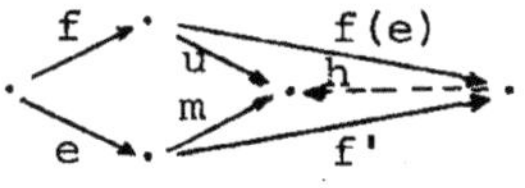

commutes. Now $[1,1,h] : P_{\mathcal{E}}(e,f) \to (f,u)$ turns out to be $P_{\mathcal{E}}$-couniversal.

(iii) $\Rightarrow$ (i) Given such a $P_{\mathcal{E}}$-couniversal arrow with $u=1$ one can easily show that $(e,hf(e))$ is a locally orthogonal $\mathcal{E}$-factorization of f.

(ii) $\Rightarrow$ (iii) Let $[j,k,h]:P_{\mathcal{E}}(e,g) \to (f,u)$ be $P_{\mathcal{E}}$-couniversal. It suffices to show that j and k are isomorphisms. The Fact K-morphism $[1,1,u]:P_{\mathcal{E}}(1,f)=(f,1) \to (f,u)$ can be written as $[1,1,u]= [j,k,h]\ P_{\mathcal{E}}\ \{u,r,t\}$ by a unique $\{u,r,t\}:(1,f) \to (e,g)$. In particular we have $jr=1$ and $kt=1$. To prove the equations $rj=1$ and $tk=1$ one looks at the two $\mathrm{Rel}_{\mathcal{E}}K$-morphisms

$$\{e,1,1\},\{erj,rj,tk\} : (1,g) \to (e,g)\ .$$

Their $P_{\mathcal{E}}$-images are equalized by $[j,k,h]$, so they are equal.□

We mention the dual concept of locally orthogonal $\mathcal{E}$-factorizations: For a class M of morphisms of K, a *locally coorthogonal* M*-factoriation* (e,m) is a locally orthogonal M^{op}-factori -

zation (m,e) in K^{op}, that is, one has $m \in M$ and the diagonalization property pictured by

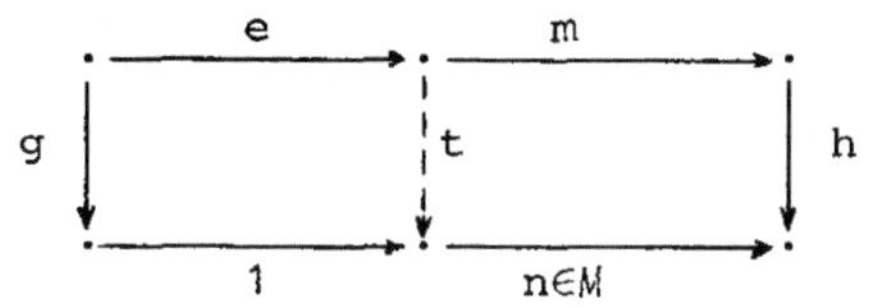

2. Examples

We restrict ourselves to examples of locally orthogonal E-factorizations which are not orthogonal a priori. The first one is general: the considered factorizations exist for all morphisms of any category with kernelpairs and coequalizers.

2.1 *Regular factorizations* (Isbell[6], Kelly[7]). For two morphisms f and g with common domain we write $\ker f \leq \ker g$, if, for all x and y, fx= fy implies gx =gy. An epimorphism e is *regular*, iff every g with $\ker e \leq \ker g$ factors as g=he. A factorization (e,m) of f is *regular*, iff e is a regular epimorphism with $\ker f \leq \ker e$. Now, for E the class of all regular epimorphisms, one easily shows that *every regular factorization is a locally orthogonal E-factorization* (hence unique up to an isomophism) and that *a locally orthogonal E-facotrization is regular, if K has coequalizers*. Furthermore, *if K admits regular factorizations, the following assertions are equivalent:*

(i) *In every regular factorization* (e,m), m *is a monomorphism.*

(ii) *Regular epimorphisms are closed under composition.*

(iii) K *admits orthogonal E-factorizations for all morphisms.*

The category Cat of small categories admits regular factorizations; these are locally orthogonal but not orthogonal since compositions of regular epimorphisms need not be regular in Cat. The same holds in the dual of the category of semigroups.

2.2 *Monotone-light factorizations.* Arising from Eilenberg's [4] investigations many authors have found generalizations of monotone-light factorizations and extended their range to all T_1-spaces. But there is no objection to considering all spaces as long as one does not expect to obtain orthogonal but just locally orthogonal factorizations. So let E be the class of quotient maps in *Top* with connected fibres. Then *Top* admits a locally orthogonal E-factorization for every continuous mapping $f:X \to Y$: call two points x_0, x_1 in X equivalent, iff there is a $y \in Y$ and a component C in $f^{-1}y$ with $x_0, x_1 \in C$, and factorize f over the quotient map. Since E is not closed under composition, this factorization is not orthogonal in general. Completely analogous results are obtained for pathwise connectedness instead of connectedness. That in this case E is not closed under composition may be seen from the first projection p of the subspace $\{(x,\sin\frac{1}{x}): x\in\mathbb{R}, x\neq 0\}\cup\{(0,y): -1\leq y\leq 1\}\subset\mathbb{R}^2$ onto $\mathbb{R}$ and the succeeding map onto a singleton space .

In a more general context one can prove without any difficulty (cf. Herrlich-Salicrup-Vázquez[5] for terminology): Let K be an initially complete concrete category and let A be a connection subcategory of K, that is A is a full subcategory of K which contains a non-void object and which is closed under epimorphic images and unions of centered families of subobjects. Then K *admits locally orthogonal* MA-*factorization for* MA *the class of all quotient maps of* K *with fibres in* A. Theorem 2.3 of [5] is an immediate consequence of this and of Proposition 1.2 above.

For more general investigations the interested reader is referred to Börger's recent thesis [2]. We also mention that corresponding considerations are possible for Strecker's [11] *submonotone-superlight factorizations;* for instance, for E the class of those quotient maps such that each fibre is contained in a path component, Top admits locally orthogonal but not necessarily orthogonal E-factorizations.

2.3 *Eilenberg-Moore factorization.* In a category Cat of categories and functors we consider the (non-full) subcategory Cat_{ra} of all right adjoint functors and its subclass M of all monadic functors. We claim that in Cat_{ra}, *every right adjoint functor* $U:A\to X$ *has a locally coorthogonal* M-*factorization, provided* A *and the Eilenberg-Moore category* X^T (T is a monad induced by U) *have coequalizers.*

Proof: It is known that the existence of coequalizers in A makes the comparison functor $K:A\to X^T$ have a left adjoint. Now, if we consider the (outer) commutative diagram

$$\begin{array}{ccccc} A & \xrightarrow{K} & X^T & \xrightarrow{U^T} & X \\ {\scriptstyle P}\downarrow & & \downarrow{\scriptstyle V} & & \downarrow{\scriptstyle R} \\ Y^S & \xrightarrow{Id} & Y^S & \xrightarrow{U^S} & Y \end{array}$$

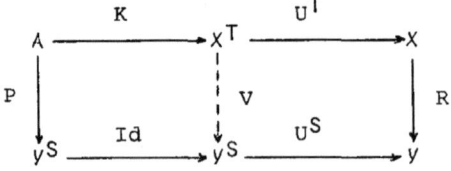

in Cat_{ra} for any monad $S=(S,\eta^S,\mu^S)$, then there is a unique functor V with $VK=P$ and $U^SV=RU^T$; we restrict ourselves to show uniqueness: for a T-algebra (X,x), $V(X,x)$ must be written as (RX,y) with an S-structure $y: SRT\to RX$, and the T-homomorphism $x:(TX,\mu X)\to(X,x)$ is transformed into an S-homomorphism $Rx\colon V(TX,\mu X) \to (RX,y)$. Since

$$V(TX,\mu X) = VKFX = PFX = (U^S PFX\,,\, U^S \varepsilon^S PFX)$$

(where F is left adjoint to U) one necessarily obtains

$$y = y \cdot SRx \cdot SR\eta X = Rx \cdot U^S \varepsilon^S PFX \cdot SR\eta X.$$

(One notices that one does not use the right adjointness of P and R to show the uniqueness of V; it is, however, used in the existence proof.) Since X^T has coequalizers V has a left adjoint by adjoint triangle arguments (apply, for instance, Corollary 7 of [12] to $U^S V = RU^T$).

2.4 *New factorizations from old by adjoint functors.* Let $U\colon A \to X$ be a functor with left adjoint F. For a class E of X-morphisms, let E' be the pushout closure of FE in A, that is, E' consists of all morphisms which are pushouts of a morphism Fe with $e \in E$. One then obtains the following lifting theorem which does not have an analogue in the orthogonal case:

Theorem 2.5: *Let A have pushouts and let E consists of X-epimorphisms only. Then, if X admits locally orthogonal E-factorizations, A admits locally orthogonal E'-factorizations.*

Proof: In order to find an E'-factorization of the A-morphism $f\colon A \to B$ we form the locally orthogonal E-factorization (e,m) of Uf and then the pushout diagram

$$\begin{array}{ccc} FUA & \xrightarrow{\;Fe\;} & FZ \\ {\scriptstyle \varepsilon A}\downarrow & & \downarrow{\scriptstyle h} \\ A & \xrightarrow{\;e'\;} & C \end{array}$$

There is a unique $m'\colon C \to B$ with $m'e' = f$ and $m'h = \varepsilon B \cdot Fm$. We claim that $(e'm')$ is a locally orthogonal E'-factorization of f. So we consider a commutative diagram

$$\begin{array}{ccccc} FX & & \xrightarrow{\;Fp\;} & & FY \\ {\scriptstyle u}\downarrow & & & & \downarrow{\scriptstyle v} \\ D & & \xrightarrow{\;p'\;} & & E \\ {\scriptstyle a}\downarrow & & & & \downarrow{\scriptstyle b} \\ A & \xrightarrow{\;e'\;} & C & \xrightarrow{\;m'\;} & B \end{array}$$

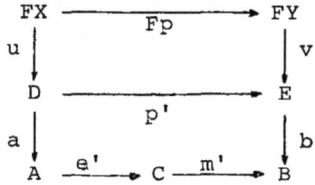

where the upper part is a pushout with $p \in E$. We have morphisms $x\colon X \to UA$, $y\colon Y \to UB$ with $\varepsilon A \cdot Fx = au$, $\varepsilon B \cdot Fy = bv$. From $\varepsilon B \cdot FUf \cdot Fx = fau = bv \cdot Fp = \varepsilon B \cdot Fy \cdot Fp$ one derives $fx = yp$. Hence there is a $t: Y \to Z$ with $tp = ex$, $mt = y$, and, by the pushout property, an $s: E \to C$ with $sp' = e'a$, $sv = h \cdot Ft$. Since E consists of epimorphisms, also E' consists of epimorphisms; so from $m'sp' = bp'$ one gets $m's = b'$, and s is unique. □

Remark: The analogue of Theorem 2.5 is valid for regular factorizations provided U is faithful. This was first shown (in the dual situation) by Mitchell [10], Lemma 2.2.

3. Locally orthogonal (E,λ)-factorizations of morphisms

We next consider factorizations of a morphism into possibly an infinite number of factors. To this end we identify the ordinal number $\lambda \geq 1$ with the ordered set of ordinals less than λ, considered as a category, and let $\alpha\delta$ be the unique morphism from α to δ for $\alpha \leq \delta$. Each limit number $\kappa \leq \lambda$ is the colimit of the set of morphims $\gamma\delta(\gamma < \kappa; \delta = \gamma+1)$. A pair (e_λ, C) is the *λ-composite* (of the morphisms $e^{\alpha\beta}(\alpha < \lambda; \beta = \alpha+1))$, if $C: \lambda+1 \to K$ is a functor with $C(0\lambda) = e_\lambda$ (and $C(\alpha\beta) = e^{\alpha\beta}$) which preserves the above colimits. The composite (e_λ, C) is said to be *in* E if $e^{\alpha\beta} \in E$ $(\alpha < \lambda; \beta = \alpha+1)$.

A *λ-factorization* (of a morphism f) is a pair $((e_\lambda, C), m)$ consisting of a composite (e_λ, C) and a morphism m such that me_λ exists (and $f = me_\lambda$). It is called an (E,λ)-factorization, if in addition (e_λ, C) is in E. A λ-composite (p_λ, D) is *orthogonal to a λ-factorization* $((e_\lambda, C), m)$, if for each commutative diagram

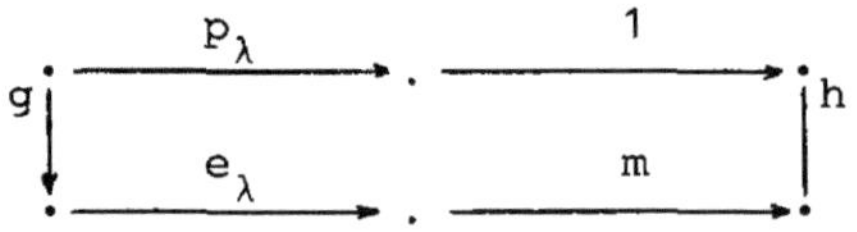

there is a unique sequence $t_1, t_2, .., t_\alpha, ..., t_\lambda$ such that $mt_\lambda = h$ and, letting $t_0 = g$, $t_\beta D(\alpha\beta) = C(\alpha\beta) t_\alpha$ whenever $0 \leq \alpha < \beta \leq \lambda$ and either $\beta = \alpha + 1$ or β a limit ordinal, as pictured by the diagram

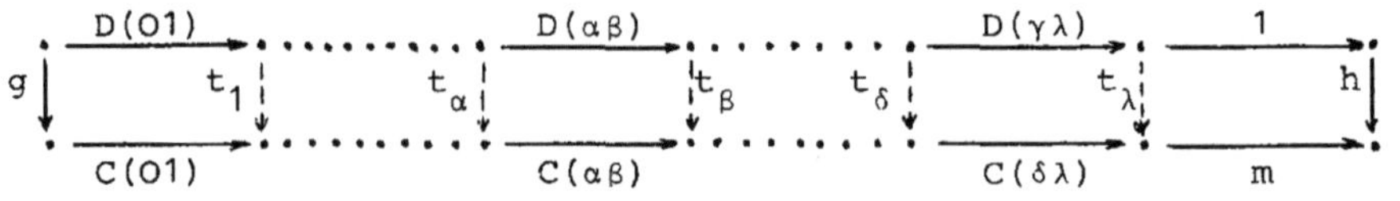

This means: there is a unique natural transformation $t: D \to C$ with $t_0^0 = g$ and $mt_\lambda^\lambda = h$. We write $(p_\lambda, D) \perp ((e_\lambda, C), m)$ in this case. An (E,λ)-factorization $((e_\lambda, C), m)$ is called *locally orthogonal*, if

$(p_\lambda, D) \perp ((e_\lambda, C), m)$ for all (p_λ, D) in E. The (locally orthogonal) $(E,1)$-factorizations may be identified with the (locally orthogonal) E-factorizations of Section 1.

Analogously to the case $\lambda = 1$ one easily proves that locally orthogonal (E,λ)-factorizations are *rigid*, that is, one necessarily has $t = 1$ whenever $D = C$, $g = 1$, and $h = m$ in the above diagram. As a consequence of this one has that two locally orthogonal (E,λ)-factorizations $(e_\lambda, C), m)$, $((e'_\lambda, C'), m')$ of the same morphism are canonically isomorphic, i.e. there is a (unique) natural equivalence $t : C \to C'$ with $t_o = 1$ and $m't_\lambda = m$.

We did not introduce a concept of orthogonal (E,λ)-factorizations, for a simple reason: We will prove first that every locally orthogonal (E,λ)-factorization is of the form

$$\cdot \xrightarrow{e} \cdot \xrightarrow{1} \cdot \xrightarrow{1} \cdot \cdots\cdots \xrightarrow{1} \cdot \xrightarrow{m} \cdot$$

if K admits orthogonal $E = (E,1)$-factorizations. Therefore, with respect to Proposition 1.2, locally orthogonal (E,λ)-factorizations may be interesting only if E is <u>not</u> closed under composition.

<u>Proposition 3.1</u>: *An orthogonal* E-*factorization* (e,m) *may be regarded as a (locally) orthogonal* (E,λ)-*factorization* $((e,C),m)$ *for any fixed ordinal* $\lambda \geq 1$ *by letting* $C: \lambda + 1 \to K$ *be defined by* $C(01) = e$ *and* $C(\alpha\beta) = 1$ *for* $1 \leq \alpha \leq \beta \leq \lambda$.

Proof: We must show that $(p_\lambda, D) \perp ((e,C),m)$ for all (p_λ, D) in E. That is, we must show that, whenever $hp_\lambda = meg$, there is a unique sequence $t_1, t_2, \ldots, t_\lambda$ such that in the diagram

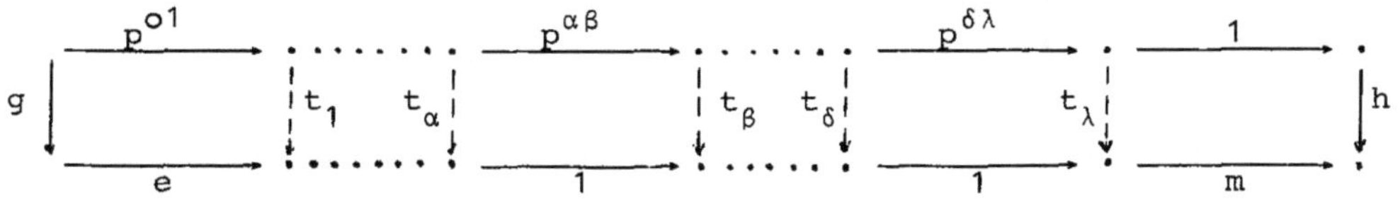

we have $eg = t_1 p^{01}$, $mt_\lambda = h$, and $t_\alpha = t_\beta p^{\alpha\beta}$ for $0 < \alpha < \beta \leq \lambda$ and $\beta = \alpha + 1$ or β = limit ordinal. The orthogonality of (e,m) implies that $p^{01} \perp (1,m)$ since $p^{01} \in E$. Thus there is a unique t_1 rendering the following diagram commutative

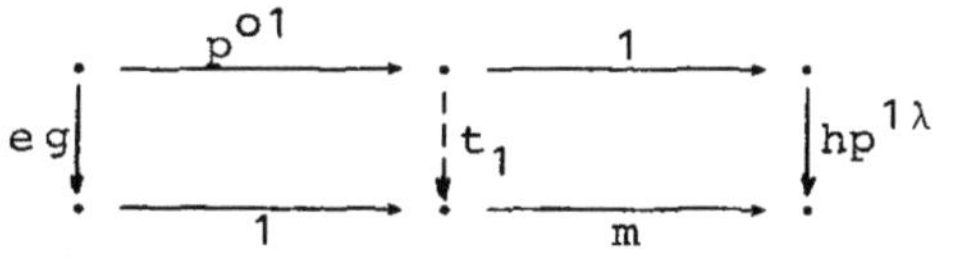

Let β be an ordinal with $1 < \beta \leq \lambda$ and suppose inductively that for $1 < \alpha < \beta$ we have $t_\alpha p^{\delta\alpha} = t_\delta$ and $mt_\alpha = hp^{\alpha\lambda}$ for a unique t_α when $\alpha = \delta + 1$ or when α is a limit ordinal and $1 \leq \delta < \alpha$. If $\beta = \alpha + 1$ then there is a unique t_β rendering the following diagram commutative

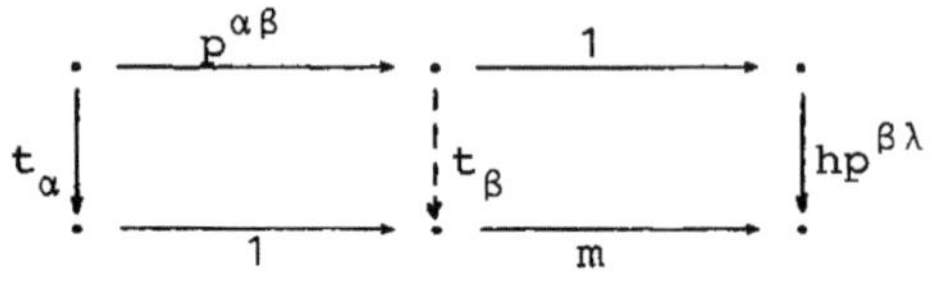

since $p^{\alpha\beta} \perp (1,m)$. If β is a limit ordinal, by the colimit property in K, there is a unique t_β such that $t_\beta p^{\delta\beta} = t_\delta$ for $1 \leq \delta < \beta$. But in the addition $mt_\beta = hp^{\beta\lambda}$ since $mt_\beta p^{\delta\beta} = hp^{\beta\lambda}p^{\delta\beta}$ for all $\delta < \beta$. This completes the inductive step. Uniqueness can be left to the reader. □

Proposition 3.2: *A locally orthogonal* (E,λ)*-factorization* $((e_\lambda, C), m)$ *may be regarded as a locally orthogonal* (E,δ)*-factorization* $((C(0\delta), \overline{C}), mC(\delta\lambda))$ *for each* δ *with* $0 < \delta < \lambda$ *where* $\overline{C}$ *is the restriction of* C *to* $\delta + 1$.

Proof: We must show that $(p_\delta, \overline{D}) \perp ((C(0\delta), \overline{C}), m\,C(\delta\lambda)))$ for all $(p_\delta, \overline{D})$ in E. The functor $\overline{D}: \delta + 1 \to K$ trivially extends to $D: \lambda + 1 \to K$ with $D(\alpha\mu) = 1$ for all $\delta \leq \alpha \leq \mu \leq \lambda$. Assume $me_\lambda g = hp_\delta$. Since $(p_\lambda = p_\delta, D) \perp ((e_\lambda, C), m)$ there is a unique sequence $t_1, t_2, \ldots, t_\delta, \ldots, t_\lambda$ rendering the following diagram commutative

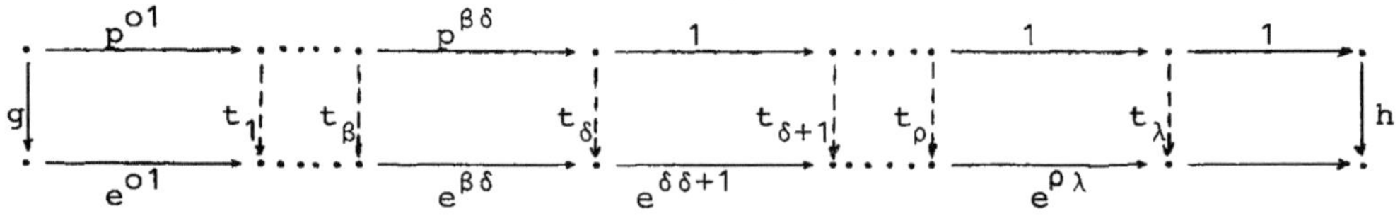

But all the morphisms t_δ, $\delta < \rho \leq \lambda$, are uniquely determined by t_δ since we have $t_\rho = e^{\delta\rho} t_\delta$, and this immediately proves the assertion.□

The main result is:

Theorem 3.3: *Suppose that every morphism* f *in* K *has a locally orthogonal* E*-factorization and that colimits of chains exist*[1] *in* K.

[1] More precisely we assume that each functor $\overline{C}: \lambda \to K$ such that $\overline{C}$ preserves colimits of chains and $\overline{C}(\alpha,\beta) \in E$ for $0 \leq \alpha < \beta < \lambda$, $\beta = \alpha + 1$, λ a limit ordinal, has a colimit in K; then $\overline{C}$ may be extended to a functor $C: \lambda + 1 \to K$ preserving colimits of chains.

Then for each ordinal $\lambda > 0$ *every morphism* f *has a locally orthogonal* (E,λ)*-factorization.*

Proof: Let $((e_1,C_1),m_1)$ be a locally orthogonal $(E,1)$-factorization of f. Suppose inductively that for all $1 < \delta < \lambda$ $((e_\delta,C_\delta),m_\delta)$ is a locally orthogonal (E,δ)-factorization of f whose restriction to a locally orthogonal (E,α)-factorization of f given by Proposition 3.2 is $((e_\alpha,C_\alpha),m_\alpha)$ for $0 < \alpha < \delta$. Diagramatically we have

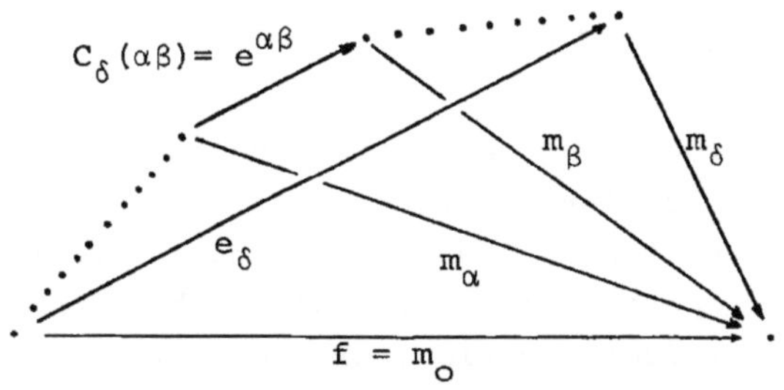

If λ is of the form $\delta+1$ we take a locally orthogonal E-factorization $(e^{\delta\lambda},m)$ of m_δ and define a functor $C\colon \lambda+1 \to K$ by $C = C_\delta$ on the restriction to λ and $C(\alpha\lambda) = e^{\delta\lambda}C_\delta(\alpha\delta)$ for $\alpha \le \delta$ and let $e_\lambda = e^{\delta\lambda}e_\delta$. If λ is a limit ordinal define $\overline{C}\colon \lambda \to K$ by $\overline{C}(\beta\mu) = C_\mu(\beta\mu)$ for $\beta \le \mu < \lambda$. This functor $\overline{C}$ preserves colimits of chains since each C_μ does. Hence by hypothesis there is an extension $C\colon \lambda+1 \to K$ with (e_λ,C) in E. We let m be the unique morphism such that $mC(\alpha\beta) = m_\alpha$ for all $\alpha \le \lambda$, which exists since $C(\lambda)$ is the object value of the colimit of $\overline{C}$.

Finally we show that $((e_\lambda,C),m)$ as just defined is a locally orthogonal (E,λ)-factorization of f. In other words, we must show that $(p_\lambda,D) \perp ((e_\lambda,C),m)$ for all (p_λ,D) in E. So let $hp_\lambda = me_\lambda g$. From the inductive hypothesis it follows that there is a unique sequence $t_1,t_2,\ldots,t_\delta$ such that $me^{\delta\lambda}t_\delta = hp^{\delta\lambda}$ and $t_\beta p^{\alpha\beta} = e^{\alpha\beta}t_\alpha$ whenever $0 \le \alpha < \delta$ and $\beta = \alpha+1$ or β a limit ordinal $\le \delta$. If $\lambda = \delta+1$ then there exists a unique t_λ such that $mt_\lambda = h$ and $t_\lambda p^{\delta\lambda} = e^{\delta\lambda}t_\delta$ since $(e^{\delta\lambda},m)$ is a locally orthogonal E-factorization. If λ is a limit ordinal then t_λ is the induced map on colimits and is the unique map with $t_\lambda p^{\alpha\lambda} = e^{\alpha\lambda}t_\alpha$ for all $\alpha < \lambda$. Thus $t_1,\ldots,t_\lambda$ satisfies the required equations and is, of course, unique. □

By Theorem 3.3 it is clear that the locally orthogonal E-factorizations of Section 2 can be extended to locally orthogonal (E,λ)-factorizations provided that all colimits exist. More precisely it is sufficient that colimits of chain-colimit preserving functors $C: \delta \to K$ with $C(\alpha,\alpha+1) \in E$ exist for $0 \leq \alpha < \delta$ and δ a limit ordinal $\leq \lambda$, or, dually, in the case of the Eilenberg-Moore-factorization. In the latter case we obtain the adjoint tower construction described in [8] where $C: \delta^{op} \to Cat_{ra}$ with $C(\alpha,\alpha+1): X^{\alpha+1} \to X^{\alpha} \in M$ = monadic functors. In this case the limit of C consists of the category X^{δ} and projections $\overline{C}(\alpha\delta): X^{\delta} \to X^{\alpha}$ for $\alpha < \delta$ where an object A of X^{δ} is a collection $\{A_{\alpha}\}_{\alpha<\delta}$ of objects A_{α} in X^{α} with $C(\alpha\mu)(A_{\mu}) = A_{\alpha}$ for $\mu < \delta$ and a morphism $f: A \to B$ of X^{δ} is a collection $\{f_{\alpha} \mid f_{\alpha}: A_{\alpha} \to B_{\alpha}\}_{\alpha<\delta}$ of morphisms f_{α} with $C(\alpha\mu)(f_{\mu}) = f_{\alpha}$ for $\mu < \delta$.

We remark that the preceeding shows that the locally coorthogonal (M,λ)-factorization, or adjoint tower construction, for a functor U in Cat_{ra} for a given ordinal λ and M = monadic functors in Cat_{ra} has universal properties exactly dual to those possessed by the locally orthogonal (E,λ)-factorization, or "regular tower" construction, for a morphism f in a category K for a given ordinal λ and E = regular epimorphisms in K.

References

1. H.Applegate and M.Tierney, Iterated cotriples, in: Lecture Notes in Math. 137 (Springer-Verlag, 1970), pp. 56 - 99.

2. R.Börger, Kategorielle Beschreibungen von Zusammenhangsbegriffen, Thesis (Hagen 1981).

3. B.Day, On adjoint functor factorization, in: Lecture Notes in Math. 420 (Springer-Verlag, 1974), pp. 1 - 19.

4. S.Eilenberg, Sur les transformations continues d'espaces métriques compacts, Fund.Math. 22 (1934), 292 - 296.

5. H.Herrlich, E.Salicrup, and R.Vázquez, Light factorization structures, Quaest.Math. 3 (1979), 189 - 213.

6. J.R. Isbell, Structure of categories, Bull. Amer. Math. Soc. 72 (1966), 619 - 655.

7. G.M. Kelly, Monomorphisms, epimorphisms, and pull-backs, J. Austral. Math. Soc. 9 (1969), 124 - 142.

8. J.MacDonald and A.Stone, The tower and regular decomposition, Cahiers Topologie Géom. Différentielle (to appear).

9. J.MacDonald and A.Stone, Essentially monadic adjunctions, in: Proc.Int.Conf.Category Theory Gummersbach 1981 (to appear).

10. B.Mitchell, The dominion of Isbell, Trans.Amer.Math.Soc. 167 (1972), 319 - 331.

11. G.E.Strecker, Component properties and factorizations, Math. Centre Tracts 52 (1974), 123 - 140.

12. W.Tholen, Adjungierte Dreiecke, Colimites und Kan-Erweiterungen, Math.Ann. 217 (1975), 121 - 129.

13. W.Tholen, Semi-topological functors I, J.Pure Appl. Algebra 15 (1979), 53 - 73.

John MacDonald
Mathematics Department
University of British Columbia
Vancouver, B.C.
Canada V6T 1Y4

Walter Tholen
Fachbereich Mathematik
Fernuniversität
5800 Hagen 1
Fed.Rep. of Germany

REMARKS ON RADICALS IN CATEGORIES

L. Márki and R. Wiegandt

1. Introduction

In order to develop a general radical theory in a category, one of the usual ways is to deal with categories like that of rings. In such a category radicals can be defined by functors or by assignments or by properties. In this note we intend to clarify the relation between various systems of axioms imposed on the category and to compare the various ways of defining radicals. Thus in section 2 we exhibit that the system of axioms introduced by Holcombe and Walker [5] is practically equivalent to that used by Andrunakievič and Rjabuhin [2]. It turns out in section 3 that a radical functor in the sense of Carreau [3] is nothing but a Hoehnke radical [4], and we prove that a radical functor is complete and idempotent if and only if the corresponding Hoehnke radical is an ideal-radical. We also exhibit the equivalence of the latter radicals with those of Šul'geĭfer [11] and Amitsur [1], and describe radical and semisimple classes by intrinsic properties.

A general reference for terminology is Mitchell [9]. As usual, monomorphisms will be denoted by $\rightarrowtail$. For a normal monomorphism and a normal epimorphism we shall use the symbols $\vartriangleright\!\!\rightarrow$ and $\rightarrow\!\!\vartriangleright$, respectively. If $B \vartriangleright\!\!\rightarrow A \rightarrow\!\!\vartriangleright C$ is an exact sequence, then the factor object C will be denoted by A/B.

2. Axioms on categories

Several systems of axioms have been imposed on categories in order to enable the development of a general radical theory. Many of these systems of axioms aim at obtaining categories similar to those of rings or Ω-groups - similarity means that one can use techniques, in particular the isomorphism theorems in their familiar forms, like in the latter categories ([1], [2], [3], [5], [6], [11], [12], [13], etc). Some other authors, on the other hand, tried to develop general radical theory in a more general setting; accordingly they could not get too far without making further restrictions either on the category or on the radicals under consideration (for recent investigations in this direction we refer to [14]).

Here we want to give a clear picture on the present situation in ring-like categories. Almost all the systems of axioms used in them for the purposes of radical theory, are slighter modifications of that of Šul'geĭfer [11]. We shall use the form presented by Andrunakievič and Rjabuhin [2]: it is supposed that the considered category $\mathcal{C}$ satisfies

(AR1) C has zero object,

(AR2) every object possesses a representative set for its normal subobjects,

(AR3) for each object its normal subobjects form a complete sublattice under the natural partial order of subobjects,

(AR4) every morphism has a kernel,

(AR5) every morphism has a normal epimorphic image,

(AR6) normal epimorphisms carry normal subobjects into normal subobjects.

We remark that (AR2) is imposed in [2] in the stronger form that all subobjects of any object form a partially ordered set, but the present weaker form suffices both for the investigations in [2] and for our purposes.

Holcombe and Walker [5] used a seemingly different system of axioms. Here we exhibit, however, that the two systems are equivalent. First of all, let us have a closer look at the categories considered by Holcombe and Walker. Part of their theory makes use of the second isomorphism theorem without a clear formulation in the categories for the latter to hold. They considered a category C satisfying the following axioms:

(HW1) C has a zero object, kernels and cokernels,

(HW2) in C intersections of arbitrary sets of normal subobjects of an object exist, further, every object possesses a representative set for its normal subobjects.

The intersection of normal subobjects is again normal ([5] Lemma 1.5).

(HW3) Every morphism of C factors through a cokernel followed by a monomorphism.

Under these axioms C has inverse images of normal subobjects which are again normal ([5] Lemma 1.4) and has epimorphic images ([5] Lemma 1.9).

(HW4) For every cokernel $t: X \longrightarrow Y$ and kernel $Z \rightarrow Y$ we have $t(t^{-1}(Z))=Z$.

The first isomorphism theorem is proved to be a consequence of these axioms ([5] Theorem 1,11).

Finally we assume

(HW5) Any set of normal subobjects of any object possesses a union which is again a normal subobject.

Let us notice that in the presence of the systems of axioms (HW1)-(HW5) the normal subobjects of any object form a complete lattice under $\cap$ and $\cup$.

For short, the systems of axioms (AR1)-(AR6) and (HW1)-(HW5) will be denoted by (AR) and (HW), respectively.

THEOREM 1. (HW) *implies* (AR), *and if any two subobjects of objects in* C *have an intersection, then the converse also holds.*

Proof: In view of what has been said above, it is clear that the system (HW) implies (AR1)-(AR5). In order to prove (AR6), let $t: X \rightarrow Y$ be a cokernel with kernel $\ker t: Z \rightarrow X$, and N be a normal subobject of X . By [5] Lemma 1.10

$t(N\cup Z)$ is a normal subobject of Y. Since $t(Z)=0\leq t(N)$, by the definition of union we have $t(N\cup Z)\leq t(N)$, hence by [5] Lemma 1.4 $t(N\cup Z)$ is a normal subobject of $t(N)$. Conversely, by the definition of the image $t(N)\leq t(N\cup Z)$, implying that $t(N)=t(N\cup Z)$ is a normal subobject of Y.

Conversely, notice that (AR) ⇒ (HW1)-(HW3) is obvious, (HW4) is a special case of (AR6) and all that needs proof in (HW5) is that the lattice-theoretical union of normal subobjects is category-theoretical union. Notice also that by (HW1)-(HW4) we already know that our category C has images and inverse images of normal subobjects. Let us consider normal monomorphisms $\alpha_i : A_i \rightarrowtail A$, $i\in I$, and let A' be the smallest normal subobject of A containing each A_i, $i\in I$, $f: A \to B$ a morphism with canonical decomposition $A \xrightarrow{f'} f(A) \rightarrowtail B$, and C be the smallest normal subobject of $f(A)$ containing all the $f(A_i)$. Suppose that there is a $\beta : B' \rightarrowtail B$ through which each $f\alpha_i$ factorizes. Then, provided that $f(A)\cap B'$ exists, it contains all the $f(A_i)$, and since the $f(A_i)$ are normal subobjects of $f(A)$ by (AR6), they are normal in $f(A)\cap B'$, too, whence $C\leq f(A)\cap B'\leq B'$.

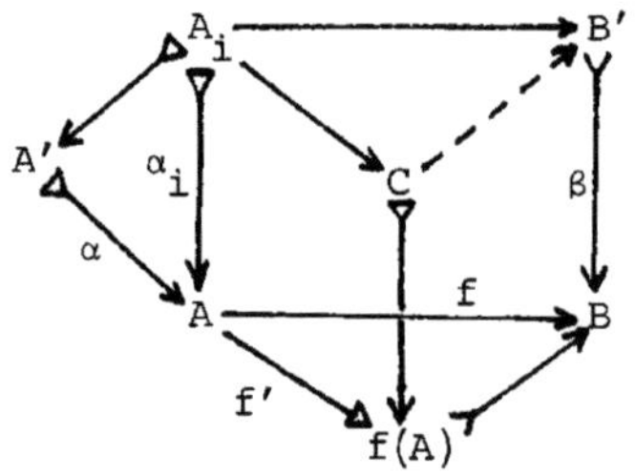

On the other hand, we obviously have $f(A')\geq C$, and conversely, $(f')^{-1}(C)\geq A_i$ for all $i\in I$, hence $(f')^{-1}(C)\geq A'$, thus $C=f'(f')^{-1}(C)\geq f'(A')=f(A')$. So we conclude that $f(A')$ and C are equivalent subobjects, whence $f\alpha$ also factorizes through β, and we are done.

Remark 1. By [2] p. 397 we know that the second isomorphism theorem holds under (AR), thus it holds under (HW1)-(HW5) as well.

Remark 2. Krempa and Terlikowska [6] and then Terlikowska-Osłowska [12], [13] introduced a self-dual system of axioms which is satisfied in the categories of associative or alternative rings but not in that of not necessarily associative ones. Hence this system of axioms cannot be equivalent to (HW) or (AR), nevertheless there is a strong connection between them. Suppose we are given a category C satisfying the system of axioms (AR). Consider the subcategory C' consisting of all objects of C and those morphisms which are kernels or cokernels or their compositions. (For establishing the basic results in the general radical theory of rings, only these morphisms are really needed.) Now it is straightforward to check that C' satisfies the system of axioms A1-A6, A6* and A7* but not A7 of Terlikowska-Osłowska [12]. Conversely, A1-A7, A6*, A7* of [12] imply most of (AR) but (AR4) and (AR5) only in a weaker form.

In the rest of the paper we shall always work in a category C satisfying the axioms (AR).

3. Radicals

In his paper [3] Carreau presented an elegant treatment of radicals in categories. What he did in genuine categorical terms is, expressed in the classical language, that radicals can be defined both by means of a function and of a semisimple class. The same idea is basic in Hoehnke's earlier development [4] of radicals in categories of universal algebras. In the category of associative rings Michler [8] introduced a notion of radical at the same time as Hoehnke did for universal algebras; these two notions are equivalent for rings.

Now we present Carreau's definition of a radical functor in the slightly modified but equivalent version given by Holcombe and Walker [5].

By the cokernel subcategory $E(C)$ of C we mean the subcategory whose objects are the objects of C and whose only morphisms are the cokernels of C. (In Carreau's terminology this is a special coextensive subcategory.) A covariant functor $\rho: E(C) \rightarrow C$ is called a radical functor, if

(i) ρ is a subfunctor of the inclusion functor $\iota: E(C) \rightarrow C$,

(ii) for all $C \in C$, $\rho(C)$ is a normal subobject in C,

(iii) $\rho(C/\rho(C))=0$ for all $C \in C$.

A radical in the sense of Hoehnke is a mapping ρ assigning to each $C \in C$ a normal subobject $\rho(C)$ in C satisfying (iii) and

(iv) $\varphi(\rho(C)) \leq \rho(\varphi(C))$ for any cokernel φ from C.

Theorem 2.2 of Carreau [3] states exactly that every radical functor defines a radical and conversely.

The most important radical functors are complete (which means that if $\rho(B)=B$ for some normal subobject B of A, then $B \leq \rho(A)$) and idempotent ($\rho(\rho(A))=\rho(A)$ for all $A \in C$).

By Carreau [3] Corollary 2.12 we know that complete idempotent radical functors are exactly the radicals in the sense of Šul'geĭfer [11]. The latter are defined, following Kuroš, as follows. Consider a function ρ assigning to each $C \in C$ a normal subobject $\rho(C)$ of C, and call C a ρ-object, if $\rho(C)=C$. This function ρ is said to be a radical function, if it satisfies:

(1) any normal epimorphic image of a ρ-object is itself a ρ-object,

(2) for any $C \in C$, if each nonzero normal epimorphic image of C has a nonzero normal subobject which is a ρ-object, then C is a ρ-object.

The other traditional definition of radical, which goes back to Amitsur [1] even for categories, is also defined by such a function ρ satisfying (1), but instead of (2) the following two properties are required: (iii) and

(α) every object $C \in \mathcal{C}$ contains a unique maximal normal subobject $\rho(C)$ which is a ρ-object.

In our category we can use the same methods as in the category of rings, and by standard reasonings we arrive at the following results.

PROPOSITION 2. Radicals in the sense of Šul'geĭfer are the same as those in the sense of Amitsur. (This completes the discussion of Amitsur radicals in Carreau [3].)

By Andrunakievič and Rjabuhin [2] V, §2, Theorem 1 we know that a function ρ satisfying (1) and (α) defines a radical if and only if the class of all ρ-objects is closed under extensions, that is,

(β) if B is a normal subobject of A and both B and A/B are ρ-objects, then so is A.

PROPOSITION 3. Under (1) and (β), (α) is equivalent with the inductivity condition

(γ) for any ascending chain (I_λ) of normal subobjects of an object $C \in \mathcal{C}$ which are all ρ-objects, also $\bigcup_\lambda I_\lambda$ is a ρ-object.

Next, we shall see that the radicals discussed above, have also been distinguished in Hoehnke's theory [4]. In fact, consider the relation

$$B \lhd A \iff B \text{ is a non-zero normal subobject of } A .$$

$\lhd$ is a "nice" M-relation in the sense of Hoehnke [6]: clearly $\lhd$ is reflexive for non-zero objects, and if A is a subdirect product of A_i, $i \in I$, and $B \lhd A$, then there is $j \in I$ such that the j-th canonical projection takes B onto a non-zero normal subobject of A_j by (AR6). According to the terminology of Hoehnke [6], a radical ρ is called an $\lhd$-radical, if it satisfies

(M1) $\rho(A)=0$ and $B \lhd A$ imply $\rho(B) \neq B$,

(M2) if $\rho(B) \neq B$ for all $B \lhd A$, then $\rho(A)=0$.

PROPOSITION 4. A radical functor ρ is complete if and only if the corresponding radical satisfies condition (M1).

Proof: Suppose that (M1) holds, and let $B \lhd A$, $\rho(B)=B$, $B \nleq \rho(A)$. Now we have

$$B'' =: B/(\rho(A) \cap B) = B' =: (\rho(A) \cup B)/\rho(A) \lhd A/\rho(A).$$

By (iii) and (M1) it follows that $\rho(B') \neq B'$, whereas considering the cokernel morphism φ of $(\rho(A) \cap B) \rightarrowtail B$, by (iv) we obtain that

$$\rho(B'') \geq \rho(B)/(\rho(A) \cap B) = B/(\rho(A) \cap B) = B'' ,$$

whence $\rho(B'')=B''$, a contradiction. The converse implication is obvious.

PROPOSITION 5. If a radical functor ρ is idempotent, then the corresponding radical satisfies (M2).

The assertion is obvious. Moreover, under the validity of (M1) the converse implication is also true.

THEOREM 6. Every complete and idempotent radical functor defines an $\triangleleft$-radical and conversely.

Proof: In view of Propositions 4 and 5 all we have to prove is that the radical functor ρ defined by an $\triangleleft$-radical is idempotent. Suppose that ρ is not idempotent. Now there is an $A \in C$ such that $\rho(A)/\rho(\rho(A)) \neq 0$. By (M2) there exists a $B/\rho(\rho(A))$ such that $B/\rho(\rho(A)) \triangleleft \rho(A)/\rho(\rho(A))$ and $\rho(B/\rho(\rho(A))) = B/\rho(\rho(A))$. By condition (iii), however, we have $\rho(\rho(A)/\rho(\rho(A)))=0$ and so for $B/\rho(\rho(A)) \triangleleft \rho(A)/\rho(\rho(A))$ condition (M1) yields $\rho(B/\rho(\rho(A))) \neq B/\rho(\rho(A))$, a contradiction.

As usual, to a radical ρ we assign two classes

$$\mathbb{R}_\rho = \{A \in C : \rho(A) = A\}$$

and

$$\mathbb{S}_\rho = \{A \in C : \rho(A) = 0\},$$

called the radical class and the semisimple class of ρ , respectively. Knowing the equivalence of the previous definitions of radicals, the connection between radical and semisimple classes as described in Andrunakievič and Rjabuhin [2], V, §2, Theorem 3, yields exactly Theorem 3.10 of Holcombe and Walker [5] and its converse. (The latter is the same as [5] Theorem 3.11; in fact the sufficient condition given in the note after this theorem, is always satisfied in view of (AR6).) Thus an $\triangleleft$-radical ρ on any object A can be determined both from below and from above:

$$\cup(B \triangleleft A : B \in \mathbb{R}_\rho) = \rho(A) = \cap(C \triangleleft A \text{ or } C = 0 : A/C \in \mathbb{S}_\rho) .$$

Till now we have characterized an $\triangleleft$-radical by means of the radical assignment (radical functor) and the radical class. It can also be characterized in terms of the semisimple class and by the pair of radical and semisimple classes, respectively. Such characterizations for not necessarily associative rings or Ω-groups exist in plenty (see e.g. [7] and [10]), and using the tools we already have in our category, their proofs can be carried out word by word in our case, too. Here we pick out just one characterization of each of the latter two types.

THEOREM 7 (Mlitz [10] Theorem 4). A class $\mathbb{S}$ of objects is the semisimple class of an $\triangleleft$-radical if and only if $\mathbb{S}$ satisfies the following three conditions:

(a) *if* $B \triangleleft A \in \mathbb{S}$, *then* B *has a non-zero factor object in* $\mathbb{S}$,

(b) $\mathbb{S}$ *is closed under subdirect products,*

(c) *for all* $A \in C$, $((A)\mathbb{S})\mathbb{S} = (A)\mathbb{S}$ *where* $(A)\mathbb{S} = \cap (B \triangleleft A$ *or* $B = 0 : A/B \in \mathbb{S})$.

THEOREM 8 (Mlitz [10] Theorem 2). *The classes* $\mathbb{R}$ *and* $\mathbb{S}$ *are the radical and semisimple classes of an* $\triangleleft$*- radical if and only if*

(A) $\mathbb{R} \cap \mathbb{S}$ *consists of zero objects,*

(B) $A \in \mathbb{R}$ *and* $A/B \neq 0$ *imply* $A/B \notin \mathbb{S}$,

(C) $A \in \mathbb{S}$ *and* $B \triangleleft A$ *imply* $B \notin \mathbb{R}$,

(D) *for any* $A \in C$ *there is a normal subobject* B *of* A *such that* $B \in \mathbb{R}$ *and* $A/B \in \mathbb{S}$.

References

[1] S. A. AMITSUR, A general theory of radicals, II, Radicals in rings and bicategories, *Amer. J. Math.* 76 (1954), 100-125.

[2] V. A. ANDRUNAKIEVIČ and Ju. M. RJABUHIN, *Radicals of algebras and structure theory* (Russian), Nauka, Moscow, 1979.

[3] F. CARREAU, Sous-catégories réflexives et la théorie générale des radicaux, *Fund. Math.* 71 (1971), 223-242.

[4] H.-J. HOEHNKE, Radikale in allgemeinen Algebren, *Math. Nachr.* 32 (1966), 347-383.

[5] M. HOLCOMBE and R. WALKER, Radicals in categories, *Proc. Edinburgh Math. Soc.* 21 (1978), 111-128.

[6] J. KREMPA and B. TERLIKOWSKA, Theory of radicals in self-dual categories, *Bull. Acad. Polon. Sci. Sér. Sci. Math. Astronom. Phys.* 22 (1974), 367-373.

[7] L. C. A. van LEEUWEN and R. WIEGANDT, Radicals, semisimple classes and torsion theories, *Acta Math. Acad. Sci. Hungar.* 36 (1980), 37-47.

[8] G. MICHLER, Radikale und Sockel, *Math. Ann.* 167 (1966), 1-48.

[9] B. MITCHELL, *Theory of categories*, Academic Press, 1965.

[10] R. MLITZ, Radicals and semisimple classes of Ω-groups, *Proc. Edinburgh Math. Soc.* 23 (1980), 37-41.

[11] E. G. ŠUL'GEĬFER, General theory of radicals in categories (Russian), *Mat. Sb.* 51 (1960), 487-500.

[12] B. TERLIKOWSKA-OSŁOWSKA, Category with self-dual set of axioms, *Bull. Acad. Polon. Sci. Sér. Sci. Math. Astronom. Phys.* 25 (1977), 1207-1214.

[13] B. TERLIKOWSKA-OSŁOWSKA, Radical and semisimple classes of objects in categories with a self-dual set of axioms, *Bull. Acad. Polon. Sci. Sér. Sci. Math. Astronom. Phys.* 26 (1978), 7-13.

[14] S. VELDSMAN, *A general radical theory in categories*, Ph. D. Thesis, University of Port Elizabeth, S. A., 1980.

ON THE STRUCTURE OF FACTORIZATION STRUCTURES

by

A. Melton and G. E. Strecker

For any category $\underline{K}$ we investigate the family of all factorization structures on $\underline{K}$. In particular, for each such structure, (E,M), we investigate the complete lattice of all factorization structures on $\underline{K}$ with left factor a subclass of E; this investigation is based on a Galois connection between all such structures and the lattice of all full isomorphism-closed subcategories of $\underline{K}$. The Galois-closed families are precisely all the E-reflective subcategories of $\underline{K}$ and all the (E,M)-dispersed factorization structures of Herrlich, Salicrup and Vazquez.

AMS (1980) subject classifications: Primary 18A20, 18A32, 18A40;
Secondary: 06A15, 18A22

§0 Introduction

The importance of factorization structures on categories is by now well appreciated. Over the years the conditions that have been considered necessary for an "(E,M)-factorization structure" to carry that name have evolved from those requiring E and M to be sufficiently nice dual-like classes of epimorphisms and monomorphisms such that each single morphism has an essentially unique (E,M)-factorization, through various stages until the current generally accepted criteria that (among other things) E be a class of morphisms and M be a conglomerate of sources such that each class-indexed source (even empty or proper class indexed) has an (E,M)-factorization, and, in the category, (E,M)-diagonalization holds. To emphasize that we require diagonalizations as well as factorizations we call such entities "diafactorization structures."

The two major references for this paper are [HSV] and [Ho], both of which made significant contributions to the clarification of the nature of (dia)factorization structures.

In [HSV] Herrlich, Salicrup and Vazquez introduced a new type of diafactorization structure called <u>dispersed</u> and proceeded to show that there is a bijection between all E-reflective subcategories of an (E,M)-category $\underline{K}$ and all (E,M)-dispersed diafactorization structures on $\underline{K}$. This was a generalization of the result that for nice categories such a correspondence exists between the epireflective subcategories of $\underline{K}$ and all perfect factorizations (cf. [He_1],[He_2],[Na],[Ne], [S_1], [S_2], [S_4]). It also improved and put into the proper context much of the earlier work on quotient reflective subcategories, connectedness properties, and corresponding factorizations (cf. [C], [P], [SV_1], [SV_2], [S_3]).

In §1, via a modification of the main result of Hoffmann [Ho] (cf. also Harvey

[Ha]), we show that the development classes of a category $\underline{K}$ (see [Ne]) are precisely those classes, E, for which there exists an M such that (E,M) is a diafactorization structure on $\underline{K}$ (Th.1.3). This answers the outstanding open problem of [HSV]. The proof of Theorem 1.3 also provides an alternative proof of the fact that for any (E,M) diafactorization structure, E must be a class of epimorphisms (cf. [HS_2], [T]). As a by-product of this theorem we also have, for any (E,M)-category $\underline{K}$, an internal characterization of all those C contained in E for which there exists a D such that (C,D) is a dispersed diafactorization structure (Th.1.9). It is interesting to note that such classes are (to within existence of the colimits) the "standard" classes of E-morphisms introduced in [S_1] and investigated further in [S_2].

In §2 we describe and investigate a Galois connection that makes precise the nature of the bijection discovered in [HSV]. Namely, the E-reflective subcategories of an (E,M)-category $\underline{K}$ and the (E,M)-dispersed diafactorization structures are precisely the Galois-closed classes and are complete lattices (in a suitably large universe) that are anti-isomorphic with each other (2.6(2)). General Galois results, as well as special properties involved, are used to investigate in more detail the structure of the complete lattice Q of all diafactorization structures (C,D) on $\underline{K}$ with C a subclass of E. In particular, it is shown that Q is partitioned into a family of complete lattices (called levels) (2.6(1)(i)) and that Q can also be viewed as a union of complete lattices (called images) all of which have a point in common and none of which meets any level non-trivially (2.6(1)(ii)).

§1. Characterization of Diafactorization Structures

1.1 Definitions and Notation

(1) In all that follows $\underline{K}$ will denote a category, and Mor $\underline{K}$, Iso $\underline{K}$ and Epi $\underline{K}$ will denote the classes of all morphisms, all isomorphisms and all epimorphisms of $\underline{K}$. All subcategories will be assumed to be full and isomorphism-closed.

(2) A $\underline{K}$-source with domain X is a pair $(X,(f_i)_I)$ where I is a class (possibly empty and possibly proper) and for each i in I f_i is a K-morphism with domain X.

(3) $\underline{K}$ is called an (E,M)-<u>category</u> and (E,M) is called a <u>diafactorization structure</u> on $\underline{K}$ provided that E is a class of $\underline{K}$-morphisms closed under composition with $\underline{K}$-isomorphisms and M is a conglomerate of $\underline{K}$-sources closed under composition with isomorphisms such that:

(a) $\underline{K}$ has the (E,M)-<u>factorization property</u>; i.e., every $\underline{K}$-source $(X,(f_i)_I)$ has a factorization $X\xrightarrow{f_i}Y_i = X\xrightarrow{e}Z\xrightarrow{m_i}Y_i$ where e belongs to E and $(Z,(m_i)_I)$ belongs to M, and

(b) $\underline{K}$ has the (E,M)-<u>diagonalization property</u>; i.e., whenever e and f are $\underline{K}$-morphisms and $(X,(m_i)_I)$ and $(Z,(h_i)_I)$ are $\underline{K}$-sources such that e is in E, $(X,(m_i)_I)$ is in M and for each i in I, $h_i e = m_i f$, then there exists a unique morphism $d:Z \to X$ such that $f = de$, and for each i in I, $h_i = m_i d$.

(*) 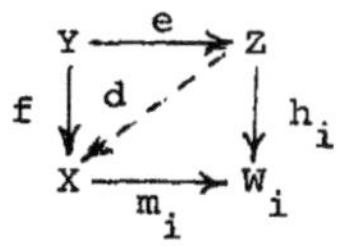

[If only (a) is satisfied, (E,M) is called a factorization structure on $\underline{K}$.]

(4) ([HSV]) If $\underline{K}$ is an (E,M)-category, then a diafactorization structure (C,D) on $\underline{K}$ is called (E,M)-dispersed iff there exists a subcategory $\underline{A}$ of $\underline{K}$ such that C is precisely all the $\underline{A}$-extendible morphisms in E;i.e.,$c:X \to Y$ is in C iff c is in E and for each $\underline{K}$-morphism $f:X \to A$ with A in $\underline{A}$ there is some $\underline{K}$-morphism $g:Y \to A$ such that $f = gc$.

(5) Let $E \subseteq \mathrm{Mor}\, \underline{K}$ then:

(a) $\alpha(E)$ will denote the conglomerate of all sources $(X,(f_i)_I)$ having the property that if $X \xrightarrow{f_i} Y_i = X \xrightarrow{e} Z \xrightarrow{h_i} Y_i$ is a factorization with e in E, then e is an isomorphism.

(b) $\Lambda(E)$ will denote the conglomerate of all sources $(X,(m_i)_I)$ having the property that if e is in E and if for each i in I the square (*) commutes, then there exists a unique $d:Z \to X$ such that for each i in I (*) commutes. (cf. $[S_1]$ and $[S_4]$)

(c) a $\underline{K}$-object Y is called an E-injective object iff for each $e:X \to Z$ in E and each $\underline{K}$-morphism $f:X \to Y$, there exists a $\underline{K}$-morphism $g:Z \to Y$ such that $f = ge$.

(d) $\underline{K}_{\underline{E}}$ is the category whose objects are members of E and whose morphisms $\hom_{\underline{K}_{\underline{E}}}(e,\hat{e})$ are pairs (f,g) where $ge = \hat{e}f$.

(e) $\Delta_0 : \underline{K}_{\underline{E}} \to \underline{K}$ is the functor defined by $\Delta_0(f,g) = f$. (cf. [Ho]).

(6) Let C and E be subclasses of $\mathrm{Mor}\, \underline{K}$ and let f, g and h be any $\underline{K}$-morphisms for which $h = gf$. Then C is said to be:

(a) iso-compositive iff h belongs to C whenever $\{f,g\} \subseteq C \cup \mathrm{Iso}\, \underline{K}$;

(b) left cancellative w.r.t. E iff whenever h belongs to C and f belongs to E, then f must belong to C;

(c) pushout prone iff

(i) every $\underline{K}$-source $(X,(c_i)_I)$ with each c_i in C has a multiple pushout $X \xrightarrow{c_i} Y_i \xrightarrow{d_i} Z$ with d in C; and

d

(ii) every 2-indexed $\underline{K}$-source, $(X,(k,c))$, with c in C has a pushout

$$\begin{array}{ccc} X & \xrightarrow{k} & Y \\ {\scriptstyle c}\downarrow & & \downarrow{\scriptstyle \hat{c}} \\ Z & \xrightarrow[\hat{k}]{} & W \end{array} \qquad \text{with } \hat{c} \text{ in } C.$$

(d) a <u>development class</u> (cf. [Ne][†]) iff

(i) $C \subseteq \mathrm{Epi}\ \underline{K}$,

(ii) C is iso-compositive, and

(iii) C is pushout prone;

(e) an E-<u>standard class</u> (cf. $[S_1]$, $[S_2]$[††]) iff

(i) C is a development class of E-morphisms, and

(ii) C is left cancellative w.r.t. E.

1.2 <u>Remark</u>. The following are some well-known properties of any diafactorization structure (E,M) on $\underline{K}$ that we will use in the sequel.

(1) E is iso-compositive.

(2) E and M determine each other: in fact $M = \Lambda(E)$.

We next obtain an improved version of the main theorem of Hoffmann [Ho] in that no conditions whatsoever are put on the category $\underline{K}$ or the class of $\underline{K}$-morphisms E. Some major steps of the proof, however, closely follow analogous steps in [Ho].

1.3 <u>Characterization Theorem for Diafactorization Structures</u>

For any category $\underline{K}$ and any class E of $\underline{K}$-morphisms, the following are equivalent:

(1) There exists a conglomerate M of $\underline{K}$-sources for which (E,M) is a diafactorization structure on $\underline{K}$.

(2) E is a development class.

(3) $(E,\Lambda(E))$ is a factorization structure on $\underline{K}$.

(4) $(E,\alpha(E))$ is a diafactorization structure on $\underline{K}$.

(5) The following hold: (a) E is iso-compositive;

(b) $\Delta_0 : \underline{K}_E \to \underline{K}$ is a topological functor[†††].

(†) In [Ne] Nel defined development classes somewhat less generally. Our formulation avoids his smallness condition.

(††) In $[S_1]$ and $[S_2]$ standard classes of epimorphisms are defined more generally, without the requirement of the existence of (multiple) pushouts in what corresponds to (6)(c).

(†††) A functor $F:\underline{A}\to\underline{X}$ is called <u>topological</u> iff each F-source $(X \xrightarrow{g_i} FA_i)_I$ has a factorization $(X \xrightarrow{g_i} FA_i) = (X \xrightarrow{r} FA \xrightarrow{Ff_i} FA_i)$ where r is an $\underline{X}$-isomorphism and $(A \xrightarrow{f_i} A_i)$ is an F-initial $\underline{A}$-source -- or, equivalently, every F-sink has an (F-final $\underline{A}$-sink, isomorphism)-factorization. (cf. $[He_3]$).

Proof: (4) $\Longrightarrow$ (1) and (1) $\Longleftrightarrow$ (3). Clear.

(1) $\Longrightarrow$ (5). 5(a) follows from 1.2(1). To show 5(b) let $S = (X \xrightarrow{f_i} \Delta_0 e_i)_I$ be a Δ_0-source. Then $(X, (e_i f_i)_I)$ is a $\underline{K}$-source which by (1) has an (E,M)-factorization $(X \xrightarrow{e} Y \xrightarrow{m_i} Z_i)_I$. Then $X \xrightarrow{1_X} \Delta_0 e \xrightarrow{\Delta_0(f_i, m_i)} \Delta_0 e_i$ is the (isomorphism, initial source)-factorization of S. (Initiality comes from (E,M)-diagonalization.)

(5) $\Longrightarrow$ (2). Since a topological functor must be faithful (see [He_3]), and since Δ_0 is topological, it is easily shown that $E \subseteq \text{Epi}\ \underline{K}$. It remains to be shown that E is pushout prone.

Let $(X, (e_i)_I)$ be a nonempty $\underline{K}$-source with each e_i in E and let $f: X \to Y$ be a $\underline{K}$-morphism. If for each i in I we let $f_i = f$, then we have the Δ_0-sink $(\Delta_0(e_i) \xrightarrow{f_i} Y)_I$. Let $\Delta_0(e_i) \xrightarrow{\Delta_0(g_i, h_i)} \Delta_0(e) \xrightarrow{r} Y$ be its (final $\underline{K}_E$-sink, isomorphism)-factorization. Then for each i in I the diagram

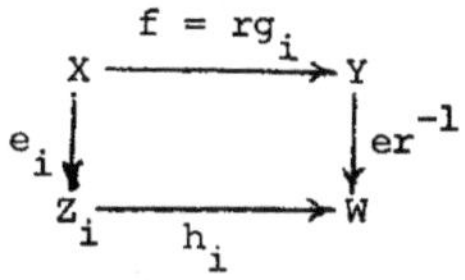

commutes; and by the finality of the sink $((g_i, h_i)_I, e)$, it follows that $((h_i)_{I \cup \{j\}}, W)$ is a colimit, where $h_j = er^{-1}$. By 5(a), er^{-1} is in E. Thus E is pushout prone.

(2) $\Longrightarrow$ (4). Suppose that $(X, (f_i)_I)$ is any $\underline{K}$-source. Since E is pushout-prone, the $\underline{K}$-source $(X, (e_j)_J)$ consisting of all those E-morphisms e_j which are first factors of each f_i, has a multiple pushout

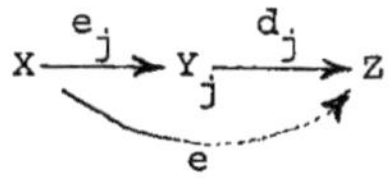

with e in E. Thus for each i in I there exists some m_i such that $f_i = m_i e$. Since E is iso-compositive and $E \subseteq \text{Epi}\ \underline{K}$, $(Z, (m_i)_I)$ is in $\alpha(E)$. Thus we have $(E, \alpha(E))$-factorizations. Suppose that e is in E, $(m_i)_I$ is in $\alpha(E)$, and f and h_i are such that for each i in I $h_i e = m_i f$. Since E is pushout-prone, the pushout $\hat{f}e = \hat{e}f$ can be formed with $\hat{e}$ in E. Thus there is a family of $\underline{K}$-morphisms k_i such that for each i the following diagram commutes. By the definition of $\alpha(E)$, $\hat{e}$ must be an isomorphism, so that $\hat{e}^{-1}\hat{f}$ is the needed diagonal. It is unique since e is an epimorphism.

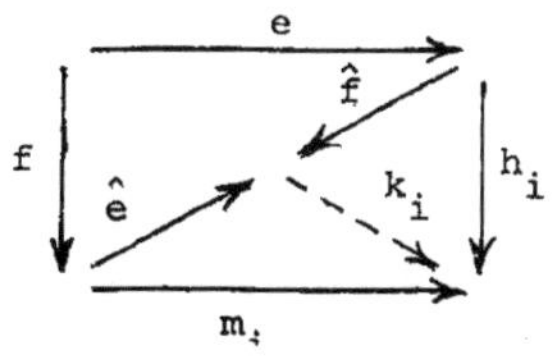

1.4 Corollary ([HS_2], [T]) If $\underline{K}$ is an (E,M)-category, then E must consist of epimorphisms in $\underline{K}$.

1.5 Corollary (cf. [Ho]) The conglomerate of all left factors of diafactorization structures on $\underline{K}$ is closed under arbitrary nonempty intersections. Thus if it has a largest member, it is a complete lattice (under the inclusion order).

1.6 Corollary. For any (E,M)-category $\underline{K}$ the subclasses C of E for which there exists a D such that (C,D) is a diafactorization structure are precisely the development subclasses of E.

1.7 Corollary. For any (E,M)-category $\underline{K}$ the conglomerate of all subclasses of E which are left factors of diafactorization structures on $\underline{K}$ is a complete lattice (under the inclusion order).

1.8 Remarks. (a) Corollary 1.4 shows that the requirement that the diagonal be unique can be deleted from the definition of diafactorization structure (1.1(3)).

(b) Corollary 1.6 answers the open problem 2.13 of [HSV].

(c) It should be noted that Theorem 3(3) of [S_4] is a forerunner of part of the following characterization theorem.

1.9 Characterization Theorem for Dispersed Diafactorization Structures

For any (E,M)-category $\underline{K}$ and $C \subseteq E$, the following are equivalent:

(1) There exists a conglomerate of $\underline{K}$-sources D for which (C,D) is an (E,M)-dispersed diafactorization structure.

(2) C is an E-standard class of morphisms.

(3) Every (C-injective)-extendible E-morphism belongs to C.

(4) C is the class of $\underline{A}$-extendible morphisms in E for some subcategory $\underline{A}$ of $\underline{K}$.

(5) The following hold:

  (a) The subcategory of C-injective objects is C-reflective.

  (b) C is left cancellative w.r.t. E.

(6) The following hold:

  (a) C is iso-compositive

  (b) C is left cancellative w.r.t. E.

  (c) $\Delta_0 : \underline{K}_C \longrightarrow \underline{K}$ is topological.

Proof: The equivalence of all but (2) and (6) is shown in [HSV] (Theorem 2.11).
(1) and (5) $\Longrightarrow$ (2). By (1) and Theorem 1.3, C is a development class; by (5) it is left cancellative w.r.t. E; and, thus, C is E-standard (1.1(6)(e)).
(2) $\Longleftrightarrow$ (6). Immediate from Theorem 1.3 and Definitions 1.1(6)(c), (d) and (e).

(2) $\Longrightarrow$ (5). By Theorem 1.3 we know that there exists some D such that (C,D) is a diafactorization structure. For any object X, let $X \xrightarrow{c} \hat{X} \xrightarrow{d_i} Y_i$ be the (C,D)-factorization of the source of all morphisms from X to C-injective objects. Diagonalization shows that $\hat{X}$ is C-injective. Thus, since $C \subseteq \text{Epi}\ \underline{K}$, $X \xrightarrow{c} \hat{X}$ is the C-reflection.

1.10 Corollary. For any (E,M)-category the conglomerate of all subclasses of E which are left factors of an (E,M)-dispersed diafactorization structure is a complete lattice (under the inclusion order).

1.11 Remark. In 2.6(2) an "external" proof of Corollary 1.10 is obtained.

§2 The Structure of All Diafactorization Structures

In this section, for any (E,M)-category $\underline{K}$, we investigate the structure of the conglomerate of all diafactorization structures (C,D) with $C \subseteq E$.

2.1 Definitions and Notation. If $(\mathcal{P},\leq)$ and $(\mathcal{Q},\precsim)$ are partially ordered classes and $F:\mathcal{P}\to\mathcal{Q}$ and $G:\mathcal{Q}\to\mathcal{P}$ are functions such that for each a in $\mathcal{P}$ and c in $\mathcal{Q}$ $c \precsim Fa$ iff $a \leq Gc$, then $((\mathcal{P},\leq), (\mathcal{Q},\precsim), F, G)$ is called a Galois connection. For each b in $G[\mathcal{Q}]$, $G^{-1}[b]$ is called the b-level of $\mathcal{Q}$, and if $(\mathcal{Q},\precsim)$ is a lattice, for each c in $\mathcal{Q}$ $F_c[\mathcal{P}] = \{Fa \wedge c \mid a \in \mathcal{P}\}$ is called the c-image of $\mathcal{P}$ in $\mathcal{Q}$.

Since for any diafactorization structure (C,D), $D = \Lambda(C)$ (1.2(2)), then by Corollary 1.6 for us to investigate the structure of all diafactorization structures with $C \subseteq E$ it is equivalent for us to investigate the structure of the conglomerate of all development subclasses of E. Throughout this section we will denote either of these conglomerates by Q. When thought of as development classes Q will be ordered by inclusion and when thought of as diafactorization structures it will be ordered by inclusion on the first elements of the pairs (C,D). We will also let P denote the conglomerate of all subcategories of $\underline{K}$, and define functions F and G as follows: For each subcategory $\underline{A}$ of $\underline{K}$: $F\underline{A}$ = all $\underline{A}$-extendible morphisms in E, and for any C in Q: GC = the full subcategory of $\underline{K}$ generated by all C-injective objects. N.B. In [HSV] $F\underline{A}$ is called the class of $\underline{A}$-concentrated morphisms and in [Ho] GC is called the germ of the diafactorization structure $(C,\Lambda(C))$.

2.2 Theorem. $((P,\subseteq), (Q,\subseteq), F, G)$ is a Galois connection on complete lattices.[†]

(†) Thus each pair $(C,\underline{A})$, where $GC = \underline{A}$ and $F\underline{A} = C$, is a regular injective structure of $\underline{K}$, in the sense of Maranda [Ma].

Proof: $(P,\subseteq)$ is clearly a complete lattice and $(Q,\subseteq)$ is one by Corollary 1.7. For any $\underline{A}$ in P, since E is a development class, it is straightforward to show that $F\underline{A}$ is a development class. Thus, by Theorem 1.3, $F\underline{A}$ belongs to Q. (That $F\underline{A}$ is a left factor has also been shown in [HSV] (Th 2.3).) Hence $F:P \to Q$. That $G:Q \to P$ is clear, and one immediately sees that for any subcategory $\underline{A}$ of $\underline{K}$ and class C of $\underline{K}$-morphisms, C is contained in the class of all $\underline{A}$-extendible E-morphisms iff each $\underline{A}$-object is C-injective. Thus the definition is satisfied.

2.3 Remark. In view of the Galois connection of Theorem 2.2 the general properties of Galois connections can be interpreted as corollaries. Before we do this we wish to first establish some special properties of the Galois connection at hand.

2.4 Proposition. For the Galois connection of Theorem 2.2:

(1) $F[P]$ = all E-standard classes of $\underline{K}$ (or, equivalently, all (E,M)-dispersed diafactorization structures of $\underline{K}$). Each development class of morphisms has a smallest E-standard class containing it. Equivalently, each (E,M)-diafactorization structure has a smallest (E,M)-dispersed diafactorization structure larger than it. The process of obtaining the "E-standard hull" (or "dispersed hull") is the Galois closure operator FG.

(2) $G[Q]$ = all E-reflective subcategories of $\underline{K}$. Each subcategory of $\underline{K}$ has an E-reflective hull and the process of obtaining E-reflective hulls is the Galois closure operator GF.

(3) The $\underline{K}$-level of Q has only one member, namely Iso $\underline{K}$, and this is the smallest member of Q and is in every image, $F_C[P]$, of P in Q.

(4) Each level in Q has a smallest member and is a complete lattice.

Proof: (1) That each $F\underline{A}$ is in left cancellative w.r.t. E is immediate. Thus each is an E-standard class (1.1(6)(e)) and so the left factor of an (E,M)-dispersed diafactorization structure (Th 1.9). The remainder follows from general Galois theory. ([MS]).

(2) For any $\underline{K}$-object X let $r_X:X \to \hat{X}$ be the multiple pushout of the source of all C-morphisms with domain X. With respect to GC, r_X is a reflection morphism in E; thus GC is E-reflective. Conversely, suppose that $\underline{B}$ is E-reflective in $\underline{K}$ and Y is a $GF\underline{B}$-object. Since the $\underline{B}$-reflective E-map $r_Y:Y \to \hat{Y}$ is in $F\underline{B}$, there is a g such that $gr_Y = 1_Y$. Thus r_Y is an isomorphism, and Y belongs to $\underline{B}$. Hence $GF\underline{B} \subseteq \underline{B}$. Since, as always, $\underline{B} \subseteq GF\underline{B}$, $\underline{B}$ is in $G[Q]$. So $G[Q]$ = all E-reflective subcategories of $\underline{K}$, and since for any $\underline{A}$ in P, $GF\underline{A}$ is the smallest member of $G[Q]$ containing $\underline{A}$, it must be the E-reflective hull.

(3) Clearly Iso $\underline{K}$ is the smallest development class (1.1(6)(d)) and $F_C\underline{K}$ = Iso $\underline{K}$. Suppose that GC = $\underline{K}$. Then every member of C must be a section. But since

$C \subseteq E \subseteq \text{Epi } \underline{K}$ (Th. 1.3), we have that $C \subseteq \text{Iso } \underline{K}$.

(4) For any E-reflective subcategory $\underline{B}$ and any object X in $\underline{K}$ let r_X be its $\underline{B}$-reflection map. Then by the construction of the reflection maps, (2), r_X is in C for each C in the $\underline{B}$-level. Thus the meet of the $\underline{B}$-level belongs to it. So the $\underline{B}$-level is complete.

2.5 Remarks. (a) It should be mentioned that a Galois connection in a more restrictive setting ([He_2], [S_1] and [S_2]) has previously been used to obtain epireflective hulls as Galois closures.

(b) That each $\underline{B}$-level has a smallest member has also essentially been shown by Hoffmann [Ho] (Prop. 2.5). Notice that general Galois theory gives a largest member for each level. This is Th. 2.3 of [Ho].

A summary of some of the properties that follow from the results of this section as well as the general Galois theory ([MS]) follows.

2.6 Summary and Sample Results. (1) Q is a complete lattice that can be viewed as a union of complete lattices in two ways:

(i) as the disjoint union of complete lattices, called levels such that in each level the join or meet of any nonempty family is its join or meet in Q, and such that the level with the smallest member of Q is a singleton.

(ii) as the union of complete lattices, called images, all of which have the smallest member of Q in common and are such that when they intersect they coincide from any common point on down. Furthermore all non-empty meets in images are the same as the corresponding meets in Q.

(2) The conglomerates of all E-reflective subcategories of $\underline{K}$ and all (E,M)-dispersed diafactorization structures on $\underline{K}$ are anti-isomorphic complete lattices. For each development class $C \subseteq E$, we have C-relativizations of all above results (C-reflective hulls, etc.). Thus, in particular, we have the result from [HSV] that, the E-reflective subcategories of K are in one-to-one correspondence with the (E,M)-dispersed diafactorization structures.

(3) Each subcategory of $\underline{K}$ has an E-reflective hull, each development class has an E-standard hull, each (E,M)-diafactorization structure has an (E,M)-dispersed hull, and each conglomerate of development subclasses of E has a development hull (the join of it in Q).

(4) The dispersed hull of any (C,D) in Q is the largest (E,M)-diafactorization structure $(\hat{C},\hat{D})$ for which the $\hat{C}$-injective objects are the C-injective objects.

(5) Let H be the family of all subcategories of $\underline{K}$ the E-reflective hulls of which are $\underline{B}$. Then $H \cup \{\bigcap H\}$ is a complete lattice with largest member $\underline{B}$ and smallest member $\bigcap H$.

(6) If C and $\hat{C}$ are development classes with $C \subseteq \hat{C}$, then there is a unique $\hat{C}$-standard class, C*, such that the C-injective objects and the C*-injective

objects coincide.

(7) In Q the $\underline{B}$-level and the C-image intersect iff $\underline{B}$ is a C-reflective subcategory of $\underline{K}$. If they do intersect, their intersection is a singleton whose member is $(C,\Lambda(C))$-dispersed.

2.7 Remark. Finally we provide a sketch of what the complete lattice Q might look like according to some of its properties that we have obtained. Note that it can be a very large lattice indeed. In the case $\underline{K} = \underline{\text{Top}}$ and $E = \text{Epi}\ \underline{K}$, it is non-legitimate ([HSV] Th 3.2). All of the nearly-horizontal lines represent levels; these are ordered by their top points, which represent the dispersed diafactorization structures (or E-standard classes). All of the nearly-vertical lines represent images. Images that meet in a point are identical from that point down and they all meet at the $\underline{K}$-level, the lowest level, which is a singleton. Points of Q only occur at the intersections of levels and images, and each such intersection has at most one point.

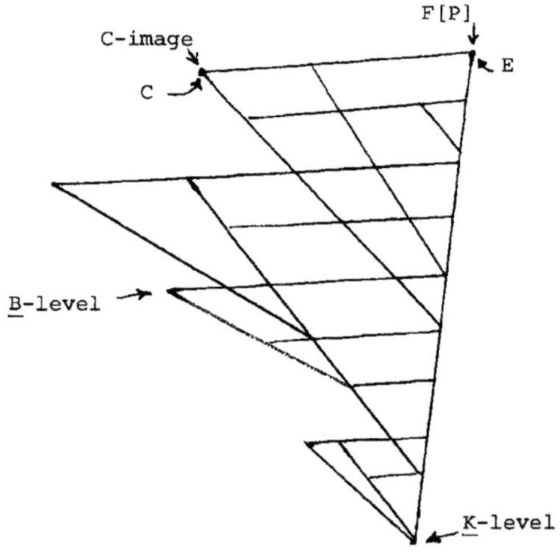

REFERENCES

[C] P. J. Collins, Concordant mappings and the concordant-dissonant factorization of an arbitrary continuous function, Proc. Amer. Math. Soc. 27 (1971), 587-591.

[Ha] J. M. Harvey, Topological functors from factorization (Proc. Int. Conf. Berlin 1978), Springer Lecture Notes in Math. 719 (1979) 102-111.

[He_1] H. Herrlich, A generalization of perfect maps (Proc. Third Prague Topological Symposium. 1971), General Topology and Its Relations to Modern Analysis and Algebra III, Academia, Prague (1972) 187-191.

[He_2] __________, Perfect subcategories and factorizations (Proc. Colloq. Karzthely, Topics in Topology). Colloq. Math. Soc. Janos Bolyai, 8, North Holland, Amsterdam (1974) 387-403.

[He_3] __________, Topological functors, General Topology and Appl. 4 (1974) 125-142.

[HSV] H. Herrlich, G. Salicrup, and R. Vázquez, Dispersed factorization structures, Can. J. Math. 31 (1979) 1059-1071.

[HS_1] H. Herrlich and G. E. Strecker, Category Theory, 2nd ed., Berlin: Heldermann-Verlag 1979.

[HS_2] ______________________________, Semi-universal maps and universal initial completions, Pacific J. Math. 82 (1979) 407-728.

[Ho] R.-E. Hoffmann, Factorization of cones, Math. Nachr. 87 (1979) 221-238.

[Ma] J. M. Maranda, Injective structures, Trans. Amer. Math. Soc. 110 (1964) 98-135.

[Me] A. Melton, Which dispersed diafactorization structures on Top are hereditary?, General Topology and Modern Analysis, Academic Press, New York (1981) 281-290.

[MS] A. Melton and G. E. Strecker, Structures of Galois connections, preprint.

[Na] R. Nakagawa, Relations between two reflections, Sci. Rep. Tokyo Kyoiku Daigaku Sect. A, 12 (1973) 80-88.

[Ne] L. D. Nel, Development classes: An approach to perfectness, reflectiveness and extension problems (Proc. Second Pittsburgh Internat. Conf., TOPO 72, General Topology and its Applications), Springer Lecture Notes in Math. 378 (1974) 322-340.

[P] G. Preuss, On factorization of maps in topological categories, preprint.

[SV_1] G. Salicrup and R. Vázquez, Reflexividad y coconexidad en Top, An. Inst. Mat. Univ. Nac. Autónoma Mexico, 14 (1976) 159-230.

[SV_2] __________________________, Reflectivity and connectivity in topological categories, preprint.

[S_1] G. E. Strecker, Epireflection operators vs perfect morphisms and closed classes of epimorphisms, Bull. Austral. Math. Soc. 7 (1972) 359-366.

[S_2] _________, On characterizations of perfect morphisms and epireflective hulls (Proc. Second Pittsburgh Internat. Conf., TOPO 72, General Topology and its Applications), Springer Lecture Notes in Math. 378 (1974) 468-500.

[S_3] ________, Component properties and factorizations, Topological Structures, Math. Centre Tract 52, Mathematisch Centrum, Amsterdam (1974) 123-140.

[S_4] ________, Perfect sources (Proc. First Categorical Topology Symposium), Springer Lecture Notes in Math. 540 (1976) 605-624.

[T] W. Tholen, Semi-topological functors I, J. Pure Appl. Algebra 15 (1979) 53-73.

A. Melton
Department of Computer Science
Wichita State University
Wichita, Kansas 67208
U.S.A.

G. E. Strecker
Department of Mathematics
Kansas State University
Manhattan, Kansas 66506
U.S.A.

A REMARK ON SCATTERED SPACES

Adam Mysior, Gdańsk, Poland

1. Introduction

A topological space is called <u>scattered</u> if each of its subsets has a relatively isolated point. It is clear that every scattered space is <u>hereditarily</u> <u>disconnected</u> i.e. it does not contain any non-trivial connected subspaces. Z.Semadeni [9] raised a long open problem whether every scattered completely regular space is zerodimensional. This problem was solved negatively by R.C.Solomon [10] . It is well known that a topological space is zerodimensional if and only if it can be homeomorphically embedded into some topological power of a two-point discrete space. Therefore a natural generalization of Semadeni's problem is the following one :

Is there a single scattered completely regular space E such that every scattered completely regular space can be embedded into some topological power of the space E ?

It turns out that the answer is negative. Namely, we get the following

THEOREM. For every cardinal $\mathfrak{m}$ there is a scattered completely regular space which cannot be embedded into any topological product of hereditarily disconnected spaces of cardinality $\leqslant \mathfrak{m}$.

2. Proof

Let $\mathfrak{m}$ be an arbitrary cardinal.

We construct an example of a scattered completely regular space X containing two distinct points p and q such that $f(p) = f(q)$ for every hereditarily disconnected space E of cardinality $\leqslant \mathfrak{m}$ and every continuous function $f : X \to E$. It is clear that such a space X cannot be embedded into any topological product of hereditarily disconnected spaces of cardinality $\leqslant \mathfrak{m}$.

The construction of the space X is based on the following lemma which was proved in [7] .

Lemma. Let M be a metrizable space with weight w and cardinality 2^w. Then there exists a space $\hat{M}$ such that

(i) M and $\hat{M}$ have the same underlying sets and the topology of $\hat{M}$ if finer (= more open sets) then the topology of M,

(ii) $\hat{M}$ is zerodimensional and moreover, each point of $\hat{M}$ has a countable compact clopen neighborhood,

(iii) for any $A, B \subset \hat{M}$, if $cl_{\hat{M}} A \cap cl_{\hat{M}} B = \emptyset$, then $|cl_M A \cap cl_M B| < 2^w$.

Observe that it follows from (ii) that the space $\hat{M}$ is scattered and completely regular.

To construct the space X take an arbitrary cardinal w such that $m \leqslant w$ and $w^{\aleph_0} = 2^w$ (see [10] Ch.8.15) and a Hilbert space H with an orthonormal base of cardinality w. H is a metric space with weight w and cardinality 2^w, all non-empty open subsets of H have cardinality 2^w and H cannot be disconnected by a set of cardinality less than 2^w. The proof of the latter is very much like for the fact that Euclidean plane cannot be disconnected by a set of cardinality less than $2^{\aleph_0}$.

Choose now two distinct points $p, q \in H$ and put $M = H - \{p,q\}$. Define a new topology on H generated by all sets open in H and all subsets of M open in $\hat{M}$. Denote the obtained space by X.

The space X is the required example. It is easy to check that X is scattered and completely regular. Let E be an arbitrary hereditarily disconnected space of cardinality $\leqslant m$ and let $f : X \to E$ be an arbitrary continuous function. We prove that $f(p) = f(q)$.

Denote by $\mathcal{Z}$ the family of all closed subspaces Z of X such that $|X - Z| < 2^w$.

Observe that all open in X neighborhoods of p or q have cardinality 2^w. Hence $p, q \in Z$ for every $Z \in \mathcal{Z}$. We prove that $f^{-1}(f(p)) \in \mathcal{Z}$. It follows that $q \in f^{-1}(f(p))$ and $f(p) = f(q)$.

The following property of the space X is the crucial point of our argumentation.

(*) If $Z \in \mathcal{Z}$ and U is a clopen in Z neighborhood of p then $U \in \mathcal{Z}$.

It is clear that the set U is closed in X. We prove that $|X - U| < 2^w$. Since $|X - Z| < 2^w$ it suffices to prove that $|Z - U| < 2^w$. Since $cl_{\hat{M}}(U \cap M) \cap cl_{\hat{M}}(Z \cap M - U) = \emptyset$ it follows from Lemma that

$|cl_M(U \cap M) \cap cl_M(Z \cap M - U)| < 2^{m}$. Hence $|cl_H U \cap cl_H(Z - U)| < 2^{m}$. On the other hand - since every non-empty open subset of H has cardinality 2^{m} - $cl_H U \cup cl_H(Z - U) = cl_H Z = H$. But the space H cannot be disconnected by a set of cardinality less than 2^{m}. Hence

$$cl_H U = H \quad \text{and} \quad |Z - U| < 2^{m}$$

or

$$cl_H(Z - U) = H \quad \text{and} \quad |U| < 2^{m}.$$

The latter is impossible - it follows from the definition of the topology of X that there is an open in H neighborhood V of p such that $V \cap Z \subset U$; since $|V| = 2^{m}$ and $|X - Z| < 2^{m}$ we have $|V \cap Z| = 2^{m}$ and $|U| = 2^{m}$. Hence $|Z - U| < 2^{m}$ and the property (*) is proved.

Define now, by transfinite recursion, a family $\{E_\alpha : \alpha < m^+\}$ of closed subspaces of E such that $E_0 = E$ and for every $\alpha < m^+$ $E_\alpha = \bigcap\{U : f(p) \in U$ and U is a clopen subset of some E_β with $\beta < \alpha\}$. Since $|E| \leqslant m$ and the family $\{E_\alpha : \alpha < m^+\}$ is decreasing there is an ordinal $\alpha_0 < m^+$ such that $E_\alpha = E_{\alpha_0}$ for every ordinal α with $\alpha_0 \leqslant \alpha < m^+$. It follows from the hereditary disconnectedness of E that $E_{\alpha_0} = \{f(p)\}$. Therefore to prove that $f^{-1}(f(p)) \in \mathcal{X}$ it suffices to show that $f^{-1}(E_\alpha) \in \mathcal{X}$ for every ordinal $\alpha < m^+$.

We proceed by transfinite induction. It is obvious that $f^{-1}(E_0) \in \mathcal{X}$. Assume that $f^{-1}(E_\beta) \in \mathcal{X}$ for all ordinals $\beta < \alpha$. We prove that $f^{-1}(E_\alpha) \in \mathcal{X}$. Since $|E| \leqslant m$ there is a family $\mathcal{U}$ of subsets of E such that $|\mathcal{U}| \leqslant m$, $E_\alpha = \bigcap\{U : U \in \mathcal{U}\}$ and every $U \in \mathcal{U}$ is a clopen subset of some E_β with $\beta < \alpha$. It follows from the continuity of f and the property (*) that $f^{-1}(U) \in \mathcal{X}$ for every $U \in \mathcal{U}$. Since $f^{-1}(E_\alpha) = \bigcap\{f^{-1}(U) : U \in \mathcal{U}\}$ and $|\mathcal{U}| \leqslant m$ we have $f^{-1}(E_\alpha) \in \mathcal{X}$. This ends the proof.

3. Categorical conclusions

Denote by Top the category of all topological spaces and continuous functions. It is known (see e.g. [5] Ch.X.37) that a full isomorphism-closed subcategory of Top is epi-reflective if and only if it is closed with respect to the formation of product and subobjects. Every subcategory $\underline{E}$ is contained in a smallest epi-reflective subcategory, its epi-reflective hull, which we denote by Top$\underline{E}$.

An epi-reflective subcategory $\underline{A}$ is said to be simple provided that there is a single space E such that $\underline{A}$ = Top$\{E\}$.

Among simple subcategories are for example the categories of all topological spaces, T_0-spaces, completely regular spaces, zerodimensional spaces and the category Top developable spaces ([3], [8]). On the other hand there is a large number of theorems stating that various subcategories are not simple (see e.g. [1], [2], [4], [6], [8]). Our THEOREM can be considered as a new result in this direction. Namely, it yields the non-simplicity of the category Top{scattered completely regular spaces} and - moreover - the non-simplicity of any subcategory lying between Top{scattered completely regular spaces} and the category of all hereditarily disconnected spaces.

References

1. D.W. Hajek and A. Mysior, On non-simplicity of topological categories, Proc. Categorical Topology Conf. Berlin 1978, Springer Lecture Notes in Math. 719 (1979) 84-93.

2. D.W. Hajek and R.G. Wilson, The non-simplicity of certain categories of topological spaces, Math.Z. 131 (1973) 357-359.

3. N.C. Heldermann, The category of $\mathcal{D}$-completely regular spaces is simple, Trans. Amer. Math. Soc. 262 (1980) 437-446.

4. H. Herrlich, Wann sind alle stetigen Abbildungen in Y konstant ?, Math. Z. 90 (1965) 152-154.

5. H. Herrlich and G.E. Strecker, Category theory, Boston 1973.

6. S. Mrówka, On universal spaces, Bull. Acad. Polon. Sci. 4 (1956) 479-481.

7. A. Mysior, The category of all zerodimensional realcompact spaces is not simple, Gen. Topology Appl. 8 (1978) 259-264.

8. A. Mysior, Two remarks on $\underline{D}$-regular spaces, Glasnik Mat. 15 (1980) 153-156.

9. Z. Semadeni, Sur les ensembles clairisemés, Dissertationes Math. 19 (1959) 1-39.

10. W. Sierpiński, Cardinal and ordinal numbers, Warsaw 1965.

11. R.C. Solomon, A scattered space that is not 0-dimensional, Bull. London Math. Soc. 8 (1976) 239-240.

Institute of Mathematics
University of Gdańsk
80952 Gdańsk
Poland

Bornological L_1-functors as Kan extensions and Riesz-like representations

L.D. Nel

The setting for this study is formed by the following closed categories:

C = (complete bornological vector spaces, bounded linear maps),

B = (Banach spaces, bounded linear maps),

A = (Banach spaces, bounded linear maps with norm at most 1).

X will be a measure space which is the union of subspaces with finite measure. The functor $L_1(X,-) : A \to A$ was shown recently [7] to have an A-right adjoint $M(X,-)$. Our main purpose here is to extend this to a C-adjunction

$$(*) \qquad [L_1(X,E), F] \simeq [E, M(X,F)] \qquad (E,F \text{ in } C),$$

thus to obtain a far reaching generalization of the classical Riesz representation for operators on L_1-spaces. For orientation we mention that $M(X,B)$ reduces to the better known Banach space $L_\infty(X,B)$ if and only if the Banach space B has the Radon-Nikodym property [7].

Our stepping stones toward the above goal evolved into results of independent interest. We establish that every functor $\Phi:B \to C$ has a left Kan extension $\mathrm{Lan}\Phi$: $C \to C$ along the inclusion $T:B \to C$. Surprisingly, the Lan of a composite $T\Omega$ with B-right adjoint Ω preserves the Lan of every such B-right adjoint composite. Using these facts, we show that every B-adjunction $\Gamma \dashv \Omega:B \to B$ implies a C-adjunction $\mathrm{Lan}T\Gamma \dashv \mathrm{Lan}T\Omega : C \to C$. The promised adjunction (*) becomes a special case.

We also obtain a related C-adjunction $\ell_1(X,-) \dashv m(X,-) : C \to C$ by applying the procedures of [7] to the C-adjunction $\ell_1(V,-) \dashv \ell_\infty(V,-) : C \to C$ that we derived recently in [6] (here V is a bornological set).

About C and categories of completant disks

For an introduction to complete bornological vector spaces (cbv-spaces) and their importance in Functional Analysis, see [3]. The category C of these spaces was shown in [6] to be autonomously algebraic over the cartesian closed topological category Bo (bornological sets). As such, C is complete, cocomplete, tensored and cotensored over Bo, has well behaved (regular epi, mono)-factorizations and it carries a Bo-related closed symmetric monoidal structure ($[-,-]$, $\otimes$, K, ...) [2]

2

where the monoidal unit K is of course the scalar field (real or complex).

Every Banach space B has an underlying cbv-space TB. The obvious associated functor

$$T: \mathcal{B} \to \mathcal{C}$$

is full and faithful; moreover, it preserves and reflects the closed monoidal structure. One thinks of $\mathcal{B}$ as a full subcategory of $\mathcal{C}$, closed under formation of hom-objects and tensor products (i.e. projective norm completed tensor products).

Recall [3] that every cbv-space E comes equipped with a specified family E-$\mathcal{Disk}$ of *completant disks* i.e. disks D contained in E such that the vector subspace E_D spanned by D becomes a Banach space when normed with the gauge of D. These D are then the basic bounded sets of E. Let us turn E-$\mathcal{Disk}$ into a category by taking the D as objects and all inclusion maps D → D' among them as morphisms. It is easy (in terms of elementary facts [3]) to verify that E-$\mathcal{Disk}$ is *small, filtered* and a *preorder* (definitions in [5]). The functor

$$E_{(-)}: E\text{-}\mathcal{Disk} \to \mathcal{C}$$

is defined to carry (D → D') to the inclusion ($TE_D \to TE_{D'}$). The important known fact that every cbv-space E is the "inductive limit" of its associated Banach spaces E_D, can now be restated in the following technically precise manner:

1. Remark. The inclusion maps $TE_D \to E$ build a colimit cone for the functor $E_{(-)}$.

2. Lemma. Suppose $\Lambda: \mathcal{P} \to \mathcal{Q}$ is a functor such that $\mathcal{P}$ is filtered and for every object Q in $\mathcal{Q}$ the comma category $(Q{\downarrow}\Lambda)$ is a non-empty preorder (i.e. for some P' there exists $Q \to \Lambda P'$ and for every P in $\mathcal{P}$ there is at most one morphism $Q \to \Lambda P$). Then Λ is a final functor (in the sense of [5]).

Proof. Routine verification.

3. Lemma. The functor

$$\mathrm{Span}: E\text{-}\mathcal{Disk} \to (T{\downarrow}E)$$

defined to carry (D → D') to ($TE_D \to TE_{D'} \to E$) (all inclusion maps), is a final functor.

3

Proof. An object $(f:TA \to E)$ of $(T\downarrow E)$ has, qua $\mathcal{C}$-morphism, a (regular epi, mono)-factorization [5], say $TA \to E' \to E$. Since the domain of f underlies a Banach space, so does its image E' (proposition 3:1(2) in [3]). Accordingly, $E' = TA'$ (say) and we can rewrite the above factorization as $TE_D \to TE_{D'} \to E$, where D and D' are the closed unit disks of A and A'. This factorization represents a $(T\downarrow E)$-morphism $(f:TA \to E) \to \mathrm{Span}D'$. Thus $(f\downarrow \mathrm{Span})$ is non-empty and it is clearly a preorder. Lemma 2 ends the proof.

Left Kan Extensions

The left Kan extensions to be considered will always be along T, so we will write $\mathrm{Lan}\Phi$ for $\mathrm{Lan}_T\Phi$. For background we refer to [5] and [1]. The theorem to follow calls for $\mathcal{C}$-enrichment for its own sake; however, for the present purpose of extending a $\mathcal{B}$-adjunction to a $\mathcal{C}$-adjunction, the given $\mathcal{S}et$-based version is adequate because $\mathcal{C}$-adjoint functors are automatically $\mathcal{C}$-functors.

4. Theorem. Every functor $\Phi:\mathcal{B} \to \mathcal{C}$ has a left Kan extension $\mathrm{Lan}\Phi$ such that $(\mathrm{Lan}\Phi)T = \Phi$.

Proof. Once existence is proved, the equation $(\mathrm{Lan}\Phi)T = \Phi$ rather than extension up to natural isomorphism will hold because T is full and faithful. For existence it is enough to show that for every E in $\mathcal{C}$ the composite

$$(T\downarrow E) \xrightarrow{T_E} \mathcal{B} \xrightarrow{\Phi} \mathcal{C}$$

has a colimit where T_E is the underlying functor $(TA \to TB \to E) \mapsto (TA \to TB)$. Since Span is final (lemma 3), this can be achieved by showing that the composite $\Phi T_E\,\mathrm{Span}$ has a colimit. But the latter is clear because $E\text{-}\mathcal{D}isk$ is small and $\mathcal{C}$ is cocomplete.

5. Remark. Since the composite

$$E\text{-}\mathcal{D}isk \xrightarrow{\mathrm{Span}} (T\downarrow E) \xrightarrow{T_E} \mathcal{B} \xrightarrow{T} \mathcal{C}$$

is just $E_{(-)}$, we have

$$\mathrm{Lan}\Phi E = \mathrm{colim}\,\Phi T_E = \mathrm{colim}\,\Phi E_{(-)}.$$

This colimit cone can be obtained by applying $\mathrm{Lan}\Phi$ to the cone of inclusions $TE_D \to E$ (D in $E\text{-}\mathcal{D}isk$).

Since right adjoints do not usually preserve colimits or left Kan extensions, the following useful fact comes as a pleasant surprise.

6. **Theorem.** The left Kan extension $\Lambda = \mathrm{Lan}T\Psi$, where Ψ is a $\mathcal{B}$-right adjoint $\mathcal{B} \to \mathcal{B}$ preserves the left Kan extension of $T\Omega$ for every $\mathcal{B}$-right adjoint $\Omega : \mathcal{B} \to \mathcal{B}$ i.e.

$$\Lambda \mathrm{Lan}T\Omega \simeq \mathrm{Lan}\Lambda T\Omega .$$

In particular, for every Banach space B and every cbv-space E we have (cf. remark 5):

(6a) $$[TB,\ E] \simeq \mathrm{colim}[TB,\ E_{(-)}] \quad \text{and}$$

(6b) $$[TB,\ \mathrm{Lan}T\Omega-] \simeq \mathrm{Lan}[TB,\ T\Omega-].$$

Proof. We first establish just the special case (6a). Define the functor $[B,-)$: $E\text{-}\mathcal{D}isk \to [TB,\ E]\text{-}\mathcal{D}isk$ by putting

$$[B,D) = \{h \in [TB,\ E] \mid h(x) \in D \text{ whenever } |x| \leq 1\}.$$

That the disk $[B,D)$ is completant is clear frcm the fact that it spans the Banach space $[TB,TE_D]$. As cbv-space $[TB,\ E]$ carries the natural bornology [3], for which the disks $[B,D)$ form a base, by definition. This means every object Q in $[TB,E]\text{-}\mathcal{D}isk$ admits at least one morphism $Q \to [B,D)$. Hence, by lemma 2, $[B,D)$ is a final functor. Observe now that we have a commutative diagram

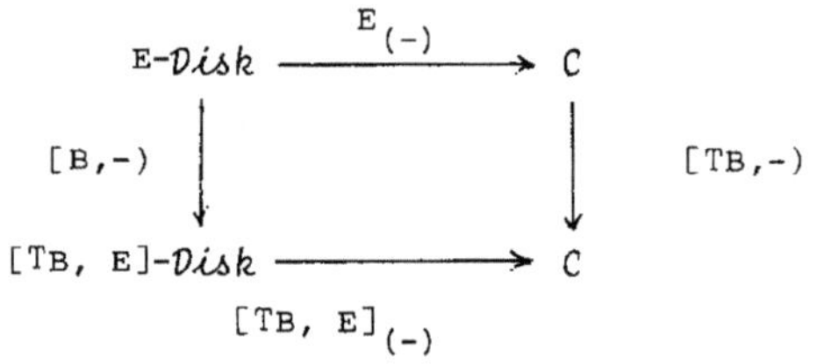

Using remark 1 and finality of $[B,-)$ we conclude that $[TB,\ E] \simeq \mathrm{colim}[TB,\ E]_{(-)} \simeq \mathrm{colim}[TB,\ E]_{(-)} \circ [B,-) \simeq \mathrm{colim}[TB,-] \circ E_{(-)}$. Thus we have (6a). Let us now pursue special case (6b). Using (6a) and remark 5, we conclude

(6c) $$[TB,\ -] \simeq \mathrm{Lan}[TB,T-].$$

Choose Γ so that $\mathcal{B}[\Gamma A,\ B] \simeq \mathcal{B}[A,\ \Omega B]$ holds in $\mathcal{B}$, natural in A and B. Then for $A \simeq K$ (the monoidal unit) we have $\Omega \simeq \mathcal{B}[\Gamma K,-]$, hence

5

(6d) $$T\Omega = [T\Omega K,\ T-].$$

We are now able to calculate as follows:

$$\begin{aligned}
[TB,\ \mathrm{Lan}T\Omega-] &= [TB,\ \mathrm{Lan}[T\Gamma K,\ T-]] && \text{(6d)}\\
&\simeq [TB,\ [T\Gamma K,\ -]] && \text{(6c)}\\
&\simeq [TB \otimes T\Gamma K,\ -\ ] && \text{(exponential law)}\\
&\simeq [T(B \otimes \Gamma K),\ -] && \text{(good nature of } T\text{)}\\
&\simeq \mathrm{Lan}[T(B \otimes \Gamma K),\ T-] && \text{(6c)}\\
&\simeq \mathrm{Lan}[TB,\ [T\Gamma K,\ T]] && \text{(retracing above steps)}\\
&\simeq \mathrm{Lan}[TB,\ T\Omega-] && \text{(6d)}.
\end{aligned}$$

This gives us (6b). Finally, we observe that every functor of the form $\Lambda = \mathrm{Lan}T\Psi$ is naturally isomorphic to a functor of the form $[TB,\ -]$, by (6d) and (6c). Thus the general statement of the theorem follows from the special case (6b).

Extending $\mathcal{B}$-adjunctions to $\mathcal{C}$-adjunctions

7. **Theorem.** If $\Gamma:\mathcal{B} \to \mathcal{B}$ is $\mathcal{B}$-left adjoint to Ω, then then $\mathrm{Lan}T\Gamma:\mathcal{C} \to \mathcal{C}$ is $\mathcal{C}$-left adjoint to $\mathrm{Lan}T\Omega$.

Proof. By hypothesis we have for every A in $\mathcal{B}$ isomorphic functors

$$[T\Gamma A,\ T-] \simeq [TA,\ T\Omega-].$$

Taking left Kan extensions and using theorem (6b), we obtain

$$\mathrm{Lan}[T\Gamma A,\ T-] \simeq [T\Gamma A,\ \mathrm{Lan}T-] \simeq [T\Gamma A,\ -] \qquad \text{and}$$

$$\mathrm{Lan}[TA,\ T\Omega-] \simeq [TA,\ \mathrm{Lan}T\Gamma-].$$

Hence $$[T\Gamma A,\ F] \simeq [TA,\ (\mathrm{Lan}T\Gamma)F]$$

natural for A in $\mathcal{B}$ and F in $\mathcal{C}$. Since the functors $[-,\ F]$ and $[-,\ (\mathrm{Lan}T\Omega)F]$ are left adjoint functors, they preserve left Kan extensions. It follows that

$$[(\mathrm{Lan}_T\Gamma)E, F] \simeq [E, (\mathrm{Lan}_T\Omega)F]$$

natural for E and F in $\mathcal{C}$, which is the desired $\mathcal{C}$-adjunction.

Let us now turn to the special $\mathcal{C}$-adjunction promised at the beginning. We define the functors

$$L_1(X,-) : \mathcal{C} \to \mathcal{C} \quad \text{and} \quad M(X,-) : \mathcal{C} \to \mathcal{C}$$

respectively as the left Kan extensions of the composites

$$\mathcal{B} \xrightarrow{L_1(X,-)} \mathcal{B} \xrightarrow{T} \mathcal{C} \quad \text{and} \quad \mathcal{B} \xrightarrow{M(X,-)} \mathcal{B} \xrightarrow{T} \mathcal{C}$$

8. Theorem. $[L_1(X,E), F] \simeq [E, M(X,F)]$ in $\mathcal{C}$, naturally in E and F.

Proof. The $\mathcal{A}$-adjunction of $L_1(X,-)$ and $M(X,-)$ established in [7] extends trivially, as any $\mathcal{A}$-adjunction does, to a $\mathcal{B}$-adjunction, since every $\mathcal{B}$-morphism can be scaled down to an $\mathcal{A}$-morphism and scaled up to itself again. This enables us to deduce the result at once from theorem 7.

In [7] we gave an explicit realization of the space $M(X,B)$ (B in $\mathcal{B}$) and also an explicit formula for the natural isomorphism $[L_1(X,A), B] \simeq [A, M(X,B)]$, which was shown to generalize the classical Riesz formula. These extend in a straight forward manner to the present setting, since everything can ultimately be reduced to Banach spaces. The space $L_1(X,E)$ introduced above, for example, is the filtered colimit of the Banach spaces $L_1(X,E_D)$ (cf. remark 1). Similarly, $M(X,F)$ is the filtered colimit of the Banach spaces $M(X,F_D)$ and $[L_1(X,E), F]$ that of the Banach spaces $[L_1(X,E_D), F_{D'}]$.

In view of the characterization of enriched right adjoints as cotensor-preserving ordinary right adjoints [4] and dually, the following is an immediate consequence of theorem 8.

9. Corollary. $L_1(X, E \otimes F) \simeq L_1(X,E) \otimes F$ and $M(X, [E,F]) \simeq [E, M(X,F)]$ in $\mathcal{C}$, naturally in E and F.

This corollary gives a further illustration of the useful functorial calculus which is inherent in enriched category theory.

7

Another Riesz-like representation

It was shown in [7] that $L_1(X,A)$ can be represented as a filtered colimit in A of a system of spaces $\ell_1(Q,A)$, where Q varies in a certain category Tes of tessellations arising from the measure space X, while $M(X,B)$ is a projective limit of an associated system of spaces $\ell_\infty(Q,B)$. In [6] we introduced cbv-spaces $\ell_1(V,E)$ and $\ell_\infty(V,F)$ which generalize the Banach spaces $\ell_1(S,A)$ and $\ell_\infty(S,B)$; actually, $\ell_1(V,E)$ is the tensor product of the bornological set V with E in C while $\ell_\infty(V,E)$ is the cotensor of such V with E, so that

$$[\ell_1(V,E),\ F] \simeq [E,\ell_\infty(V,F)]$$

naturally for E and F in C. By departing from this natural isomorphism and making V vary through the tessellations of [7] , considered as indiscrete bornological sets, the procedure of [7] can be carried out mutatis mutandis in the present setting. Accordingly, one defines $\ell_1(X,E)$ as the filtered colimit of the $\ell_1(Q,E)$ and $m(X,F)$ as the associated projective limit of the $\ell_\infty(Q,F)$. In this way one arrives at the following result.

10. Theorem. $[\ell_1(X,E),F] \simeq [E,\ m(X,F)]$
naturally for E and F in C.

The spaces $\ell_1(X,E)$ consists, as does $L_1(X,E)$, of (equivalence classes) of integrable functions $X \to E$; but unlike $L_1(X,E)$, these functions can attain essentially only countably many values. This comes about because, as a vector space, $\ell_1(X,E)$ is just the union of spaces of the form $\ell_1(Q,E)$. Thus the filtered colimit of the system $\ell_1(Q,A)$, when formed in A gives us $L_1(X,A)$ and when formed in C gives us the rather different cbv-space $\ell_1(X,A)$ which is not a Banach space even though A is. Thus theorem 10 also illustrates that not every C-left adjoint is the extension of a B-left adjoint.

References

1 E. Dubuc, Kan extensions in Enriched Category Theory, Springer Lecture Notes in Math., 145 (1970).

2 S. Eilenberg and G.M. Kelly, Closed Categories, Proc. Conf. Categorical Algebra, La Jolla 1965, Springer, Berlin (1966) 421 - 562.

3 H. Hogbe-Nlend, Bornologies and Functional Analysis, North Holland, Amsterdam (1977).

4 G.M. Kelly, Adjunction for Enriched Categories, Reports of the Midwest Category Seminar III, Springer Lecture Notes in Math. 106 (1969) 166-177.

5 S. Mac Lane, Categories for the Working Mathematician, Springer, New York

5 S. Mac Lane, Categories for the Working Mathematician, Springer, New York (1971).

6 L.D. Nel, Enriched algebraic categories with applications in Functional Analysis, Proc. Conf. Categorical Aspects of Topology and Analysis, Carleton Univ. 1980 (to appear).

7 L.D. Nel, Riesz-like representations for operators on L_1 by categorical methods, Advances in Math. (to appear).

Carleton University, Ottawa, Canada.

NSERC aided.

EXACTNESS AND PROJECTIVITY*

S. B. Niefield
Union College
Schenectady, New York 12308

1. Introduction

Let $\underline{V}$ be a symmetric monoidal category. An object M of $\underline{V}$ is exact if the functor $M\otimes -:\underline{V}\to\underline{V}$ has a left adjoint.

In [6], we obtain the following characterization of exact R-modules and commutative R-algebras over a commutative ring R. A module M is exact if and only if it is finitely generated and projective if and only if it is finitely presented and flat. In particular, one can show that the left adjoint to $M\otimes -$ is given by $\mathrm{Hom}_R(M,-)$. An algebra A is exact if and only if the underlying module is exact. The construction of the left adjoint is obtained by applying the following general lemma with $\underline{M}$ and $\underline{A}$ the categories of modules and algebras, respectively, T the forgetful functor, S the symmetric algebra functor, $F = A\otimes -$, and $G = TA\otimes -$.

Lemma 1 Consider the following diagram of categories

$$F \circlearrowright \underline{A} \underset{S}{\overset{T}{\rightleftarrows}} \underline{M} \circlearrowleft G$$

where $TF = GT$, and S is left adjoint to T with the counit $\varepsilon_A:STA\to A$ a regular epimorphism for all A. If G has a left adjoint, then so does F.

Proof It suffices to show that for every M in $\underline{M}$, $\underline{A}(SM,FA)$ is a representable functor of A, for every object A' of $\underline{A}$ admits a presentation $SM \rightrightarrows SN \to A'$ since $\varepsilon_{A'}$ is a regular epi. But, if $G' \dashv G$, we have

$$\underline{A}(SM,FA) \cong \underline{M}(M,TFA) \cong \underline{M}(M,GTA) \cong \underline{M}(G'M,TA) \cong \underline{A}(SG'M,A)$$

and the desired result follows. ▮

Let Loc denote the category whose objects are locales (i.e. complete lattices A satisfying the distributive law $a \wedge \bigvee S = \bigvee\{a \wedge s \mid s \in S\}$ for all $a \in A$ and $S \subseteq A$), and morphisms are functions which preserve finite meets and arbitrary sups. In 1978, Martin Hyland [3] showed that a locale A is exact if and only if it is a continuous lattice. What follows is an outgrowth an attempt to obtain an abstract construction of the left adjoint to $A\otimes -$ for a continuous locale A, that is, a construction like the one for commutative algebras (Hyland's construction uses theories).

Now, Joyal and Tierney have obtained many results for locales by considering the category Sl of sup lattices (i.e. complete lattices and sup preserving maps) as abelian groups, and locales as commutative rings. After several discussions we decided that one should be able to obtain Hyland's theorem using a method much like

* This research was supported by a Killam Postdoctoral Fellowship at Dalhousie University.

that of the module/algebra results. Unfortunately, this doesn't quite work. Sup-lattices behave like modules, but locales are too different from algebras to apply lemma 1. But there is a way around this problem, as we shall see in section four.

2. Exact sup lattices

We begin with some basic facts about sup lattices. First, we note that Sl is a $*$-autonomous category in the sense of Barr [1]. The internal hom is given by $\mathrm{Hom}(M,N) = \underline{\mathrm{Sl}}(M,N)$, where the sup of a family $\{f_i : M \to N\}$ is defined by $(\bigvee f_i)(m) = \bigvee\{f_i m\}$, the unit I is complete lattice $\mathbf{2}$, and the dual of M is the opposite lattice M^o. As usual, $\mathrm{Hom}(M,N) = \mathrm{Hom}(N^o,M^o)$ and $M \otimes N = \mathrm{Hom}(M,N^o)^o$. Note that we shall reserve the notation M^* for the dual $\mathrm{Hom}(M,I)$.

If $\{M_i\}$ is a family of sup lattices, their product $\prod M_i$ in Sl is given by the cartesian product with pointwise sup. Now, $\prod M_i$ is also the coproduct of the M_i in Sl since $(\ )^o$ is a complete duality. Moreover, every sup lattice can be expressed as a quotient $\mathbf{2}^X \xrightarrow{e} M$, where X is a set, namely take $X = M$ and $e = \bigvee$, using the identification of $\mathbf{2}^X$ with the power set $\mathcal{P}(X)$. Note that $\mathbf{2}^X$ is both the product and coproduct of X copies of $\mathbf{2}$.

Theorem 2.1 The following are equivalent for a sup lattice M

a) $M \otimes -$ has a left adjoint

b) $\mathrm{Hom}(M,-)$ has a right adjoint

c) M is flat

d) M is projective

e) $M \otimes - \cong \mathrm{Hom}(M^*,-)$

f) M^* is projective and $M^{**} \cong M$

g) $M^* \otimes - \cong \mathrm{Hom}(M,-)$

h) $\mathrm{Hom}(M,-) \dashv M \otimes -$

Proof We shall show that a) ⇒ c) ⇒ d) ⇒ e) ⇒ f) ⇒ g) ⇒ d), e) ⇒ h) ⇒ a), and h) ⇒ b) ⇒ d).

a) ⇒ c) If $M \otimes -$ has a left adjoint, then M is flat since every right adjoint preserves monomorphisms.

c) ⇒ d) If M is flat, then $\mathrm{Hom}(M,-)$ preserves epimorphisms since $M \otimes -$ preserves monomorphisms, $\mathrm{Hom}(M,-) = (M \otimes -^o)^o$, and $(\ )^o$ interchanges monics and epis.

d) ⇒ e) If $M = \mathbf{2}^X$, then the canonical map $M \otimes N \to \mathrm{Hom}(M^*,N)$ is an isomorphism for it is the composite of the isomorphisms

$$\mathrm{Hom}(\mathbf{2}^X,N^o)^o \longrightarrow [\mathrm{Hom}(\mathbf{2},N^o)^X]^o \longrightarrow (N^{oX})^o \longrightarrow N^X \longrightarrow \mathrm{Hom}(\mathbf{2},N)^X \longrightarrow \mathrm{Hom}(\mathbf{2}^X,N) \longrightarrow \mathrm{Hom}(\mathbf{2}^{X^*},N)$$

Note that $\mathbf{2}^X$ is projective since it is the free sup lattice on the set X. Now if M is any projective sup lattice, then M is a retract of $\mathbf{2}^X$ for some X, and the desired result follows.

e) ⇒ f) Clearly, $M^{**} \cong M$. Also, M^* is projective since $M \otimes -$ preserves epis being a left adjoint.

f) $\Rightarrow$ g) If M^* is projective, then applying d) $\Rightarrow$ e) we have $M^* \otimes - \cong \mathrm{Hom}(M^{**},-)$, and the latter is isomorphic to $\mathrm{Hom}(M,-)$.

g) $\Rightarrow$ d) Clearly, M is projective since $M^* \otimes -$ preserves epis.

e) $\Rightarrow$ h) If $M \otimes - \cong \mathrm{Hom}(M^*,-)$, then $M^* \otimes - \dashv M \otimes -$. But, we also know that $M^* \otimes - \cong \mathrm{Hom}(M,-)$ by e) $\Rightarrow$ g).

h) $\Rightarrow$ a) and h) $\Rightarrow$ b) are clear.

b) $\Rightarrow$ d) M is clearly projective since $\mathrm{Hom}(M,-)$ has a right adjoint. ∎

After seeing this theorem, Mike Barr remarked that the equivalence of a) and e) is valid in any $\star$-autonomous category. In fact, we have the following proposition

Proposition 2.2 The following are equivalent for an object M in a $\star$-autonomous category

a) $M \otimes -$ has a left adjoint

b) $M^* \otimes - = \mathrm{Hom}(M,-)$

c) $\mathrm{Hom}(M,-)$ has a right adjoint

d) $M \otimes - \cong \mathrm{Hom}(M^*,-)$

e) $\mathrm{Hom}(M,-) \dashv M \otimes -$

Proof a) $\Rightarrow$ b) If $L \dashv M \otimes -$, then

$$\mathrm{Hom}(M,N)^{o} \cong M \otimes N^{o} \cong \mathrm{Hom}(I, M \otimes N^{o}) \cong \mathrm{Hom}(LI, N^{o}) = (LI \otimes N)^{o}$$

and

$$(LI)^{o} \cong \mathrm{Hom}(I,(LI)^{o}) \cong \mathrm{Hom}(LI, I^{o}) \cong \mathrm{Hom}(I, M \otimes I^{o}) \cong M \otimes I^{o} \cong \mathrm{Hom}(M,I)^{o}$$

Therefore, $\mathrm{Hom}(M,-) \cong LI \otimes - \cong M^* \otimes -$.

b) $\Rightarrow$ c) is clear

c) $\Rightarrow$ d) If R denotes the right adjoint to $\mathrm{Hom}(M,-)$, then

$$M \otimes N \cong \mathrm{Hom}(M,N^{o})^{o} \cong \mathrm{Hom}(I,\mathrm{Hom}(M,N^{o})^{o}) \cong$$

$$\mathrm{Hom}(\mathrm{Hom}(M,N^{o}),\ ^{o}) \cong \mathrm{Hom}(N^{o},R(I^{o})) \cong \mathrm{Hom}(R(I^{o})^{o},N)$$

and

$$R(I^{o}) \cong \mathrm{Hom}(I,R(I^{o})) \cong \mathrm{Hom}(\mathrm{Hom}(M,I),I^{o}) \cong \mathrm{Hom}(M^*,I^{o}) \cong \mathrm{Hom}(I,M^{*o}) \cong M^{*o}$$

Therefore, $M \otimes - \cong \mathrm{Hom}(R(I^{o})^{o},-) \cong \mathrm{Hom}(M^*,-)$

d) $\Rightarrow$ a) and e) $\Rightarrow$ a are clear

b) $\Rightarrow$ e) If $M^* \otimes - \cong \mathrm{Hom}(M,-)$, then $\mathrm{Hom}(M,-) \dashv \mathrm{Hom}(M^*,-)$. Now, we have already shown that b) $\Rightarrow$ d). Therefore, $\mathrm{Hom}(M^*,-) \cong M \otimes -$, and the proof is complete. ∎

3. Projective sup lattice

One is tempted at this point (in analogy with the module/algebra theorem) to attempt to prove that an exact locale is flat or projective as a sup lattice. This is not the case. We know (from Hyland's theorem) that the exact locales are the continuous lattices, and a more careful analysis of projective sup lattices will show that these lattices are completely distributive, whereas there are clearly continuous lattice which are not completely distributive, e.g. the opens of the reals.

Recall that a complete lattice M is completely distributive if for every family $\{m_{ij} | i \in I, j \in J(i)\}$ we have

$$\bigwedge_{i\in I}(\bigvee_{j\in J(i)} m_{ij}) = \bigvee_{\substack{f:I\to\bigcup_i J(i)\\ \text{over } I}}(\bigwedge_{i\in I} m_{if(i)})$$

A complete lattice M is completely continuous if for all $m \in M$ we have $m = \bigvee\{n | n \ll m\}$, where $n \ll m$ (read n is completely below m) if whenever $m \le \bigvee S, S \subseteq M$, then $n \le s$, for some $s \in S$. Note that if we require that the subsets S of M be directed, then we have precisely the definition of a continuous lattice. Thus, a completely continuous lattice is necessarily continuous.

Theorem 3 The following are equivalent for a sup lattice M

a) M is projective

b) M is completely distributive

c) M is completely continuous

Proof a) $\Rightarrow$ b) Suppose M is projective. Then the map $\bigvee_M : \mathcal{P}(M) \to M$ admits a left adjoint right inverse. Indeed, $\bigvee_M$ has a right inverse i, and replacing i by the map $m \mapsto \{n | n \le s$, for some $s \in i\, m\}$ gives the desired map. Therefore, $\bigvee_M$ preserves all sups and infs. Thus, M is completely distributive since $\mathcal{P}(M)$ is.

b) $\Rightarrow$ c) Suppose $m \in M$. Let $I = \{S \subseteq M | m \le \bigvee S\}$, $J(S) = S$, and $m_{S,j} = j$. Then by complete distributivity we have

$$\bigwedge_{S\in I} \bigvee_{j\in S} m_{S,j} = \bigvee_{\substack{f:I\to\bigcup_S S\\ \text{over } I}} (\bigwedge_{S\in I} m_{S,f(S)})$$

Now, the left-hand side is clearly equal to m, for $\{m\} \in I$. We claim that if $f:I \to \bigcup_{S\in I} S$ over I, then $n = \bigwedge_{S\in I} m_{S,f(S)} \ll m$. If $m \le \bigvee S$, then $n \le m_{S,f(S)} = f(S) \in S$.

c) $\Rightarrow$ a) It suffices to show that $\bigvee_M : \mathcal{P}(M) \to M$ admits a right inverse, for $\mathcal{P}(M)$ is clearly projective. If $m \in M$, define $i(m) = \{n | n \ll m\}$. Then, $\bigvee_M i = 1_M$, since M is completely continuous. To see that i is sup preserving we note that (in a completely continuous lattice) if $n \ll \bigvee S$, then $n \ll s$ for some $s \in S$, since $\bigvee S = \bigvee\{t | t \ll s$, for some $s \in S\}$. ∎

Remarks 1. There is an analogue of b) $\Rightarrow$ c) for continuous lattices [2;p.58] in which complete distributivity is replaced by a weaker notion, i.e. the sets $\{m_{ij} | j \in J(i)\}$ are required to be directed.

2. Note that projective locales (i.e. projective in Loc) are necessarily projective as sup lattices, (but not conversely) for a local A is projective if and only if the map $\bigvee_A : \mathcal{P}(A) \to A$ admits a right inverse in Loc. Moreover, it is not difficult to show that A is projective in Loc if and only if A is completely continuous,

the completely below relation preserves finite meets and $1 \ll 1$, where 1 denotes the top of A. An example of a locale that is projective in Sl but not Loc is the lattice $\mathbf{2} \times \mathbf{2}$. It is clearly projective in Sl (it is the coproduct of projective sup lattices), but $(1,1) \not\ll (1,1)$ since $(1,1) = (1,0) \vee (0,1)$.

4. Exact locales

Now we return to the original problem, i.e. an abstract proof of Hyland's theorem. At first, it seems that the locale/commutative algebra analogy breaks down, but André Joyal observed that this is not really the case. He noticed that, in the proof of lemma 1, we do not use the full power of the existence of the left adjoint to G, we only use the fact that the functor $\underline{M}(M,GT-) : \underline{A} \to \underline{\text{Sets}}$ is representable.

To apply (this amended form of) lemma 1, we need to know that the forgetful functor $\underline{\text{Loc}} \to \underline{\text{Sl}}$ has a left adjoint. But, every locale is a quotient of a free locale $\mathbf{2}^X$ on a set X, and

$$\underline{\text{Sl}}(\mathbf{2}^X, A) \cong \underline{\text{Sets}}(X,A) \cong \underline{\wedge\text{-slat}}(K(X),A) \cong \underline{\text{Loc}}(\downarrow\text{Cl}(K(X)),A)$$

where $\wedge$-slat denotes the category of meet semi-lattices and meet preserving maps, K denotes the free meet semi-lattice functor, and $\downarrow\text{Cl} : \underline{\wedge\text{-slat}} \to \underline{\text{Loc}}$ denotes the free locale functor, i.e. $\downarrow\text{Cl}\,M$ is the locale of downward closed subsets of M.

Theorem 4 The following are equivalent for a locale A

a) A is exact in Loc

b) A is a continuous lattice

c) $A \otimes - : \underline{\text{Loc}} \to \underline{\text{Sl}}$ preserves monomorphisms

d) $A \otimes - : \underline{\text{Loc}} \to \underline{\text{Sl}}$ preserves equalizers

e) $A \otimes - : \underline{\text{Loc}} \to \underline{\text{Sl}}$ is representable.

Remark The following proof of a) $\Rightarrow$ b) is due to Joyal and is outlined in the introduction to [4]. Any attempt to improve upon this beautiful argument would be futile.

Proof of theorem 4 a) $\Rightarrow$ b) If $L : \underline{\text{Loc}} \to \underline{\text{Loc}}$ denotes the left adjoint to $A \otimes -$, then L preserves projectives, since $A \otimes -$ preserves epimorphisms. In particular, if S denotes the Sierpinski locale (i.e. the free locale on the sup lattice $\mathbf{2}$), then $L(S)$ is projective. But, the lattice $\underline{\text{Loc}}(P,\mathbf{2})$ of points of a projective locale is necessarily continuous, and

$$\underline{\text{Loc}}(L(S),\mathbf{2}) \cong \underline{\text{Loc}}(S, A \otimes \mathbf{2}) \cong \underline{\text{Loc}}(S,A) \cong \underline{\text{Sl}}(\mathbf{2},A) \cong A$$

Therefore, A is continuous.

b) $\Rightarrow$ c) To see that $A \otimes - : \underline{\text{Loc}} \to \underline{\text{Sl}}$ preserves monomorphisms, it suffices to show that given a diagram

$$\begin{array}{ccc} & & A \\ & & \downarrow g \\ C^o & \xrightarrow{f^o} & B^o \end{array}$$

where g is sup preserving and $f:B \rightarrowtail C$ is a monomorphism of locales, there exists $h:A \to C^{o}$ such that $f^{o}h = g$.

If A is continuous, then the morphism of locales $\bigvee : \mathrm{Idl}\,A \to A$ admits a sup preserving right inverse, where $\mathrm{Idl}\,A$ denotes the locale of downward closed (upward) directed ideals of A. But $\mathrm{Idl}\,A$ (and hence any retract) in $\underline{\mathrm{Sl}}$ clearly satisfies the desired property since the map $\underline{\mathrm{Sl}}(\mathrm{Idl}\,A, M) \to \underline{\mathrm{Lat}}(A,M)$ given by

$$(f:\mathrm{Idl}\,A \to M) \longmapsto (A \xrightarrow{\downarrow \mathrm{seg}} \mathrm{Idl}\,A \xrightarrow{f} M),$$

is an isomorphism, where $\underline{\mathrm{Lat}}$ denotes the category of distributive lattices and finite meet and join preserving maps. In particular, given a diagram

$$\begin{array}{ccc} & & \mathrm{Idl}\,A \\ & & \downarrow g \\ C^{o} & \xrightarrow[f^{o}]{} & B^{o} \end{array}$$

$fg(\downarrow \mathrm{seg}):A \to C^{o}$ is finite meet and join preserving, and hence induces the desired fill-in $\mathrm{Idl}\,A \to C^{o}$.

c) $\Rightarrow$ d) This is immediate since every monomorphism in $\underline{\mathrm{Sl}}$ is an equalizer, and $A \otimes - : \underline{\mathrm{Loc}} \to \underline{\mathrm{Sl}}$ takes equalizers to monomorphisms.

d) $\Rightarrow$ e) If $A \otimes -$ preserves equalizers, then $A \otimes -$ preserves all limits, since it preserves products, in any case. Thus, it suffices to show that $A \otimes - : \underline{\mathrm{Loc}} \to \underline{\mathrm{Sl}}$ satisfies the solution set condition [5;p118].

If $f:A \to B^{o}$ is an element of $A \otimes B = \mathrm{Hom}(A,B^{o})^{o}$, then $f:A^{o} \to B$ preserves finite meets, and hence induces a locale morphism $\bar{f}:\downarrow \mathrm{Cl}(A^{o}) \to B$ such that $\bar{f} \downarrow \mathrm{seg} = f$. We claim that $1 \otimes f : A \otimes \downarrow \mathrm{Cl}(A^{o}) \to A \otimes B$ takes $A \xrightarrow{\mathrm{seg}} \downarrow \mathrm{Cl}(A^{o})^{o}$ to f, and so $\{\downarrow \mathrm{Cl}(A^{o})\}$ provides a solution set for $A \otimes -$. But, if $g:C \to B$ is any morphism of locales, $1 \otimes g : A \otimes C \to A \otimes B$ is described as follows. Composition with $g^{o}:B^{o} \to C^{o}$ induces a sup preserving map $\mathrm{Hom}(A,B^{o}) \to \mathrm{Hom}(A,C^{o})$, and hence an inf-preserving map $A \otimes B \to A \otimes C$, whose left adjoint is $1 \otimes g$. With this description, it is not difficult to show that $1 \otimes \bar{f}$ takes $\downarrow \mathrm{seg}$ to f.

e) $\Rightarrow$ a) By (the amended version of) lemma 1, it suffices to show that the functor

$$\underline{\mathrm{Sl}}(M, A \otimes -) : \underline{\mathrm{Loc}} \to \underline{\mathrm{Sl}}$$

is representable, for every M. Now, e) takes care of the case where $M = \mathbf{2}$. If $M = \mathbf{2}^{X}$ and $L(\mathbf{2})$ represents $A \otimes -$, then $\oplus_{X} L(\mathbf{2})$ represents $\underline{\mathrm{Sl}}(\mathbf{2}^{X}, A \otimes -)$. For a general M, we express M as a quotient of $\mathbf{2}^{X}$, and take $L(M)$ to be the appropriate quotient of $L(\mathbf{2}^{X})$. ∎

In conclusion, I would like to thank André Joyal for many helpful observations and suggestions.

References

1. M. Barr, "$\star$-Autonomous Categories", Springer Lecture Notes 752, (1979).

2. G. Gierz, K. H. Hofmann, et. al., A Compendium of Continuous Lattices, Springer-Verlag, 1980.

3. J.M.E. Hyland, "Function spaces in the category of locales", Continuous Lattices, Springer Lecture Notes in Mathematics, to appear.

4. P.T. Johnstone and A. Joyal, "Continuous categories and exponentiable toposes", to appear.

5. S. MacLane, Categories for the Working Mathematician, Springer-Verlag, 1971.

6. S.B. Niefield, "Cartesianness:topological spaces, uniform spaces, and affine schemes", J. Pure Appl. Alg., to appear.

Constructive Arithmetics

M. Pfender R. Reiter M. Sartorius

Introduction

We found essential parts of Arithmetic on the concept of 'Primitive Recursive Universe' which is weaker than Topos Theory with Natural Numbers Object. It just allows formalization of the schema of primitive recursion and hence can be assumed to be consistent, i.e. not to collapse. It turns out that this basic universe has a notion of equality, that Arithmetic of the natural numbers can be developped relative to this equality and that equality can be extended consistently in such a way that internally and externally defined equality between maps coincide. Altogether this shows that it is possible to develop Arithmetic of the natural numbers in a framework whose consistency is built exclusively on consistency of the schema of primitive recursion formalized map theoretically.

As far as things are now we still depend on a quite evident property of taking the integer part of a root. Elimination of this dependency is for further study.

We first introduce the general frame 'Primitive Recursive Universe' and develop within this frame the basic algebraic structure of the natural numbers and some of the logical structure.

1. Algebraic definition of Primitive Recursive Universes

A Primitive Recursive Universe is a category U (of 'sets' and 'maps') with a terminal object 1 ('one-element-set'), binary (cartesian) product and a Natural Numbers Object N in the sense of [Fr], propos. 5.21, i.e. with the following 'scheme of primitive recursion': There are maps $0\colon 1 \to N$ ('zero') and $s\colon N \to N$ ('successor') given and to any given maps $f\colon A \to B$ ('initialisation'), $g\colon B \to B$ ('step function') is associated a map $g^*f\colon A\times N \to B$ ('iteration', intuitively: $g^*f(a,n) = g^n(f(a))$), satisfying the equations

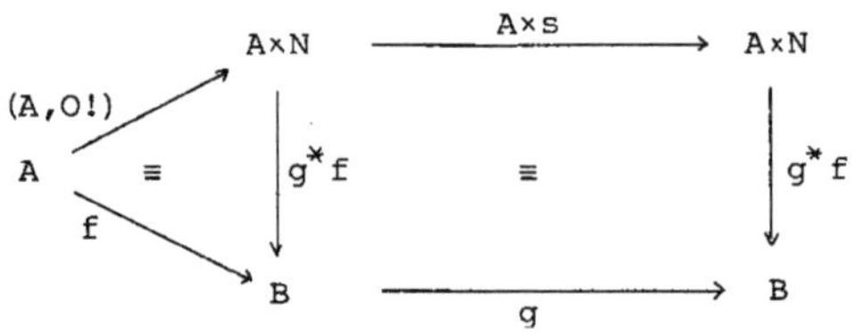

(1)

and which is unique in this regard, i.e. if $h: A\times N \longrightarrow B$ instead of g^*f satisfies (1), then $h \equiv g^*f$. ($!: A \longrightarrow 1$ is the unique map into the terminal object 1).

As seen above for the NNO-structure, the other parts of the structure too are defined algebraically, i.e. in terms of operations (e.g. $A,B \longmapsto A\times B$, $A,B \longmapsto l: A\times B \longrightarrow A$ (left projection), $A,B \longmapsto r: A\times B \longrightarrow B$, $f: C \longrightarrow A$, $g: C \longrightarrow A \longmapsto (f,g): C \longrightarrow A\times B$ (induced map)) and equations and - for the NNO-structure - an implication of equations.

2. General scheme of primitive recursion

For the classical natural numbers the (general) scheme of primitive recursion reads: To maps $g: A \longrightarrow B$ ('initialisation') and $h: A\times N\times B \longrightarrow B$ is associated a unique $f: A\times N \longrightarrow B$ such that $f(a,0) = g(a)$ and $f(a,y+1) = h(a,y,f(a,y))$. (2)
In a Primitive Recursive Universe this would read: To any $g: A \longrightarrow B$, $h: (A\times N)\times B \longrightarrow B$ is associated a unique $f: A\times N \longrightarrow B$ such that $A \xrightarrow{(A,0!)} A\times N \xrightarrow{f} B \equiv A \xrightarrow{g} B$ and $A\times N \xrightarrow{A\times s} A\times N \xrightarrow{f} B \equiv A\times N \xrightarrow{((l,r),f)} (A\times N)\times B \xrightarrow{h} B$.
Freyd shows this for topoi, his proof uses only means of Primitive Recursive Universes ([Fr], 5.22). Viewing variables as projections allows us to state (and partially also to prove) assertions like (2) litterallyas in the classical frame.

3. Construction of 'the' Primitive Recursive Universe

Because the notion of Primitive Recursive Universe is algebraic, it is possible to give a primitive recursive algorithm which produces (counts) the sets and the maps of the absolutely free structure $\underline{E}$ of type of P.R. Universe over the two empty sets of sets and of maps and which produces (counts) at the same time a congruence on that structurewhich - taken as equality $\equiv$ - makes it into 'the' (initial) Primitive Recursive Universe (the Lindenbaum-Algebra), i.e. the free one over the two empty sets of sets and of maps. By the nature of free construction, this structure $(\underline{E},\equiv)$ constitutes our theory as well as its canonical model (Herbrand-model). Terms of the theory are the sets of $\underline{E}$ (set terms) and the maps of $\underline{E}$ (map terms), theorems are all pairs $(f: A \longrightarrow B, g: A \longrightarrow B)$ in $\equiv$ (equations). The counting algorithm ('universal program') is formalizable in our system, since it is primitive recursive. This latter fact is shown in detail in [La] by writing it up as a primitive recursive PL - without-GO-TO-program in the sense of [B-L].

4. Semiring structure of the Natural Numbers Object N

After introducing $+: N\times N \longrightarrow N$ and $\cdot: N\times N \longrightarrow N$ as usual by primitive recursion one shows first $0+x \equiv x$ and $s(x+y) \equiv sx+y$ and from this that $(N,0,1,+,\cdot)$ is a unitary commutative semiring.

5. Logical structure of Primitive Recursive Universes

Freyd shows in [Fr], 5.11 that $1 \xrightarrow{0} N \xleftarrow{s} N$ constitutes a sum. We generalize this slightly to the fact that for an arbitrary set A $A \xrightarrow{(A,0!)} A\times N \xleftarrow{A\times s} A\times N$ constitutes a sum, i.e. for arbitrary $f: A \to C$, $g: A\times N \to C$ there is a unique $(f|g): A\times N \to C$ such that $(f|g)(A,0!) \equiv f$ and $(f|g)(A\times s) \equiv g$. We use N instead of a not yet available object 2 for case-distinction: Given maps $f,g: A \to C$. The map $h :\equiv (A \xrightarrow{f} C \mid A\times N \xrightarrow{l} A \xrightarrow{g} C)$ has the property $h(A,0!) \equiv f$ and $h(A\times s)(A,0!) \equiv g$ which gives in particular $h(a,0) \equiv f(a)$ and $h(a,1) \equiv g(a)$.

Let us study further the 2-like structure of N: The signum-function $sg := (0|1): N \to N$ intuitively identifies every natural number greater 0 with 1. Maps of the form $A \xrightarrow{\chi} N \equiv A \xrightarrow{f} N \xrightarrow{sg} N$ are called predicates, for A = 1 nullary predicates or truth values. Define $neg :\equiv (1|0!): N \to N$, $N\times N \xrightarrow{and} N :\equiv N\times N \xrightarrow{((0|0!)|(0|1!)1)} N$, i.e. by their 'truth-tables', similarly the other logical junctors. The boolean equations for the junctors are showed by truth-table arguments using uniqueness of the induced $(f|g)$ out of the sum. By the same method all tautologies on predicates used in the sequel can be showed.

6. Further Algebra on N

We define the predecessor $p: N \to N$ as induced out of the sum $1 \xrightarrow{0} N \xleftarrow{s} N$ by $p :\equiv (0|N)$, i.e. $p(0) \equiv 0$, $p(sx) \equiv x$, and the truncated subtraction by $N\times N \xrightarrow{\dot{-}} N :\equiv N\times N \xrightarrow{p^*N} N$, i.e. $x\dot{-}0 \equiv x$, $x\dot{-}sy \equiv p(x\dot{-}y)$. For this operation we show $0\dot{-}x \equiv 0$, $sx\dot{-}sy \equiv x\dot{-}y$ (compensation), $x\dot{-}x \equiv 0$, $p(x\dot{-}y) \equiv px\dot{-}y$, $(x+y)\dot{-}y \equiv x$ (absorption) $(x+y)\dot{-}(y+z) \equiv x\dot{-}z$, $(x\dot{-}y)\dot{-}z \equiv x\dot{-}(y+z)$ (association). These laws are important for properties of equality and order on N to be introduced below and for the proof of induction principles.

Exponentiation is defined by primitive recursion, likewise division with remainder. The equations of these operations necessary in the sequel are shown using the above lemmata. Then the notions of 'prime number' and 'greatest common divisor' can be defined as usual.

We now introduce equality on N and derive Peano's axioms from primitive recursion. We will discuss an important extension of Peano-induction, the 'diagonal induction'.

7. Predicate of equality on N and the proof of Peano's axioms

The distance between natural numbers is defined by $dist(x,y) :\equiv (x\dot{-}y)+(y\dot{-}x)$, equality by $N\times N \xrightarrow{=} N :\equiv N\times N \xrightarrow{dist} N \xrightarrow{neg} N$ (neg means equality with 0). Equality is reflexive and symmetric and equality on N implies logical equivalence.

Peano's axioms read in our framework as follows:
Natural numbers are arrows $x: 1 \to N$.

P1 O: $1 \longrightarrow N$ is a natural number

P2 (x=y) impl (sx=sy) $\equiv$ 1!

where infix notation is used: x=y instead of =(x,y), a impl b instead of impl(a,b), impl being the implication junctor.

P 2 means that the successor is well defined, i.e. that the 'map' s has this crucial (internal) property of a mapping.

P3 (sx=sy) impl (x=y) $\equiv$ 1!

i.e. s is injective as a mapping.

P4 neg(sx=O!) $\equiv$ 1!

i.e. O is not a successor.

P5 induction, here in first order form:

If for a predicate (this stands for 'property') $\chi : A\times N \longrightarrow N$

(i) χ(a,O) $\equiv$ 1! and

(ii) χ(a,x) impl χ(a,sx) $\equiv$ 1!

then χ(a,x) $\equiv$ 1! i.e. is overall true.

Specializing to A = 1 gives the classical P5.

The proofs of P1 to P4 are straight forward by the use of tautologies, the lemmata above and the cited properties of equality.

For proving P5, show that

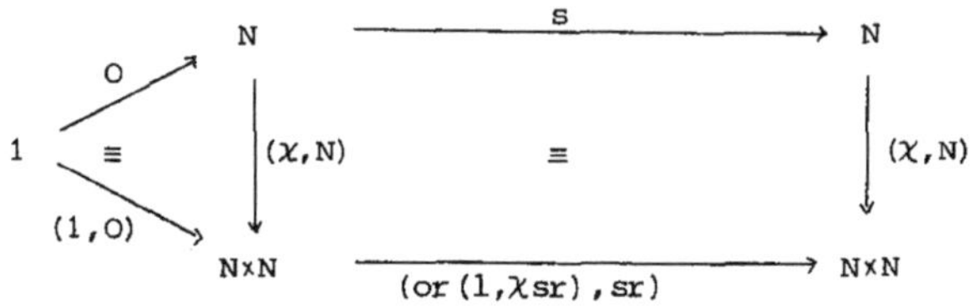

commutes, using or(1,χsr) $\equiv$ and(or(1,χsr),impl(χ,χs)). Then conclude by uniqueness of the iteration map above.

Sometimes we use a sharper induction scheme, the $\bigwedge$-induction:

If χ(a,O) $\equiv$ 1! and ($\bigwedge_{i=0}^{x}$ χ(a,i)) impl χ(a,sx) $\equiv$ 1! then χ(a,x) $\equiv$ 1!

Herein $\bigwedge_{i=0}^{x} \chi(a,i)$ is given by iteration of and: $N\times N \longrightarrow N$.

This induction principle is proved by application of the foregoing principle.

8. Diagonal induction

Only in special cases the induction principles so far discussed are appropriate for showing general truth of n-ary predicates on N. In particular they do not suffice for Elementary Arithmetics. We need another induction principle: 'diagonal induction', i.e. induction along the direction of the diagonal of N×N which reads as follows:

If a predicate $\chi : A\times(N\times N) \longrightarrow N$ of two NNO-variables satisfies

(i) $\chi(a,(x,0!)) \equiv 1!$ i.e. χ is true on the x-axis,

(ii) $\chi(a,(0!,y)) \equiv 1!$ i.e. χ is true on the y-axis, and

(iii) $\chi(a,(x,y))$ impl $\chi(a,(sx,sy)) \equiv 1!$ i.e. truth is inherited in parallel to the diagonal,

then $\chi(a,(x,y)) \equiv 1!$ i.e. χ is overall true.

Geometrically, this principle is obvious. But since we cannot be sure at the moment that N×N is what we understand geometrically by it, we have to use our Arithmetics so far developped to prove it. In order to reduce the two dimensions to one, we first define an isomorphism count: $N \longrightarrow N\times N$ in analogy to the usual meander-counting of the rationals by

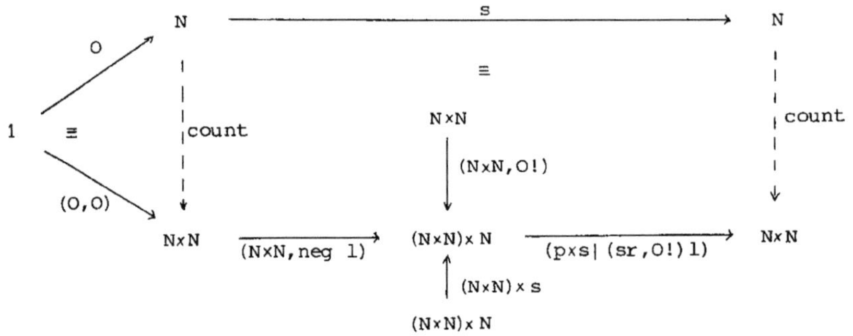

i.e. by

$$\mathrm{count}(0) \equiv (0,0)$$

$$\mathrm{count}(sn) \equiv \begin{cases} (s\ r\ \mathrm{count}(n), 0) & \text{if } 1\ \mathrm{count}(n) \equiv 0 \\ (p\ 1\ \mathrm{count}(n), s\ r\ \mathrm{count}(n)) & \text{else} \end{cases}$$

and try to show that count has as an inverse $\mathrm{count}^{-1}: N\times N \longrightarrow N$ defined by

$$\mathrm{count}^{-1}(x,y) :\equiv \mathrm{half}((x+y)\cdot(x+sy))+y$$

where half(0) :≡ 0, half(sx) :≡ half(x)+odd(x) and odd(0) :≡ 0, odd(sx) :≡ neg odd(x)

First we show that count^{-1} is a retraction for count.

For this commutativitiy of (*) in

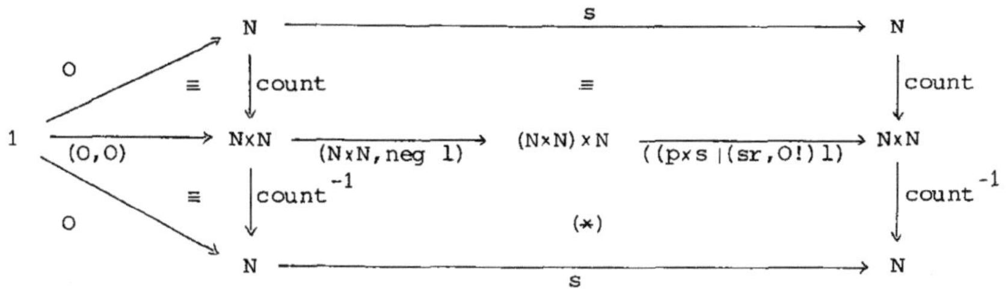

is sufficient by uniqueness of the iterated map.

By pulling back the sum $N{\times}N \xrightarrow{(N\times N,O!)} (N{\times}N)\times N \xleftarrow{(N\times N)\times s} (N{\times}N)\times N$ along $(N{\times}N,\text{neg } l)$ the upper edge of (*) becomes the induced out of the sum $N \xrightarrow{(O!,N)} N{\times}N \xleftarrow{s\times N} N{\times}N$ of (s,O!) and ((O!,s) | s×s), the latter being an induced map out of N×N decomposed the same way.

Next it is shown that the upper part of (*) is the induced out of the sum above of half((sx)·(ssx)) and the induced out of the (same) sum of half((sx)·(ssx))+sx and half((sx+sy)·s(sx+sy))+sy. Similarly one shows that the lower path is induced by half((x·sx)+sx) and the induced of half((sx·ssx)+sx) and half((ssx+y)·(ssx+sy))+sy. Corresponding inducing components of upper and lower path are equal: this is immediate for components two and three and verified by definition of half and distributivity of '·' over '+' for the component one.

It would be sufficient to show that count^{-1} is a section by giving an arbitrary retraction of it. A proof using count seems not to work very well. We follow instead [Pe], [H-B] and [Da] by using the integer (part of) root of a natural number, designated by root(x). Abbreviate

$$q(z) := \text{half}(\text{root}(8z+1)+1) \dot{-} 1$$

$$q'(z) := 2z \dot{-} (q(z))^2$$

Then, as in the literature cited

$$\widetilde{\text{count}}(z) := (q(z) \dot{-} \text{half}(q'(z) \dot{-} q(z)), \text{half}(q'(z) \dot{-} q(z)))$$

is the canditate for the wanted retraction of count^{-1}. The crucial point in Davis' proof is the equality $q(\text{count}^{-1}(x,y)) = \text{half}(\text{root}((2x+2y+1)^2+8y)+1) \dot{-} 1 = x+y$.
If we can show this equality in Primitive Recursive Universes we are done, because from this the retraction property of $\widetilde{\text{count}}$ is straight forward. The wanted equality is further reducible to

$$\text{root}((2x+2y+1)^2+8y) = \begin{cases} 2x+2y+1 & \text{if } 2y \leq 2x+1 \\ 2x+2y+2 & \text{otherwise} \end{cases}$$

This is classically 'evident', and we have a classical proof for it using nested induction.

Diagram form of the last statement:

$$N\times N \xrightarrow{((x,y),\,sg(2y \dot{-} (2x+1)))} (N\times N)\times N \xrightarrow{(2x+2y+1|\,(2x+2y+2)\,1)} N$$

$$\equiv$$

$$\text{root}((2x+2y+1)^2+8y)$$

We will call fulfillment of this the 'root property' for a Primitive Recursive Universe.

For the definition of root there are at least two possibilities. The first follows [H-B]: $\text{root}(0) \equiv 0,\quad \text{root}(sx) \equiv \text{root}(x)+\text{neg}((s\ \text{root}(x))^2 \dot{-} sx)$

The second one - used in our proof of the classical statement - is the following which is combined with the remainder by taking the integer part of the root:

$$\text{root}(0) \equiv 0,\quad \text{rrem}(0) \equiv 0$$

$$\text{root}(sx) \equiv \begin{cases} s\ \text{root}(x) & \text{if} \quad \text{rrem}(x) = 2\cdot\text{root}(x) \\ \text{root}(x) & \text{otherwise} \end{cases}$$

$$\text{rrem}(sx) \equiv \begin{cases} 0 & \text{if} \quad \text{rrem}(x) = 2\cdot\text{root}(x) \\ s\ \text{rrem}(x) & \text{otherwise} \end{cases}$$

(categorical translation by using sums)

So, from the 'root property' follows $N \cong N\times N$. From this we prove the principle of diagonal induction as follows:

By the $\bigwedge$-induction principle we show $\chi(A\times\text{count}) \equiv 1!$, count being an isomorphism yields $\chi \equiv 1!$ what we have to show.
The proof of $\chi(A\times\text{count}) \equiv 1!$ turns out to be quite a difficult one. It uses the uniqueness of the induced map out of the sum
$N \xrightarrow{\text{count}^{-1}(0!,N)} N \xleftarrow{\text{count}^{-1}(S\times N)} N\times N$,
properties of $\bigwedge$ and special induction principles for n-ary predicates.

By diagonal induction we now show transitivity of equality on N. Then we expaned equality to all objects of our initial universe and prove its properties: reflexivity, symmetry, transitivity and others. Furthermore we define the usual order on N and verify its properties of a linear ordering.

9. Extension of the notion of equality to all objects, substitutivity

All objects of 'the' universe are of the form 1 or an (iterated) cartesian product of 1 and N. Equality on1 is trivial, equality can be extended from 1 and N to cartesians products by componentwise definition. We show reflexivity of equality and - by diagonal induction - substitutivity, i.e. $(a=b)\ \text{impl}\ (f(a)=f(b)) \equiv 1!$ for all $f: A \to B$ in the universe $\underline{E}$. From substitutivity follows transitivity. Symmetry holds for this equality. So our 'equality' has all the properties of an equality.

10. Order on N

We define $\leq : N \times N \longrightarrow N$ by $x \leq y :\equiv neg(x \dot{-} y)$ and show - using diagonal induction and several arithmetic laws (again proved by diagonal induction) - reflexivity, antisymmetry, transitivity, linearity and trichotomy.

11. Relationship of arithmetic operations versus equality and versus order

We show the additive and multiplicative simplification rule, distributivity, additive and multiplicative monotony and absence of zero divisors.

12. Equality externally and internally

$f \equiv g$ implies $(f=g) \equiv 1!$ (reflexivity of equality). The converse, 'coreflexivity' seems us not to hold a priori. We 'force' it by introducing into the universe an extended notion of external equality: $f \approxeq g$ iff $(f=g) \equiv 1!$ which is compatible with all the defining operations of a Primitive Recursive Universe. We show consistency of this notion of equality by proving: If $0 \approxeq s0$ then $0 \equiv s0$.

A Primitive Recursive Universe such as the above, provided with an equality notion satisfying all the properties mentioned above inclusive coreflexivity, is called Primitive Recursive Universe with Equality. In such universes the diagonal is an equalizer of equality and $1!$, injective maps f (i.e. $(fa=fb)$ impl $(a=b) \equiv 1!$) are monic.

13. Introduction of formal quotients by equivalence predicates and of extensions of predicates

We now introduce into our universes formal quotients by equivalence predicates and formal extensions of predicates.
Sets of the extended universe are triples $(A, A \times A \xrightarrow{\rho} N, A \xrightarrow{\chi} N)$ with predicates ρ, χ and ρ an equivalence, i.e. reflexive, symmetric and transitive, all of these components in a Primitive Recursive Universe $\underline{U}$ with Equality. Maps from $A^{\rho}_{\chi} = (A, \rho, \chi)$ to B^{β}_{ψ} are $\underline{U}$-maps $f: A \rightarrow B$ satisfying ρ impl $\beta(f \times f)$ and χ impl ψf.
Notion of equality: $f \approxeq g$ iff χ impl $(f \beta g) \equiv 1!$
We have to show:

1) The new universe $\underline{V}$ is a Primitive Recursive Universe with Equality.
2) $(A, A \times A \xrightarrow{\rho} N, A \xrightarrow{1!} N)$ is the quotient A/ρ of A by ρ.
3) $(N, N \times N \xrightarrow{(neg \mid sg\ 1)} N, N \xrightarrow{1!} N)$ represents the two-element set 2 with its usual properties.
4) $\underline{V}$ has extensions $\{\chi : A\}$ of predicates $\chi: A \longrightarrow 2$.

This $\underline{V}$ has finite limits, monos are injective and $\underline{V}$ constitutes a good framework for Arithmetics of the integers and of the rationals.

For details of this paper see [Pf], [Re2], [Sa] and [Re1].

References

[B-L] W.S. Brainerd, L.H. Landweber, Theory of Computation, New York 1974

[Da] M. Davis, Computability and Unsolvability, New York, Toronto, London 1958

[Fr] P. Freyd, Aspects of Topoi, Bulletin of the Australian Mathematical Society 7 (1972), 1-76

[Ge] G. Gentzen, Die Widerspruchsfreiheit der reinen Zahlentheorie, Math.Annalen 117 (1940), 493-565

[Gö] K. Gödel, Über formal unentscheidbare Sätze der Principia Mathematica und verwandter Systeme I, Monatshefte für Mathematik und Physik 38 (1931), 173-198

[H-B] D. Hilbert, P. Bernays, Grundlagen der Mathematik, Berlin 1934

[La] M. Laßmann, Gödels Nichtableitbarkeitstheoreme und Arithmetische Universen, Diplomarbeit TU Berlin 1981

[Pe] R. Péter, Recursive functions, New York 1967

[Pf] M. Pfender, Algebraische Mengenlehre, Vorlesungsskript TU Berlin 1979

[Re1] R. Reiter, Mengentheoretische Konstruktionen in Arithmetischen Universen, Diplomarbeit TU Berlin 1980

[Re2] R. Reiter, Grundlagen einer algebraisch konstruktiven Fundierung der Arithmetik (in Vorbereitung)

[Sa] M. Sartorius, Kategorielle Arithmetik, Diplomarbeit TU Berlin 1981

ADJOINT DIAGONALS FOR

TOPOLOGICAL COMPLETIONS

Hans-E. Porst

Abstract : Given a commutative square of functors formed by two topological (i.e. initial or final) completions of some concrete category and a finest or coarsest diagonal of this diagram we construct an adjoint of this diagonal as a diagonal in a related square. Properties and applications of this construction are discussed, too.

Quite recently E. J. Dubuc [4] has given a construction of an adjoint situation between a topological category $\underline{T}$ and a category of quasispaces $\underline{Q}$ defined by a suitable subcategory $\underline{C}$ of $\underline{T}$. We will show that this construction is a special instance of a far more general one using the well known external characterizations of topological functors (cp. [2] , [3] , [7] , [9]).

Our general setting will be a commutative square of functors

$$\text{(I)} \qquad \begin{array}{ccccc} & & \underline{C} & & \\ & \swarrow^{E_1} & & \searrow^{E_2} & \\ \underline{A}_1 & & \xleftarrow{\;q\;} & & \underline{A}_2 \\ & \searrow_{T_1} & & \swarrow_{T_2} & \\ & & \underline{X} & & \end{array}$$

where the functors T_i (i=1,2) may be thought to be (proper) topological and amnestic and the functors E_i (i=1,2) to be full embeddings. Hence we might think of this basic situation as of a concrete category $\underline{C}$ over $\underline{X}$ together with two topological completions of $\underline{C}$. In general, however, we will not make use of all these assumptions.

Any diagonal $q:\underline{A_2}\longrightarrow \underline{A_1}$ of this square (there exist at least a coarsest and a finest one which will be different in general [3]) defines a new square of functors

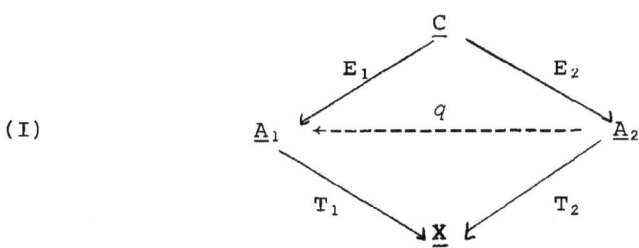

(II)

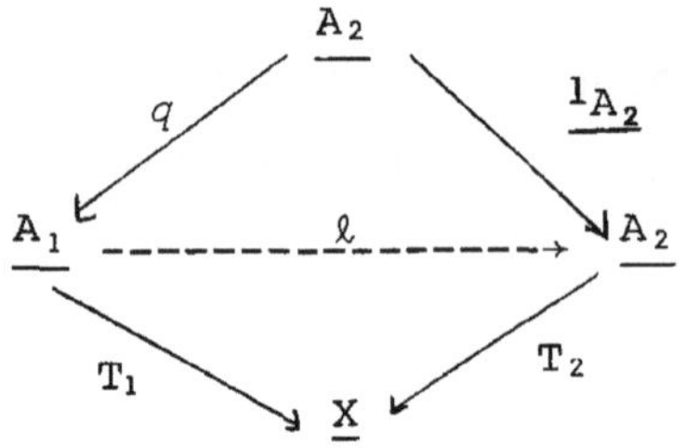

This square now admits a quasidiagonal $\ell : \underline{A}_1 \longrightarrow \underline{A}_2$, i.e. a functor ℓ such that $T_2\ell = T_1$ and $\ell_q \leq_{T_2} 1$. We shall always assume ℓ to be the T_2-coarsest quasidiagonal; if in addition q is full, then $\ell q = 1$ (cp. [2]).

It is now our main purpose to look for conditions which make ℓ a left adjoint for q. A necessary condition is given by our first proposition which is a slight generalization of a result implicitly contained in [2, Thm. 2.6].

<u>Proposition 1</u> : Let be given a commutative square

(II)

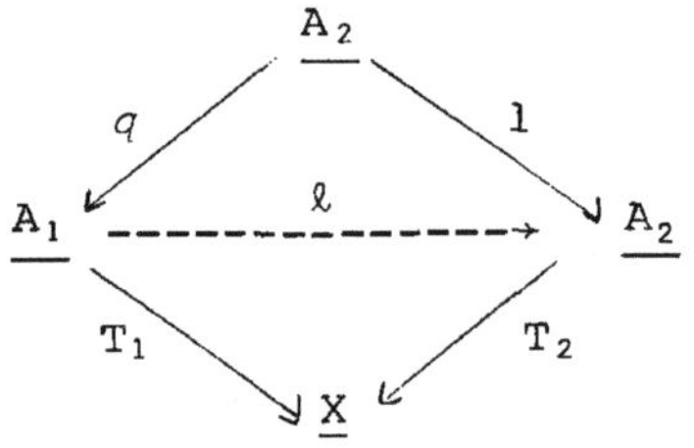

where the functor T_2 is topological. If then $\ell : \underline{A}_1 \longrightarrow \underline{A}_2$ is a left adjoint of q s.t. $T_2 \circ \ell = T_1$ then ℓ is the coarsest quasidiagonal of that diagram.

<u>Proof</u> : To show : ℓ is constructed by initial lifts, i.e.

(*) $T_2B \xrightarrow{h} T_2\ell A = T_2(B \xrightarrow{\bar{h}} \ell A)$

iff

(**) $T_2B \xrightarrow{h} T_2\ell A \xrightarrow{T_2\bar{f}} T_2B' = T_2(B \xrightarrow{h_f} B')$

for all $f : A \longrightarrow qB'$ where $(\bar{f})$ is the T_2-initial lift of $(T_1 f)$.

Hence assume (**) and $f \in A_1(A, qB')$ hence $\bar{f} \in A_2(\ell A, B')$

By Tholen's generalization [8] of Wyler's taut lift theorem [10] we can assume that the unit η of the adjunction $\ell \longmapsto q$ has underlying maps

$$T_1\eta_A = 1_{T_1A}.$$

Hence we have for maps f and $\tilde{f}$ corresponding by adjunction $T_1 f = T_2\tilde{f}$. Moreover by faithfulness of T_2 we have $\bar{f} = \tilde{f}$ (with the above notations), and given any $g : \ell A \longrightarrow B'$ it is (by adjunction) an $\bar{f}$ for some $f : A \longrightarrow qB'$. If now (**) holds then we have in particular

$$h = T_2B \xrightarrow{h} T_2\ell A \xrightarrow{T_2\bar{\eta}_A} T_2\ell A = T_2(B \xrightarrow{\bar{h}} \ell A) \quad \text{q.e.d.}$$

That ℓ acts on morphisms as claimed comes from faithfulness of T_2 and the condition $T_2\ell = T_1$. □

<u>Proposition 2</u> : Given a commutative square

(I)

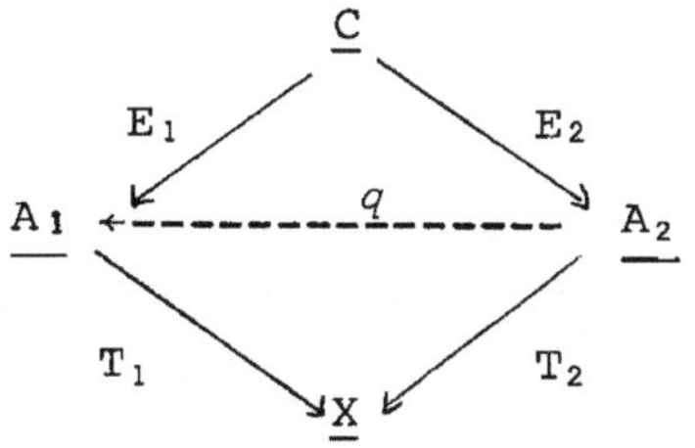

with E_1 codense (i.e. T_1-finally dense) and T_2 topological Assume that q, any diagonal of (I), exists. Let ℓ be the coarsest quasidiagonal of the corresponding square (II). Then q preserves initiality. Consequently then ℓ is a left-adjoint of q.

<u>Proof</u>: Given a T_2-initial source $(A, p_i : A \rightarrow A_i)$ in $\underline{A_2}$, and morphisms $h_i : B \rightarrow qA_i$ and $k : T_1B \rightarrow T_1qA$ s.t.

$$T_1B \xrightarrow{k} T_1qA \xrightarrow{T_1qp_i} T_1qA_i = T_1h_i$$

we have to look for a unique $h: B \to qA$ with $T_1 h = k$ and $qp_i \circ h = h_i$ for each i. Consider the following diagram where $\varepsilon: \ell q \to 1$ is given by the condition $\ell q \leq 1$ and $(m_j, B)_j$ is a T_1-final sink given by codenseness of E_1 .

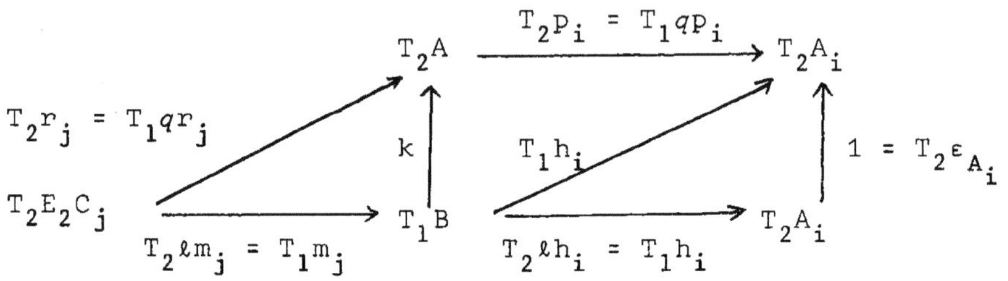

Here r_j (for each $j \in J$) is the lift of $k \circ T_1 m_j$ due to T_2-initiality of (A, p_i). Finality of the sink (m_j, B) now provides a unique $h: B \to qA$ with $T_1 h = k$ and $h \circ m_j = qr_j$. To get $qp_i \circ h = h_i$ it is sufficient to prove $qp_i \circ h \circ m_j = h_i \circ m_j$. By the following proposition we have $h_i \circ m_j = q(\tilde{m}_{ij})$ and hence $T_1 q(p_i \circ r_j) = T_1 q(\tilde{m}_{ij})$. By faithfulness of T_2 we get $p_i \circ r_j = \tilde{m}_{ij}$, hence $qp_i \circ h \circ m_j = qp_i \circ qr_j = h_i \circ m_j$.

To prove uniqueness of h, assume that for some $\bar{h}: B \to qA$ we have $T_1 \bar{h} = k$ and $qp_i \circ \bar{h} = h_i$ for each i. Then $T_1 \bar{h}$ will have the same final lift by (m_j, B) as k, by a similar argument as above, hence $\bar{h} = h$. Again by Tholen's taut lift Theorem [8] we conclude that q has a left adjoint ℓ with $T_2 \circ \ell = T_1$. Now apply proposition 1. □

We will point out later to what extent this proposition covers Dubuc's construction mentioned in the introduction.

We go on establishing some conditions equivalent to the crucial hypothesis in the preceding proposition that $\ell E_1 = E_2$.

<u>Proposition 3</u> : Let q be any diagonal of (I) and ℓ the corresponding coarsest quasidiagonal of (II). Then the following conditions are equivalent:

(i) ℓ is a diagonal of diagram (I)

(ii) $\ell E_1(C) = E_2(C)$ for each $C \in \text{ob } \underline{C}$

(iii) for every $h: E_1 C \longrightarrow qB$ there exists some $\bar{h}: E_2 C \to B$ with $T_2 \bar{h} = T_1 h$

(iv) for all $C \in \text{ob } \underline{C}$ and $B \in \text{ob } \underline{A}_2$ the maps $q: \underline{A}_2(E_2 C, B) \longrightarrow \underline{A}_1(qE_2 C, qB) = \underline{A}_1(E_1 C, qB)$ are surjective.

<u>Proof</u> : (i) and (ii) are obviously equivalent since T_2 is faithful. (iii) and (iv) are equivalent since q commutes with the underlying functors T_i which in addition are faithful.

(i) ⇒ (iv) : Given $h: E_1 C \longrightarrow qB$ consider

$$\mu_B: \ell q B \longrightarrow B \quad \text{with} \quad T_2 \mu_B = 1$$

(where (μ_B) is the natural transformation $\ell q \longrightarrow 1$ arising from the condition $\ell q \leqslant 1$ [2]).
Hence we get $T_2(\mu_B \circ \ell h) = T_1 h$ where $\mu_B \circ \ell h: \ell E_1 C = E_2 C \longrightarrow B$ according to (i).

(iv) ⇒ (i): As in the preceding proof we have natural morphisms $\mu_C: \ell q E_2 C = \ell E_1 C \longrightarrow E_2 C$ with $T_2(\mu_C) = 1$ for each $C \in \text{ob } \underline{C}$. Hence it remains to show that for each C

$$1: T_2 E_2 C \longrightarrow T_2 \ell E_1 C = T_2(j_C) \quad \text{for some} \quad j_C: E_2 C \longrightarrow \ell E_1 C.$$

By definition of ℓ this would be the case provided $T_2 E_2 C \xrightarrow{1} T_2 \ell E_1 C \xrightarrow{T_2 \bar{k}} T_2 B = T_2(E_2 C \xrightarrow{k'} B)$ where the family $(\bar{k})$ is the T_2-initial lift of the family of all maps $(T_1 k = T_1 E_1 C \longrightarrow T_1 qB)$.

Now $T_2\bar{k}\circ 1 = T_1 k = T_2 qk' = T_2 k'$ by (iv). □

Remarks

(i) The equivalent conditions of proposition 3 are obviously fulfilled if the diagonal q is full or if E_2 is (initially) dense (which will not be the case in general).

(ii) The equivalent conditions of proposition 3 are fulfilled in the following cases

(a) $\underline{X}$ = Set, $\underline{C}$ = $\underline{\text{CompT}_2}$ (compact Hausdorff spaces)

(a_1) $\underline{A}_1$ = Span (Spanier-spaces), $\underline{A}_2$ = Top (topological spaces)

(a_2) $\underline{A}_1$ = Span A_2 = Unif (Uniform spaces)

Here (i) is checked easily if q is the finest diagonal (cp [4]).

(b) $\underline{X} = \underline{1}$, $\underline{C}$ any partially ordered set considered as a concrete category over $\underline{1}$

$\underline{A}_1$ any join-dense completion of $\underline{C}$

$\underline{A}_2$ any meet-dense completion of $\underline{C}$ (cp [5]).

(iii) The conditions of proposition 3 are not fulfilled automatically as is shown by the following example in the setting of remark (iib).

Let $\underline{C}$ be the partially ordered set

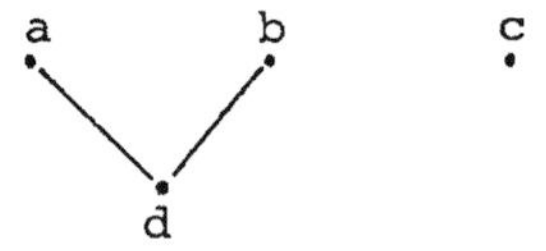

and take $\underline{A}_1$ the MacNeille completion of $\underline{C}$ and $\underline{A}_2$ the largest final completion $\underline{C}^{-1}$ in the notation of [5]. If the q is chosen as the coarsest diagonal one gets

$$q(x) = \inf_1 \{y \in \underline{C} \mid x \leq y\}$$
$$\ell(x) = \inf_2 \{y \in \underline{A}_2 \mid x \leq q(y)\}.$$

From these equations it follows easily (e.g. using the graphs of [5]) that $\ell(a) = d(\neq a)$.

(iv) The adjointness established in proposition 2 specializes in the setting of remarks (ii, b) to a pair of monotone maps $q:\underline{A}_2 \longrightarrow \underline{A}_1$, $\ell:\underline{A}_1 \longrightarrow \underline{A}_2$ which map $\underline{C}$ identically and which moreover satisfy the following relations

$x \leqslant qy$ iff $\ell x \leqslant y$
$y \leqslant \ell x$ iff $qy \leqslant x$
x, qy are incomparable iff $y, \ell x$ are incomparable.

The following relations between the functors ℓ and q are already proved to be useful in the special instances of remark (ii, a) (cp [4]).

Proposition 4 : Given any diagram (I) fulfilling the basic assumptions with $q:\underline{A}_2 \to \underline{A}_1$ an arbitrary diagonal. If then ℓ is a coarsest diagonal of (II) the following statements hold:

(i) $q\ell q \leqslant_{T_1} q$

(ii) $\ell q\ell \leqslant_{T_2} \ell$

Proof: (i) By construction of ℓ there is the canonical natural transformation

$$\mu:\ell q \to 1 \quad \text{with} \quad T_2\mu = 1.$$

Hence $q\mu:q\ell q \to q$ and $T_1(q\mu) = T_2\mu = 1$.

(ii) With μ as in the preceding proof we have
$\mu\ell:\ell q\ell \to \ell$ with $T_2(\mu\ell) = 1$. □

Proposition 5 : Under the assumptions of proposition 2 the following assertions hold

(i) $\ell q\ell = \ell$
(ii) $q\ell q = q$.

Proof : (i) is an application of [2, Thm 2.3 (ii)]
(ii) By adjunction we have a natural transformation $\nu:1 \to q\ell$

with $T_1(\nu) = 1$.

Hence $\nu q : q \to q\ell q$ is natural with $T_1(\nu q) = 1$ i.e. $q \leq_{T_1} q\ell q$. Together with proposition 5(i) this gives the result. □

The above relations between ℓ and q allow the introduction of a new common subcategory of the categories $\underline{A}_1$ and $\underline{A}_2$:

Let $\underline{K}$ be the full subcategory of $\underline{A}_2$ consisting of those objects satisfying the equation $\ell q B = B$, and let $\underline{L}$ be the full subcategory of $\underline{A}_1$ consisting of all q-images of $\underline{K}$.

Corollary : Assume the assumptions of proposition 2.
Then $\text{ob}\,\underline{K} = \ell(\text{ob}\,\underline{A}_1)$ and $\text{ob}\,\underline{L} = q(\text{ob}\,\underline{A}_2)$, and the restrictions of ℓ and q serve as an isomorphism of categories $\underline{K} \simeq \underline{L}$.
Moreover $\underline{K}$ is coreflexive in $\underline{A}_2$ with the counit ε as coreflexion map and $\underline{L}$ is reflexive in $\underline{A}_1$ with the unit η as reflexion map.

If we call $\underline{K}$ (as in [4]) the category of model generated spaces one can rephrase the last results as follows:
Under the assumptions of proposition 2 the category of model generated spaces is at the same time a full reflexive subcategory of $\underline{A}_1$ and a full coreflexive subcategory of $\underline{A}_2$. (cp. [4, Cor. 2.3]).

Finally we want to describe the category $\underline{K}$ internally.

Proposition 6 : Assume the assumptions of proposition 2, where T_1 is topological in addition. Then for each $A \in \text{ob}\,\underline{A}_2$ the following statements are equivalent

(i) $A \in \text{ob}\,\underline{K}$

(ii) the sink $(E_2 C \xrightarrow{f} A)$ of all morphisms with domain in $\underline{C}$ and codomain A is T_2-final.

Proof: To show that (i) implies (ii) assume $A = \ell q A$.
Since the sink $(E_1 C \xrightarrow{k} qA)$ of all such k is T_2-final by assumption, the sink $(\ell E_1 C \xrightarrow{\ell k} \ell q A) = (E_2 C \xrightarrow{\ell k} A)$ is T_2-final by the dual of the (generalized) taut lift theorem. This implies (ii).

To prove the converse assume that $(E_2C \xrightarrow{f} A)$ is T_2-final. It suffices to show that

$$1 : T_2A \to T_2\ell qA = T_2(A \xrightarrow{i} \ell qA)$$

which by definition of ℓ and our assumption is equivalent to

$$T_2E_2C \xrightarrow{T_2f} T_2A \xrightarrow{1} T_2\ell qA \xrightarrow{T_2\tilde{h}} T_2A' = T_2(E_2C \xrightarrow{f_h} A')$$

for each f of the above sink and each $\tilde{h}$ of the initial lift of the source of all maps $T_1h : T_1qA \to T_1qA'$.
Now the above composite is equal to

$$T_1E_1C \xrightarrow{T_1\bar{f}} T_1qA \xrightarrow{1} T_1qA \xrightarrow{T_1h} T_1qA' = T_1(h \circ \bar{f})$$

where h is as described above and $\bar{f}$ corresponds to f by adjunction.

Now condition (iii) of proposition 3 gives the result. □

With notations and assumptions as above we get in particular

Corollary 1 : $\underline{K} = \underline{A}_2$ iff E_2 is finally dense.

Corollary 2 : Let $\underline{C}$ be at the same time a full finally dense subcategory of topological categories $\underline{A}_1$ and $\underline{A}_2$ and assume that the equivalent conditions of proposition 3 hold. Then $\underline{A}_2$ is (isomorphic to) a reflective subcategory of $\underline{A}_1$.

Corollary 3 : The MacNeille completion of any concrete category $U : \underline{A} \to \underline{X}$ is reflectively contained in any finally dense topological completion of $\underline{A}$.

Using the fact that the concept of topological functors is self-dual one gets obvious dualizations of the preceding results as for example:

Corollary 3^{op} : The MacNeille completion of any concrete category $U : \underline{A} \to \underline{X}$ is a coreflective subcategory of any initially dense topological completion of $\underline{A}$.

Examples

(i) Assume in the general setting that $\underline{X} = \underline{Set}$ and $\underline{A}_1$ is the category of quasispaces determined by

$U := T_2E_2 : \underline{C} \to \underline{Set}$ (and a Grothendieck-topology J on $\underline{C}$) in the sense of [4]. Hence objects of $\underline{A}_1$ are pairs $(X, ad(C,X)_{C \in ob \underline{C}})$ where X is set and $ad(C,X)$ is a set of maps $C \to X$ where the family $(ad(C,X))_C$ of admissible maps is among others subject to the condition

(*) $\alpha \in ad(C,X)$, $f \in \underline{C}(D,C) \Rightarrow \alpha f \in ad(D,X)$.

An $\underline{A}_1$-morphism $(X, ad(C,X)_C) \to (Y, ad(C,Y)_C)$ is a map $f : X \to Y$ such that $f\alpha \in ad(C,Y)$ whenever $\alpha \in ad(C,X)$. The embedding $E_1 : \underline{C} \to \underline{A}$ with $E_1C = (UC, \underline{C}(C',C)_{C' \in ob \underline{C}})$ is easily checked to be finally dense (using (*)). The finest diagonal $q : \underline{A}_2 \to \underline{A}_1$ arising from this situation is then given by Dubuc's formula [4] $q(A) = (T_2A, \underline{A}_2(E_2C,A)_C)$ as is shown by a straightforward calculation. Hence propositions 2, 5, and it's corollary apply to this situation since moreover very obviously condition (iii) of proposition 3 is fulfilled. For the identification of categories $\underline{K}$ arising in this setting we refer to [4, Ex 2.12, 2.13].

(ii) The MacNeille completion of $\underline{CompT_2}$ which is well known to be a somewhat unhandy category is a coreflexive subcategory of the category $\underline{Prox}$ of proximity spaces since the latter is an initially dense completion of $\underline{CompT_2}$ (cp. [6]).

This examples shows how the techniques developed in this note may be used in order to get additional relations between different topological completions of a concrete category.

Acknowledgement

The author is indebted to G.C.L. Brümmer whose most valuable comments led in particular to the present general form of proposition 2. Thanks also to W. Tholen who drew the author's attention to [8].

Remarks added in proof:

1. If one drops amnesticity and the properness-condition in the definition of a topological functor, everything goes through. One only has to replace the obvious equations by equivalences (e.g. $T_2 \circ \ell \simeq T_1$, $q\ell q \simeq q$); cp. [9] for the construction of diagonals in this case.

2. In applications one is often concerned with extremal - i.e. finest or coarsest - diagonals (cp [2] or Dubuc's construction). It can be proved that the construction of ℓ to a given q changes the order.

REFERENCES

[1] Adamek, J., Herrlich, H., Strecker, G.E., Least and largest initial completions, Comment. Math. Univ. Carolinae 20 (1979), 43-77

[2] Brümmer, G.C.L., Topological functors and structure functors, Springer Lecture Notes in Math 540 (1976), 109-135

[3] Brümmer, G.C.L., Hoffmann, R.-E., An external characterization of topological functors, Springer Lecture Notes in Math. 540 (1976), 136-151

[4] Dubuc, E.J., Concrete quasitopoi, Proc. Durham Conference, Springer Lecture Notes in Math. 753 (1979), 239-254

[5] Herrlich, H., Initial completions, Math. Z. 150 (1976), 101-110

[6] Herrlich, H.,Strecker, G.E., Semi-universal maps and universal initial completions, Pacific J. Math. 82 (1979), 407-428

[7] Porst, H.-E., Characterizations of MacNeille completions and topological functors, Bull. Austral. Math. Soc. 18 (1978), 201-210

[8] Tholen, W., On Wyler's taut lift theorem, General Topol. and its Appl. 8 (1978), 197-206

[9] Tholen, W., Wischnewsky, M.B., Semitopological functors II, J. Pure and Appl. Alg. 15 (1979), 75-92

10] Wyler, O., On the categories of general topology and topological algebra, Arch. Math. (Basel) 22 (1971), 7-17

Fachbereich Mathematik
Universität Bremen
2800 BREMEN
Fed. Rep. of Germany

INTERNAL CATEGORIES AND CROSSED MODULES

Timothy Porter
School of Mathematics and Computer Science
University College of North Wales

Bangor, Gwynedd, Wales(U.K.)

This note is an attempt to indicate how one might initiate a combinatorial study of presentations in algebraic categories other than that of groups. It will concentrate attention on one construction, namely that of the crossed module associated to a presentation. This construction occurs in the study of the identities between relations in the presentation (Brown-Huebschmann [1]) as well as forming a part of the crossed resolution used by Huebschmann [3] in group cohomology. Generalising this construction to other categories of interest e.g. algebras, Lie algebras, monoids, small categories etc., poses certain problems. In fact in the last two cases the existence of associated crossed modules is in doubt. To be able to handle such cases we note a well known result on internal categories in the category of groups, namely that this category is equivalent to that of crossed modules. Thus we could have assigned to each presentation an internal category and have used this instead. The existence of free internal categories in essentially algebraic categories is known, hence it only remains to give as neat as possible a description of such a construction. In the case of small categories (over a fixed object set), and hence of monoids, such a construction was given and used by Mitchell [4]. For another large family of algebraic categories, those monadic over the category of groups (i.e. groups with operations), we give an explicit construction, which generalises that in the group case. In many of these categories of groups with operations, the equivalence of crossed modules and internal categories still holds, so one can replace the internal category by a crossed module which is smaller and hence easier to study.

The plan of the paper is as follows. In section 1, we briefly review the theory of crossed modules (in Groups) and indicate how one may define analogous objects in other settings. This is followed by an account of Mitchell's construction. In section 3, we prove the main result on the construction of free internal categories within categories of "groups with operations"; in this the important point is not their existence, but the simple and explicit nature of the construction. Finally we illustrate this with a brief discussion of the situation in associative k-algebras for k a commutative ring.

1. Presentations of groups

We consider a presentation of a group G to be a triple $(X; R, v)$ where X is a set and $v : R \longrightarrow UF(X)$ is a function taking values in the underlying set of the free group on X, such that the cokernel of the adjoint map

$$\tilde{v} : F(R) \longrightarrow F(X)$$

is G .

Classically one had v an inclusion but as $\tilde{v}$ will usually not be a monomorphism, this restriction is misleading. In fact the problem of studying identities amongst the relations is essentially that of calculating the kernel of $\tilde{v}$ or rather $\tilde{v}$ extended over a "formal normal closure" of $F(R)$ along $\tilde{v}$.

A crossed module consists of a group homomorphism $\theta : A \longrightarrow B$ together with an action of B on A (written $(a, b) \longmapsto a.b$) such that the following two properties hold:

(i) for all $a \in A$, $b \in B$

$$\theta(a.b) = b^{-1}\theta(a)b$$

(ii) for all a_1 , $a_2 \in A$,

$$a_1.\theta(a_2) = a_2^{-1}a_1a_2 \qquad \text{(the Peiffer identity)}$$

For future reference we note the following simple way of writing these two conditions:

<u>Lemma</u>

$\theta : A \longrightarrow B$ is a crossed module if and only if the following diagram commutes

$$\begin{array}{ccccccccc}
1 & \longrightarrow & A & \longrightarrow & A \ltimes A & \rightleftarrows & A & \longrightarrow & 1 \quad \ldots\ldots (1) \\
 & & \downarrow{=} & & \downarrow & & \downarrow{\theta} & & \\
1 & \longrightarrow & A & \longrightarrow & B \ltimes A & \rightleftarrows & B & \longrightarrow & 1 \quad \ldots\ldots (2) \\
 & & \downarrow{\theta} & & \downarrow & & \downarrow{=} & & \\
1 & \longrightarrow & B & \longrightarrow & B \ltimes B & \rightleftarrows & B & \longrightarrow & 1 \quad \ldots\ldots (3)
\end{array}$$

where the three rows are split exact, thus representing respectively the action of A on itself by conjugation, of B on A and of B on itself by conjugation; the central unlabelled vertical maps are the obvious maps induced by θ on the semi-direct products.

The proof is routine.

There is clearly a category of crossed modules and a forgetful functor from that to the category, $(\text{Groups})^{\mathbf{2}}$, of group morphisms. We need to construct a left adjoint to this functor.

Suppose $f : H \longrightarrow G$ is a group homomorphism, then we can form the free G-morphism on H by taking :

(a) $H_1 = \coprod_{g \in U(G)} H_g$ with isomorphisms $\theta_g : H_g \cong H$ and with the natural G-action, obtained by permuting the indices of the coproduct,

(b) $f_1 : H_1 \longrightarrow G$, given by $f_1(h_g) = g^{-1}f(\theta_g(h_g))g$ if $h_g \in H_g$.

Now it should be clear how to form the free crossed module.

Any element of the form

$$h^{-1}k^{-1}h(k.f_1(h)) \qquad h , k \in H_1 \quad \ldots. (4)$$

is in $\operatorname{Ker} f_1$. Moreover the subgroup, P , of H_1 generated by such elements is normal and G-invariant. Hence on forming the quotient group $C = H_1/P$ one finds that it has a natural G-action and that there is an induced G-morphism

$$\bar{f} : C \longrightarrow G$$

satisfying axiom (ii) : for all c_1 , $c_2 \in C$, $c_1.\bar{f}(c_2) = c_2^{-1}c_1c_2$.

In the study of identities, elements such as in (4) are always present regardless of the form of the presentation. As such they have little use in the initial stages of understanding the identities, and one loses little on dividing out by them. These elements are often called Peiffer elements and the subgroup P , the Peiffer group of f .

Although the lemma clearly allows one to generalise the definition of a crossed module to other algebraic categories than groups, the important existence of free crossed modules on morphisms in such categories is not immediate. The equational way of defining a crossed module suggests that such an existence theorem should be true, but it is not obvious initially how to provide a general categorical construction for this. To circumvent these difficulties, we replace the category of crossed modules (in Groups) by that of internal categories again in Groups. The equivalence between these two categories is fairly well known (Brown-Spencer [2]) but we give a sketch of the proof as it will be useful later.

Given any group homomorphism $\theta : A \longrightarrow B$ and a right B-action on A , one can form the semidirect product, $B \ltimes A$, and define two functions

$$B \ltimes A \underset{d_1}{\overset{d_0}{\rightrightarrows}} B$$

where $d_0(b, a) = b$, $d_1(b, a) = b\theta(a)$. d_0 is a homomorphism but d_1 is a homomorphism if and only if θ is a B-morphism, (i.e. satisfies (i) above). There is also the natural splitting $s : B \longrightarrow B \ltimes A$ and one can attempt to define a composition on "composable pairs" by

$$(b, a) \circ (b\theta(a), c) = (b, ac) \quad ..... \quad (5)$$

This is associative and with the $s(b)$ as identity elements one gets a category. However this category will not be internal unless $\circ$ is a group homomorphism and this holds if and only if (ii) holds.

Conversely given $(C, B; d_0, d_1, s, \circ)$, an internal category in Groups, one can take $A = \mathrm{Ker}\ d_0$, $\theta = d_1|A$ and the B-action on A induced by conjugation within C , to get a crossed module $\theta : A \longrightarrow B$.

It is easily checked that in many categories of "groups with operations", corresponding approximately to Orzech's "categories of interest", [5], a similar result holds. However for small categories (with fixed object set) and thus for monoids, one does not get an equivalence only an embedding of crossed modules into internal categories. In another family of important categories, including those of unitary algebras over commutative rings, the formation of kernels takes one out of the category concerned, so the construction of crossed modules from internal categories breaks down completely. In the next section we consider the first of these cases and the solution given by Mitchell to this basic problem in [4].

2. Presentations of categories and 2-categories

Let $\mathbb{C}$ be a small category with 0 its set of objects and let R be a set of relations in $\mathbb{C}$, that is, R is a set of pairs (a, b) of elements of $\mathbb{C}$ with the same domain and codomain.

Suppose $A , B : p \longrightarrow q$ in $\mathbb{C}$ we say that a symbol (x, a, b, y) is a path (of length one) from A to B if

$$A = x\,a\,y$$
$$x\,b\,y = B ,$$

and with (a, b) or $(b, a) \in R$.

Let $\mathbb{P}(p, q)$ be the set of all paths of length one between elements of $\mathbb{C}(p, q)$ and let $\varepsilon_0(x, a, b, y) = x\,a\,y$, $\varepsilon_1(x, a, b, y) = x\,b\,y$. For each $p, q \in 0$, let

$$(\Omega_0(R)(p, q) , \mathbb{C}(p, q); \varepsilon_0, \varepsilon_1, \ ; \perp)$$

be the free category on the directed graph,

$$(\mathbb{P}(p, q), \mathbb{C}(p, q); \varepsilon_0, \varepsilon_1) .$$

We call $\perp$ the "vertical composition". There is also a possible "horizontal composition":

Suppose we write $(\underline{x}, \underline{a}, \underline{b}, \underline{y})$ for an element of $\Omega_0(R)(p, q)$ where $\underline{x} = (x_1, \ldots, x_n)$ etc. and $\varepsilon_0(\underline{x}, \underline{a}, \underline{b}, \underline{y}) = x_1\, a_1\, y_1 = A$ say, $\varepsilon_1(\underline{x}, \underline{a}, \underline{b}, \underline{y}) = x_n\, b_n\, y_n = B$; similarly $(\underline{u}, \underline{c}, \underline{d}, \underline{v})$ in $\Omega_0(R)(q, r)$ with $\varepsilon_0(\underline{u}, \underline{c}, \underline{d}, \underline{v}) = C$, $\varepsilon_1(\underline{u}, \underline{c}, \underline{d}, \underline{v}) = D$. Define the "horizontal composite" of these two elements by

$$(\underline{x}, \underline{a}, \underline{b}, \underline{y}) \cdot (\underline{u}, \underline{c}, \underline{d}, \underline{v}) = (\underline{x}, \underline{a}, \underline{b}, \underline{y}C) \perp (B\underline{u}, \underline{c}, \underline{d}, \underline{v}) ,$$

where $\underline{y}C = (y_1C, y_2C, \ldots, y_nC)$ etc.

One can represent this schematically as follows:

A path of length one (x, a, b, y) is

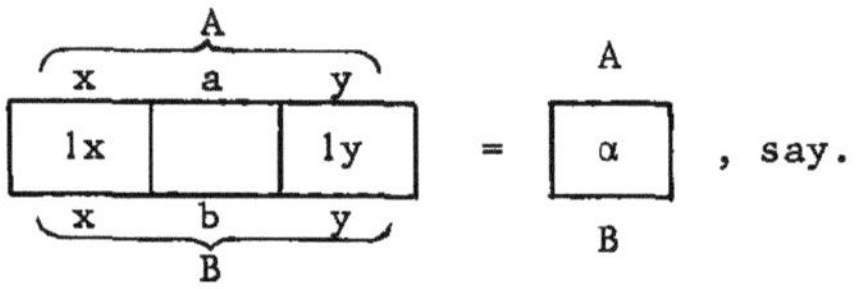

Vertical composition is the obvious :

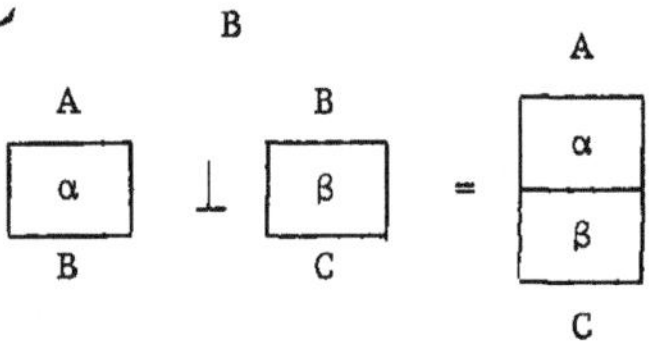

"Horizontal composition" is

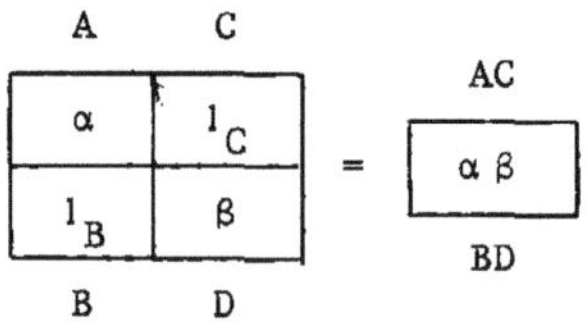

This looks fine until one checks the "Godement interchange law"

$$(\alpha \perp \beta) \; . \; (\gamma \perp \delta) = (\alpha \; . \; \gamma) \perp (\beta \; . \; \delta)$$

which will make composition into a functor and the $\Omega_0(R)$ into a 2-category - it fails.

The problem is that there is no reason why one should have the equality

$$\begin{array}{|c|c|}\hline \alpha & 1 \\ \hline 1 & \beta \\ \hline \end{array} \; = \; \begin{array}{|c|c|}\hline 1 & \beta \\ \hline \alpha & 1 \\ \hline \end{array}$$

This is the analogue of the Peiffer identity in this setting and one should probably call such pairs Peiffer pairs. On dividing out by the congruence generated by these Peiffer pairs one obtains a 2-category.

Remark

The analogy between Peiffer pairs and Peiffer elements seems very strong, but it will need more study to be certain that the rôles played by them are as close as it seems. Mitchell, [4], uses a notion of degenerate closed path to define the congruence relation. This notion is the analogue of that of a primary identity sequence in combinatorial group theory. The relationship between these latter and the Peiffer elements is explored in [1].

3. Presentations and internal categories

As suggested above, in some cases internal categories may be more easily associated to presentations than crossed modules. In fact considering the obvious definition of presentation in a general situation, this is fairly clear.

Suppose given $\mathbb{C} \underset{F}{\overset{U}{\rightleftarrows}} \mathbb{D}$, F left adjoint to U and an object C of $\mathbb{C}$. A presentation of C relative to $\mathbb{D}$ should consist of an object, X , of generators within $\mathbb{D}$ and a pair of maps

$$R \underset{d_1}{\overset{d_0}{\rightrightarrows}} UF(X)$$

such that the adjoint diagram

$$F(R) \underset{\tilde{d}_1}{\overset{\tilde{d}_0}{\rightrightarrows}} F(X)$$

has as equaliser an epimorphism $\mu : F(X) \longrightarrow C$.

Thus a presentation is essentially an internal directed graph. To construct a free internal category on such a directed graph, one can mimic the construction used in Sets, taking into account the difficulty that in general coproduct and pullbacks need not commute. This construction works in many algebraic categories and so guarantees that the basic program suggested in this note is feasible. However in a large and interesting class of categories one can say more; one can control the construction of the free category much as one could the construction of the free crossed module in Groups.

Initially let us suppose the category C has coproducts. In this case one can form, on any internal directed graph, an internal "directed graph with identities" or D.G.I., that is one with distinguished vertex loops. The construction is simple.

Given

$$C_1 \underset{d_1}{\overset{d_0}{\rightrightarrows}} C_0 \quad \text{in} \quad DG(\mathbb{C})$$

we form

$$C_0 \coprod C_1 \underset{d_1}{\overset{d_0}{\rightrightarrows}} C_0 \quad \text{in} \quad DGI(\mathbb{C})$$
$$\overset{\longleftarrow}{s}$$

with the d_0 and d_1 extended in the natural way and with s the inclusion morphism. This provides a left adjoint for the forgetful functor from $DGI(\mathbb{C})$ to $CG(\mathbb{C})$.

Now we must impose extra conditions on $\mathbb{C}$. We shall assume there is a forgetful functor, U', from $\mathbb{C}$ to Groups, which preserves "internal categories". (If U' preserves pullbacks, this suffices so, for instance, if U' has a left adjoint, the condition is satisfied.)

In groups and hence in $\mathbb{C}$, the composition in internal categories is determined by the group multiplication. More precisely the interchange law,

$$(a \circ b) \,.\, (c \circ d) = (a \,.\, c) \circ (b \,.\, d)$$

which expresses the fact that $\circ$ is a group homomorphism readily implies (cf. Brown-Spencer [2]) that

$$a \circ b = a\, s\, (d_0(b))^{-1} b \quad \ldots\ldots \quad (6)$$

Thus if $b : y \longrightarrow z$ and $c : \ell \longrightarrow m$, the interchange law implies that in any internal category the equations

$$s(y)^{-1} b\, c\, s(m)^{-1} = c\, s(m^{-1})\, s(y)^{-1} b \quad \ldots\ldots \quad (7)$$

hold.

Thus given

$$D \rightleftarrows C_0 \quad \text{in} \quad DGI(\mathbb{C})$$

we may form a congruence on D by relating the two sides of (7) plus any other equations coming from the interchange laws for possible other operations. Dividing by this congruence gives a new d.g.i.

$$D/\!\sim\ \rightleftarrows C_0$$

and now defining $\circ$ by (6), one shows that this is the underlying d.g.i. of an internal category. The universal property is easily checked.

To summarise one has:

If $\mathbb{C}$ has a left exact "forgetful" functor to Groups then in the adjunction

$$Cat(\mathbb{C}) \rightleftarrows DGI(\mathbb{C})$$

the unit is naturally an epimorphism with kernel generated by "generalised Peiffer elements" such as

$$[s(y)^{-1} b\, c\, s(m)^{-1}][c\, s(m)^{-1}\, s(y)^{-1} b]^{-1}$$

and similarly for any other operations existing on objects of $\mathbb{C}$.

It is perhaps better to think of this as a list of instructions rather than as a theorem. To illustrate it, we will examine the case of a category of unitary k-algebras for k a commutative ring.

4. Example

Let $\mathbf{C}$ = k-algebras (with 1), $U : \mathbf{C} \longrightarrow Ab \subset Groups$ preserves pullbacks so our general principle holds. We consider presentations relative to Sets i.e. for the adjoint pair

$$\text{k-alg} \underset{F}{\overset{U}{\rightleftarrows}} \text{Sets}$$

A presentation $(X; R; d_0, d_1)$ leads to a d.g.i.

$$k[X] \underset{k}{\otimes} k[R] \underset{\underset{s}{\longleftarrow}}{\overset{d_0}{\underset{d_1}{\rightrightarrows}}} k[X]$$

and hence to Peiffer elements for $+$ and $\times$. The Peiffer identity for $+$ is trivial as $+$ is commutative. That for $\times$ reduced to the expression

$$(-b\,c + s(y)(c - s(m) + d)) = (a\,d - (a + b - s(y))\,s(m)) \;\ldots\ldots \quad (8)$$

Dividing out by this congruence leaves an internal category. Simplification occurs if, forgetting the unit elements, we put $A = \text{Ker}\, d_0$, $A_0 = k[X]$, $\theta = d_1|A$. If $b \in A$ and $c \in \text{Ker}\, d_1$, (8) gives $b\,c = 0$. As for any c, $c - s\,d_1(c) \in \text{Ker}\, d_1$, $b\,c = b\,s(d_1(c))$. Similarly $c\,b = s\,d_1(c)b$. Remembering how the A_0-action on A comes from the splitting s, this gives us equations

(i) $\theta(a'\,a) = a'\theta(a)$, $\theta(a\,a') = \theta(a)a'$ for $a \in A$, $a' \in A_0$

(ii) $\theta(a)a' = a\,a' = a\theta(a')$ for $a, a' \in A$.

These correspond exactly to commutativity of the diagram (1) - (2) - (3) considered earlier and so θ is a crossed module.

One can construct this crossed module directly as follows. Let $A_0^+[R]$ be the k-algebra (without 1) of polynomials over A_0 in the (non-commuting) indeterminates, R, having zero constant term. There is a map $d_0 - d_1 : A_0^+[R] \longrightarrow A_0$, which is zero on elements of the form $\theta(a)a' - a\,a'$ etc. Let P be the ideal generated by such elements, then $A = A_0^+[R]/P$ and θ is the induced map.

Consequences of this construction will be explored in another paper.

References

1. R. Brown and J. Huebschmann, Identities among relations, to appear in: Low Dimensional Topology, Ed. R. Brown and T.L. Thickstun, London Math. Soc. Lecture Notes, C.U.P.
2. R. Brown and C.B. Spencer, G-groupoids, crossed modules and the fundamental groupoid of a topological group, Proc. Kon. Ned. Akad. v. Wet., 19 (1976) 296-302.
3. J. Huebschmann, Crossed n-fold extensions and cohomology, Comm. Math. Helv., 55 (1980) 302-314.
4. B. Mitchell, Rings with Several Objects, Adv. Math., 8 (1972) 1-161.
5. G. Orzech, Obstruction Theory in Algebraic Categories (I, II), Journ. Pure Appl. Algebra, 2 (1972) 287-314, 315-340.

SUBDIRECT IRREDUCIBILITY AND CONGRUENCES

Aleš Pultr, Prague

Subdirect irreducibility was originally defined for varieties of algebras ([1]). An object was said to be subdirectly irreducible if

(1) for each monomorphism $\mu : A \longrightarrow \prod A_i$ such that all the $p_i \mu : A \longrightarrow A_i$ are onto, at least one of the morphisms $p_i \mu$ is an isomorphism.

Equivalently, A is subdirectly irreducible if

(2) in any system $\{E_i\}$ of congruences on A such that $\bigcap E_i$ is trivial, some of the E_i is trivial.

When extending the notion to more general concrete categories, we should follow the (perhaps less elegant) form (1). It expresses explicitly what the subdirect irreducibility really means, namely that s.i. objects are those which cannot be non-trivially obtained from others when using the constructions of forming products and taking subobjects ("prime objects" with respect to this basic constructions arsenal).

Regarding the statement (2) in a more general setting we are, first, encountered with the problem to tell what we mean by a congruence. But this is easy : intuitively speaking, one has in mind an equivalence on the underlying set of an object such that the factor-object remains in the category in question (for a formal definition see 2.1 below). The trouble is that thus understood extension of (2) <u>is not</u> equivalent with the "primeness" above and, moreover, there is not even an implication in any of the directions. Consider the following simple examples : In the category of graphs (sets with binary relations) (X,R) and relation preserving mappings take $A = (2, \{(0,1),(1,0)\})$. A is not subdirectly irreducible, since A is a subobject of $B \times B$ where $B = (2, 2\times 2 \smallsetminus \{(0,0)\})$. But, an intersection of any system of non-trivial congruences on A (in this category, all equivalences are congruences) is obviously non-trivial. On the other hand, $C = (3, 3\times 3 \smallsetminus \{(0,1)\})$ is subdirectly irreducible in the sense of never being a subobject of a product unless contained as a subobject in some of the factors; here, however, one has a system of non-trivial congruences intersecting in the trivial one.

The aim of this note is to present a characterization of finite subdirectly irreducible objects in reasonably general concrete categories in terms of the behavior of congruences (Theorem 3.3 ; cf. [4] where such a characterization was proved for the special case of classes of graphs).

The restriction to finite objects, besides technical reasons (to be able to apply Theorem 3.3 from [5] without modifications), is due mainly to the fact that for the finite case it is reasonably clear what the natural generalization of the subdirect irreducibility (as the "primeness") is, while in the infinite case this question needs further analysis (cf. [6], [7]). E.g., (1) with "monomorphism" replaced by "subobject" is not the same as "if A is a subobject of $\prod A_i$, it is a subobject of some of the A_i" which may be a more suitable definition and may be not.

§1. Definitions

1.1. A concrete category $(\mathcal{K},U)$ is a category $\mathcal{K}$ together with a fixed faithful $U:\mathcal{K} \longrightarrow \mathrm{Set}$. A monomorphism $\mu:A\rightarrow B$ is said to be a subobject (in $(\mathcal{K},U)$) if

whenever $U\mu \cdot f = U\psi$ for a $\psi:C\rightarrow B$, there is a $\varphi:C\rightarrow A$ with $U\varphi = f$.

(cf. e.g. [2], [3]).

Dually, an epimorphism $\varepsilon:B\rightarrow A$ is said to be a quotient (in $(\mathcal{K},U)$) if

whenever $f\cdot U\varepsilon = U\psi$ for a $\psi:B\rightarrow C$, there is a $\varphi:A\rightarrow C$ with $U\varphi = f$.

(One sees easily that if α is a subobject resp. a quotient, $U\alpha$ is one-one resp. onto, and that in everyday-life categories subobjects are what one intuitively understands under the term : embeddings of subspaces, of induced subgraphs, etc. Similarly it is with the quotients. In varieties of algebras, all monomorphisms are subobjects.)

1.2. An object A is said to be subdirectly irreducible in $(\mathcal{K},U)$ if for every subobject

$$\mu:A\longrightarrow\prod A_i$$

such that all the $p_i\mu:A\rightarrow A_i$ are onto, at least one of the $p_i\mu$

is an isomorphism.

1.3. Let $(\mathcal{K},U)$ be a concrete category. For a set X define a preordered class (sometimes called the fibre over X)

$$\mathcal{K}UX = (\{A : UA=X\}, \prec)$$

with $A \prec B$ iff there is an $\alpha : A \to B$ with $U\alpha = 1_X$ (in such a case we will write $\alpha : A \prec B$). We write $\mathcal{K}UA$ for $\mathcal{K}U(UA)$.

1.4. The characterization theorem 3.3 below will be proved for concrete categories satisfying the conditions (i) - (vi) we will now list. The reader will certainly observe that these conditions are quite commonly satisfied in everyday-life categories.

(i) U preserves the limits, $\mathcal{K}$ is complete and cocomplete.

(ii) If X is a set and $f : X \to UA$ an invertible mapping, there is an isomorphism φ with $U\varphi = f$.

(iii) If $\varphi : A \to B$ is an isomorphism and $U\varphi = 1_X$ then $\varphi = 1_A$.

(iv) Each $\mathcal{K}UX$ is a set and it is finite if X is.

(v) Every subcategory of $\mathcal{K}$ closed with respect to products and subobjects is reflective in $\mathcal{K}$.

(vi) For every morphism $\varphi : A \to B$ there is a decomposition

$$\varphi = (A \xrightarrow{\varepsilon} C \overset{\iota}{\prec} D \xrightarrow{\mu} B)$$

with μ a subobject and ε a quotient.

1.5. Remarks : 1. (iii) and (iv) make sure that $\mathcal{K}UX$ is a poset. According to (ii), a bijection $X \to Y$ induces an isomorphism $\mathcal{K}UX \cong \mathcal{K}UY$.

2. In categories of algebras, $\mathcal{K}UX$ are antichains. On the other hand, in a category with heredity (i.e. where for each object A and each $X \subset UA$ there is a subobject $\mu : B \to A$ with $U\mu : X \subset UA$), all the $\mathcal{K}UX$ are meet-semilattices (see [5 ;1.8]).

3. Obviously, if a coequalizer γ can be written as $\mu \cdot \varepsilon$ with μ a monomorphism and $U\varepsilon$ onto, then μ is an isomorphism. Consequently, in a category satisfying (vi) all coequalizers are carried by onto maps.

1.6. Convention : An object A is said to be maximal (in $(\mathcal{K},U)$) if it is a maximal element of $\mathcal{K}UA$. Similarly, A is said to be meet-irreducible if it is such in $\mathcal{K}UA$. (Note that in categories of algebras all the objects are maximal).

§2. Congruences and critical congruences

2.1. A $\mathcal{K}$-equivalence is a reflexive symmetric transitive relation

$$E \overset{\varepsilon_1}{\underset{\varepsilon_2}{\rightrightarrows}} X$$

in a category $\mathcal{K}$ (some may prefer the more usual representation $\varepsilon : E \to X\times X$ with $p_i\varepsilon = \varepsilon_i$). It is said to be a $(\mathcal{K},U)$-congruence if U preserves the coequalizer of $\varepsilon_1, \varepsilon_2$, i.e., if we have for $\gamma = \text{coequ}(\varepsilon_1,\varepsilon_2)$

$$U\gamma(x) = U\gamma(y) \quad \text{iff} \quad \exists e \in UE,\ U\varepsilon_1(e)=x \ \&\ U\varepsilon_2(e)=y\ .$$

2.2. Remarks : 1. In varieties of algebras the fact that a set-theoretical equivalence can be represented as a $\mathcal{K}$-equivalence suffices, the second condition being satisfied automatically. On the other hand, in categories with heredity (see 1.5.2), each set-theoretical equivalence can be represented as a $\mathcal{K}$-equivalence, while the second condition is essential.

To give an example of a category where both the conditions play a role at once consider the category of non-idempotent groupoids (the binary algebras satisfying the implication $x.x = x \Rightarrow x = y$) and take the algebra with three elements a,b,c and the multiplication given by $b.b = c$, $x.y = b$ otherwise. See what happens when identifying a with b but not with c (which violates the first condition) and when identifying b with c but not with a (which violates the second condition).

2. One checks easily that if U preserves limits, the intersection of a system of $(\mathcal{K},U)$-congruences is always a $(\mathcal{K},U)$-congruence.

2.3. Convention (in categories satisfying (i)-(vi)) : Let $(\varepsilon_1,\varepsilon_2):E \to A$ be a congruence, $\varepsilon:E \to A\times A$ defined by $p_i\varepsilon = \varepsilon_i$. Then we have a set-theoretical equivalence $E' = U\varepsilon(UE)$ on UA and for $\gamma = \text{coequ}(\varepsilon_1,\varepsilon_2):A \to A'$ an obvious bijection $f:UA' \to UA/E'$. Take the isomorphism $\varphi:A' \to A''$ with $U\varphi = f$. The object A'' will be denoted by

$$A/(\varepsilon_1,\varepsilon_2)$$

and the $\text{coequ}(\varepsilon_1,\varepsilon_2)$ will be, as a rule, represented as $\varphi\gamma: A \to A/(\varepsilon_1,\varepsilon_2)$.

2.4. Lemma : Let U preserve limits, $(\varepsilon_1,\varepsilon_2) = \mathrm{difKer}\,\varphi$, let $\varphi = \mu\circ\gamma$ with μ a monomorphism and γ a quotient. Then

(1) $U\gamma = \mathrm{coequ}(U\varepsilon_1, U\varepsilon_2)$,

(2) $\gamma = \mathrm{coequ}(\varepsilon_1,\varepsilon_2)$.

Proof: Since U preserves limits, we have the pullback

$$\begin{array}{ccc} UP & \xrightarrow{U\varepsilon_1} & UA \\ \downarrow{\scriptstyle U\varepsilon_2} & & \downarrow{\scriptstyle U\varphi} \\ UA & \xrightarrow{U\varphi} & UB \end{array}$$

and hence, up to isomorphism, $UP = \{(x,y)\in UA\times UA : U\varphi(x) = U\varphi(y)\} = \{(x,y)\in UA\times UA : U\gamma(x) = U\gamma(y)\}$ and $U\varepsilon_i(x_1,x_2) = x_i$. Thus, (1) holds. Now, let $\alpha\cdot\varepsilon_1 = \alpha\cdot\varepsilon_2$ .Then $U\alpha\cdot U\varepsilon_1 = U\alpha\cdot U\varepsilon_2$ and hence there is an f with $U\alpha = f\circ U\gamma$. Since γ is a quotient, $f = U\beta$ for a β satisfying $\alpha = \beta\cdot\gamma$ and (2) follows. □

2.5. Corollary : In a concrete category satisfying (i) and (vi)

- each difference kernel is a congruence,
- the coequalizers coincide with the quotients,
- each quotient is a coequalizer of its difference kernel. □

2.6. A $(\mathcal{K},U)$-congruence is said to be critical if it is non-trivial (i.e., $\varepsilon_1 \neq \varepsilon_2$) and if

whenever for $\iota : A \precneqq B$ the couple $(\iota\varepsilon_1,\iota\varepsilon_2):E\to B$ is still a congruence, one has $A/(\varepsilon_1,\varepsilon_2) \precneqq B/(\iota\varepsilon_1,\iota\varepsilon_2)$.

2.7. Remarks : More losely speaking : a critical congruence yields a stronger-structured quotient whenever applicable on an object with a stronger structure. In particular, an equivalence which is a congruence on A but on no strictly stronger B is a critical congruence.

It is worth realizing that $(\iota\varepsilon_1,\iota\varepsilon_2):E\to B$ is still a congruence (if $(\varepsilon_1,\varepsilon_2)$ is) iff $U\mathrm{coequ}(\iota\varepsilon_1,\iota\varepsilon_2) = U\mathrm{coequ}(\varepsilon_1,\varepsilon_2)$.

§3. Subdirect irreducibility and congruences

3.1. Lemma : In a category satisfying (i) and (vi) the following two statements on an object A are equivalent :

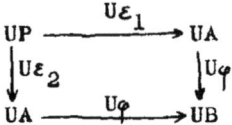

(a) There exists a system $(\mu_i : A\to B_i)$ such that $(\forall i\ \mu_i\cdot\alpha = \mu_i\cdot\beta)$ implies $\alpha = \beta$ but none of the μ_i is a monomorphism.

(b) There exists a system of non-trivial congruences on A such that its intersection is trivial.

Proof : Given the system (μ_i) consider the system of the difference kernels of the μ_i. On the other hand, given the system of congruences, consider their coequalizers. Checking the properties is a matter of an easy counting. □

3.2. Lemma : In a category satisfying (i) and (vi) the following two statements are equivalent for A with finite UA :

(a) For every $\varphi: A \to A'$ with $\mathrm{card}UA' < \mathrm{card}UA$ there is a $\iota: A \not\subseteq B$ and a $\varphi': B \to A'$ such that $\varphi' \cdot \iota = \varphi$.

(b) There is no critical congruence on A .

Proof : I. Let (a) hold and let $(\varepsilon_1, \varepsilon_2): E \to A$ be a non-trivial congruence. Consider $\gamma = \mathrm{coequ}(\varepsilon_1, \varepsilon_2): A \to A'$. We have $\mathrm{card}UA' < \mathrm{card}UA$ and hence there is a $\iota: A \not\subseteq B$ and a $\gamma': B \to A'$ such that $\gamma' \cdot \iota = \gamma$.

Let $\varphi \cdot \iota \cdot \varepsilon_1 = \varphi \cdot \iota \cdot \varepsilon_2$ for some $\varphi: B \to C$. Then there is a $\psi: A' \to C$ such that $\psi\gamma = \varphi\iota$. Since, however, $\gamma = \gamma'\iota$ and ι is an epimorphism, we obtain $\psi\gamma' = \varphi$. Thus, $\gamma' = \mathrm{coequ}(\iota\varepsilon_1, \iota\varepsilon_2)$. Consequently, $(\iota\varepsilon_1, \iota\varepsilon_2)$ is a congruence on B (we have $U\gamma' = U\gamma$ and $U(\iota\varepsilon_i) = U\varepsilon_i$) and we have $B/(\iota\varepsilon_1, \iota\varepsilon_2) = A' = A/(\varepsilon_1, \varepsilon_2)$ so that $(\varepsilon_1, \varepsilon_2)$ is not critical.

II. Let (b) hold and let $\varphi: A \to A'$ be such that $\mathrm{card}UA' < \mathrm{card}UA$. Consider a decomposition $\varphi = \mu \cdot \gamma$ with γ a quotient and μ a monomorphism. Then $\gamma = \mathrm{coequ}(\varepsilon_1, \varepsilon_2)$ where $(\varepsilon_1, \varepsilon_2)$ is the difference kernel of γ (recall 2.5).

The congruence $(\varepsilon_1, \varepsilon_2)$ is non-trivial. Since it cannot be critical, there is a $\iota: A \not\subseteq B$ such that $(\iota\varepsilon_1, \iota\varepsilon_2)$ is a congruence on B and $B/(\iota\varepsilon_1, \iota\varepsilon_2) = A/(\varepsilon_1, \varepsilon_2)$. Thus, for $\gamma' = \mathrm{coequ}(\iota\varepsilon_1, \iota\varepsilon_2)$: $B \to B/(\iota\varepsilon_1, \iota\varepsilon_2)$ we have $\gamma' \cdot \iota = \gamma$; hence, $\varphi = \mu \cdot \gamma = (\mu\gamma') \cdot \iota$. □

3.3. Theorem : Let the concrete category $(\mathcal{K}, U)$ satisfy the conditions (i)-(vi). Then a finite object A is subdirectly irreducible iff

either A is maximal and the intersection of a system of non-trivial congruences is non-trivial,

or A is non-maximal meet-irreducible and there is no critical congruence on A .

Proof : By [5; Thm 3.3] , A is subdirectly irreducible iff

either it is maximal and each system $(\mu_i : A \to B_i)$ such that $(\forall i \ \mu_i\alpha = \mu_i\beta \Rightarrow \alpha = \beta)$ contains a monomorphic element, or it is non-maximal meet-irreducible and the statement (a) from 3.2 holds true. Thus, the statement of the theorem follows from 3.1 and 3.2. □

References :

[1] G.Birkhoff, Lattice Theory (AMS Colloquium Publications Vol.25, Providence, RI, 1967)

[2] N.Bourbaki, Théorie des Ensembles, Ch.IV Structures (Hermann, Paris, 1953)

[3] M.Hušek, S-categories, Comment.Math.Univ.Carolinae 5 (1964), 37-46

[4] A.Pultr, On productive classes of graphs determined by prohibiting given subgraphs, Coll.Math.Soc.János Bolyai,18, Combinatorics (Keszthely 1976), 805-820

[5] A.Pultr and J.Vinárek, Productive classes and subdirect irreducibility, in particular for graphs, Discr.Math.20 (1977), 159-176

[6] W.Tholen, Birkhoff's Theorem for Categories, Seminarberichte Nr.8, Fernuniversität Hagen (1981), 153-159

[7] J.Vinárek, Remarks on subdirect representations in categories, Comment.Math.Univ.Carolinae 19 (1978), 63-70

ALGEBRAIC CATEGORIES OF TOPOLOGICAL SPACES

Guenther Richter

Abstract

It is well known that every epireflective, full subcategory of the category $\underline{Comp}_2$ of compact Hausdorff spaces is algebraic in the sense of HERRLICH [6,§32]. Conversely, every algebraic, epireflective, full subcategory of the category of all Hausdorff spaces is contained in $\underline{Comp}_2$. This generalizes a result of HERRLICH and STRECKER [5] and yields a complete new proof for it. The lattice of such algebraic categories is very large.
For arbitrary full subcategories $\underline{C}$ of topological (not necessary Hausdorff) spaces the following holds:
If $\underline{C}$ is algebraic, closed-hereditary, and contains the ordinal spaces $[0,\beta]$ for every limit ordinal β then each space in $\underline{C}$ is compact (not necessary Hausdorff).

AMS Subj. Class. (1980): Primary 18B30, 18C10, 54D30; Secondary 18A40

0. Introduction

In [5] HERRLICH and STRECKER proved the following nice algebraic characterization of the category $\underline{Comp}_2$ of compact Hausdorff spaces:

0.1 Theorem

If $\underline{C}$ is a full, isomorphism-closed, epireflective and nontrivial subcategory of the category $\underline{Top}_2$ of all Hausdorff spaces then the following conditions are equivalent:

(1) $\underline{C}$ is varietal (i.e. the forgetful functor $U:\underline{C} \longrightarrow \underline{Set}$ is varietal in the sense of LINTON [8])

(2) $\underline{C} = \underline{Comp}_2$.

"Nontrivial" means that there is a space in $\underline{C}$ containing at least two points. The most difficult part of the proof "(1) $\Rightarrow$ (2)" is to show that every space in $\underline{C}$ is compact:

First one obtains that every space X in $\underline{C}$ is normal because nonempty, disjoint, closed subsets $A,B \subseteq X$ become distinct points in the quotient space X/R where $R :=$ $(A \times A) \cup (B \times B) \cup \{(x,x) \mid x \in X\}$ is a closed equivalence relation on X. By assumption (1), the space X/R belongs to C, especially it is Hausdorff. A and B have disjoint neighborhoods in X/R, hence in X.

Since C is productive, each power X^I of a C-object X is normal. This implies, by a result of NOBLE [9], that each C-object is compact.

Essential in this proof is that C is closed under the formation of certain quotients X/R which are, in addition, Hausdorff spaces. This enables one to apply the result of NOBLE. Nevertheless, there is a generalization of 0.1 for which quotients, separation axioms, and the result of NOBLE are not needed. Instead of varietal categories, we consider the weaker concept of algebraic categories in the sense of HERRLICH [6,§32]:

0.2 Definition

A category C is called algebraic with respect to a functor $U:\underline{C}\to\underline{Set}$ provided that C has coequalizers, U has a left adjoint and preserves and reflects regular epimorphisms.

Each algebraic category (C,U) is uniquely (regular epi, mono)-factorizable and U preserves and reflects these factorizations. Especially, U reflects isomorphisms. Moreover, U is faithful and C is complete and cocomplete.

Note, that the existence of coequalizers in C is a much weaker condition than the closedness under the formation of coequalizers (quotients) in a bigger category like $\underline{Top}_2$ in the proof above. Up to equivalence, an algebraic category C is a quasivariety, i.e. a full subcategory of universal algebras of a certain type Ω (not necessary with rank) which is closed under the formation of subalgebras and products in the category Alg-Ω of all algebras of type Ω and all Ω-homomorphisms between them. On the other hand, a varietal category C corresponds to a variety which is, in addition, closed under the formation of quotients [2,11].

1. Results

In the sequel, C always denotes a full, isomorphism-closed subcategory of the category Top of all topological spaces and all continuous maps and $U:\underline{C}\to\underline{Set}$ denotes the forgetful functor. If (C,U) is algebraic then every bijective map in C is a homeomorphism.

Let ß be an ordinal and

$$[0,\beta] := \{\alpha \mid \alpha \leq \beta,\ \alpha \text{ ordinal}\} \quad , \quad [0,\beta[:= \{\alpha \mid \alpha < \beta,\ \alpha \text{ ordinal}\}$$

the usual ordinal spaces together with the order topology.

1.1 Lemma

If every bijective map in $\underline{C}$ is a homeomorphism, β is a limit ordinal, and $\mathcal{T}$ a topology on $[0,\beta[$ which is finer than or equal to the order topology, then

$$[0,\beta] \in \underline{C} \Rightarrow ([0,\beta[,\mathcal{T}) \notin \underline{C} \quad .$$

Proof: The map $h: [0,\beta[\to [0,\beta]$ defined by $h(0) = \beta$, $h(n) = n - 1$ for $n \in \mathbb{N}$, $n \geq 1$, and $h(\alpha) = \alpha$, otherwise, is continuous and bijective but not a homeomorphism, because $h[\{0\}] = \{\beta\}$ is not an open subset in $[0,\beta]$ (for a limit ordinal β) but $\{0\}$ is open in $([0,\beta[, \mathcal{T})$.

1.2 Proposition

If $\underline{C}$ is closed-hereditary, (surjective,injective)-factorizable, and has finite products which are preserved by U, and if $\underline{C}$ contains the two element discrete space $D_2 = \{0,1\}$ then $\underline{C}$ is closed under the formation of finite coproducts (sums) in Top.

Proof: By assumption, the topology of finite products in $\underline{C}$ is finer than the usual product topology, because the natural projections are still continuous. If X is in $\underline{C}$, then in the case of $X\prod_{\underline{C}}\{0,1\}$ the two injections of X, $x \mapsto (x,0)$, $x \mapsto (x,1)$, are continuous and, therefore, the topology on $X\prod_{\underline{C}}\{0,1\}$ is coarser than the sum topology on $X \times \{0,1\} = X \dot{\cup} X$ which coincides with the product topology in Top. This means

$$X \dot{\cup} X = X \times \{0,1\} = X\prod_{\underline{C}}\{0,1\} \in \underline{C} \; .$$

Now, consider an element x_0 of X and the constant map $c: X \to X$, $x \mapsto x_0$. This defines a map $\mathrm{id}\, X \dot{\cup} c$ in $\underline{C}$ which admits a (surjective, injective)-factorization

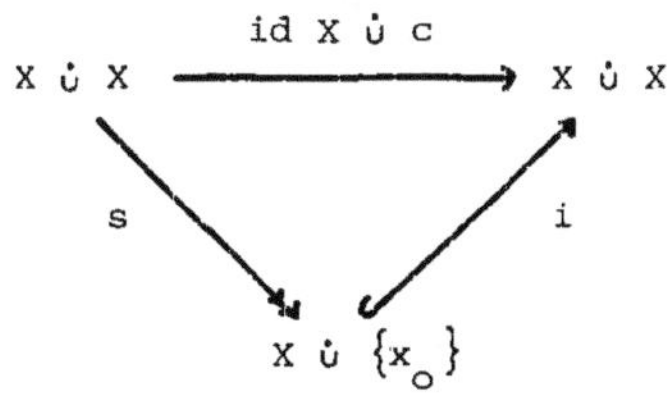

The usual coproduct topology is the only topology on $X \dot{\cup} \{x_0\}$ such that s and i become continuous. Hence, the sum $X \dot{\cup} \{x_0\}$ belongs to $\underline{C}$.

If we have two nonempty spaces $X,Y \in \underline{C}$, $x_0 \in X$, $y_0 \in Y$, then $S := (X \times \{y_0\}) \dot{\cup} (\{x_0\} \times Y)$ is a disjoint union of two closed subsets in $P := (X \dot{\cup} \{x_0\}) \prod_{\underline{C}} (Y \dot{\cup} \{y_0\})$, because these subsets are already closed in the coarser ordinary product topology. Thus, by assumption, S together with the induced topology belongs to $\underline{C}$ and, moreover, it is the topological sum of its subspaces $X \times \{y_o\}$ and $\{x_o\} \times Y$ which are isomorphic to X and Y, resp.

1.3 Lemma

If $\underline{C}$ is closed-hereditary and $\eta: I \to UFI$ U-universal then $\eta[I]$ is dense in FI.

Proof: The closure $\overline{\eta[I]}$ of $\eta[I]$ in FI belongs to $\underline{C}$. Therefore, there is a unique $\underline{C}$-morphism $s: FI \to \overline{\eta[I]}$ such that the diagram

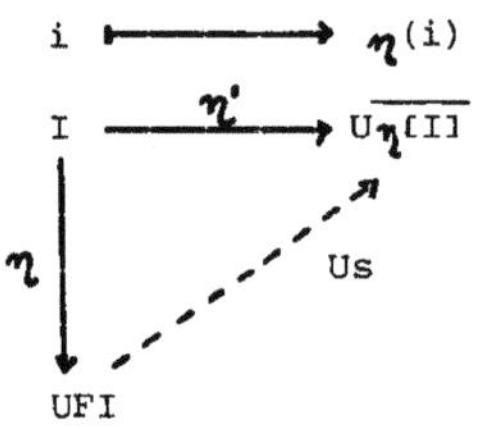

commutes. If $j: \overline{\eta[I]} \to FI$ denotes the injection we have

$$U(js)\eta = UjUs\eta = Uj\eta' = \eta = U(\text{id } FI)\eta \quad .$$

This implies $js = \text{id } FI$, hence j is surjective.

1.4 Theorem

If $\underline{C}$ contains the ordinal spaces $[0,\beta]$ for all limit ordinals β, is closed-hereditary, and algebraic with respect to the forgetful functor $U: \underline{C} \to \underline{Set}$ then every space in $\underline{C}$ is compact.

Proof: For a topological space X the following conditions are equivalent:

(1) X is compact (not necessary Hausdorff)

(2) Every decreasing family $(A_\alpha)_{\alpha \in [0,\beta[}$ of nonempty closed subsets $A_\alpha \subseteq X$, β a limit ordinal, has a nonempty intersection,

because, it is well known [1] that (1) is equivalent to

(3) Every well-ordered decreasing family $(A_i)_{i \in I}$ of closed nonempty subsets of X has a nonempty intersection.

Now, every well-ordered set $(I,\leq)$ is orderisomorphic to some $[0,\beta[$. Obviously, it suffices to consider limit ordinals β. Otherwise there is a greatest element in $[0,\beta[$, hence in $(I,\leq)$, and a smallest member in the family $(A_i)_{i \in I}$.

Therefore, it is enough to show that every decreasing family $(A_\alpha)_{\alpha \in [0,\beta[}$ of nonempty closed subsets of a space $X \in \underline{C}$ has a nonempty intersection, where β is a limit ordinal. Now, take an $x_\alpha \in A_\alpha$ for each $\alpha < \beta$ and consider the following diagram, where $\eta : [0,\beta[\rightarrow UF[0,\beta[$ is U-universal and f, g are defined by the universal property of η :

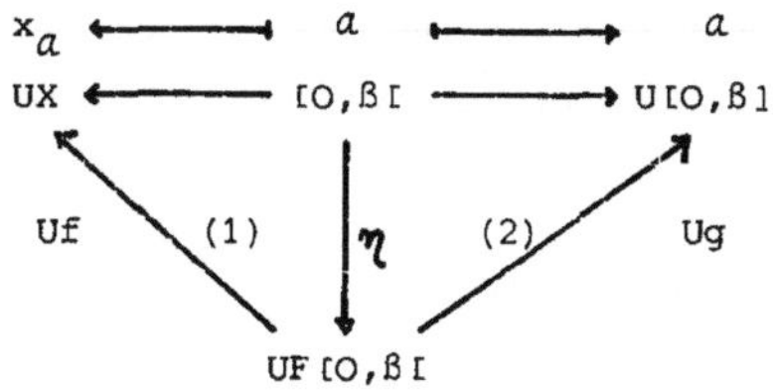

We will prove that the intersection of the family $(f^{-1}[A_\alpha])_{\alpha \in [0,\beta[}$ is nonempty. (This implies $\emptyset \neq f[\bigcap_\alpha f^{-1}[A_\alpha]] \subseteq \bigcap_\alpha f[f^{-1}[A_\alpha]] \subseteq \bigcap_\alpha A_\alpha$.) To do this, we show

a) g is surjective

b) $g^{-1}(\beta) \subseteq f^{-1}[A_\alpha]$ for all $\alpha < \beta$.

Now, by the commutativity of (2), $[0,\beta[\subseteq g[F[0,\beta[] \subseteq [0,\beta]$. If $[0,\beta[= g[F[0,\beta[]$, then the middle space of the (regular epi, mono)-factorization of g would give us a topology $\mathcal{T}$ on $[0,\beta[$ which is finer than or equal to the order topology such that $([0,\beta[,\mathcal{T})$ belongs to $\underline{C}$. This contradicts 1.1. Hence $g[F[0,\beta[]=[0,\beta]$ and g is surjective

For each $\alpha < \beta$ the set $[0,\alpha] = [0,\alpha+1[$ is clopen in $[0,\beta]$. Therefore, $[0,\beta] = [0,\alpha] \dot{\cup} [\alpha+1,\beta]$ is a topological sum. The left adjoint F of U preserves coproducts. By 1.2, we have $F[0,\beta[= F[0,\alpha] \dot{\cup} F[\alpha+1,\beta[$ ($\underline{C}$ is closed hereditary, hence D_2 belongs to $\underline{C}$), and diagram (2) splits as follows:

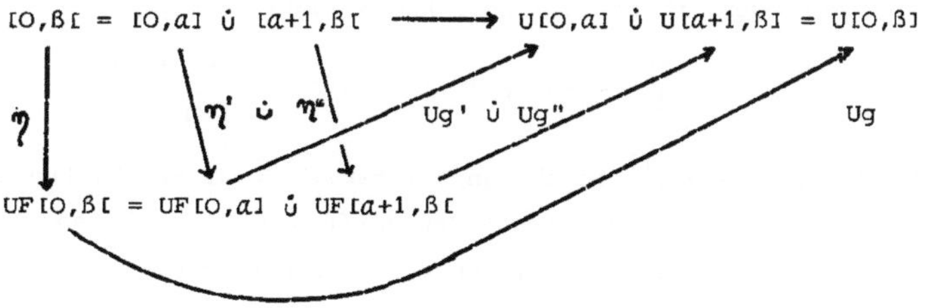

This shows

$$g^{-1}(\beta) \subseteq g^{-1}[\alpha+1,\beta] = g''^{-1}[\alpha+1,\beta] = F[\alpha+1,\beta[\quad (+) .$$

For each $\gamma \in [\alpha+1,\beta[$ we have $f\eta(\gamma) = x_\gamma \in A_\gamma$ and, therefore,

$$f^{-1}[A_\alpha] \supseteq f^{-1}[A_\gamma] \supseteq f^{-1}(x_\gamma) \ni \eta(\gamma) .$$

This implies $f^{-1}[A_\alpha] \supseteq \eta[[\alpha+1,\beta[]$ and, by 1.3 and (+)

$$f^{-1}[A_\alpha] = \overline{f^{-1}[A_\alpha]} \supseteq \overline{\gamma[[\alpha+1,\beta[]} = F[\alpha+1,\beta] \supseteq g^{-1}(\beta) \; .$$

Thus, we have b) which completes the proof of 1.4.

1.5 Corollary

If $\underline{C} \subseteq \underline{Top}_2$ is closed-hereditary and contains the ordinal spaces $[0,\beta]$ for all limit ordinals β then the following conditions are equivalent:

(1) $(\underline{C},U)$ is algebraic

(2) $\underline{C}$ is an epireflective subcategory of $\underline{Comp}_2$.

Proof: "(1) $\Rightarrow$ (2)": $\underline{C}$ fulfills the assumptions of 1.4, hence $\underline{C} \subseteq \underline{Comp}_2$. Moreover, the inclusion $\underline{C} \hookrightarrow \underline{Comp}_2$ is algebraic [6,§32]. Thus, $\underline{C}$ is a reflective subcategory of $\underline{Comp}_2$ and, in addition, epireflective, because it is closed-hereditary.

"(2) $\Rightarrow$ (1)": Every epireflective subcategory of $\underline{Comp}_2$ is regular-epi-reflective and, therefore, algebraic.

1.6 Corollary

If $\underline{C} \subseteq \underline{Top}_2$ is epireflective then the following conditions are equivalent:

(1) $(\underline{C},U)$ is algebraic (varietal)

(2) $\underline{C} \subseteq \underline{Comp}_2$ ($\underline{C} = \underline{Comp}_2$ or $\underline{C}$ is trivial)

Proof: "(1) $\Rightarrow$ (2)": $\underline{C} \subseteq \underline{Top}_2$ is epireflective iff $\underline{C}$ is closed-hereditary and productive [4]. Especially, for nontrivial $\underline{C}$ all closed subspaces of the powers $D_2^{\,I}$, i.e. all Stone spaces, are contained in $\underline{C}$, hence all ordinal spaces $[0,\beta]$. Thus, by 1.4, $\underline{C} \subseteq \underline{Comp}_2$. Moreover, there is a well known quotient map

$$s : D_2^{\mathbb{N}} \longrightarrow [0,1], \quad s(f) := \sum_{i=0}^{\infty} \frac{f(i)}{2^{i+1}} \; ,$$

from the Cantor space $D_2^{\mathbb{N}}$ to the unit interval $[0,1]$ which, therefore, belongs to $\underline{C}$ if $\underline{C}$ is varietal [5]. Every compact Hausdorff space X is homeomorphic to a closed subspace of a certain power $[0,1]^I$, hence it belongs to $\underline{C}$, too.

A careful analysis of the proofs above yields the following topological formulation of 1.6 which is comparable with DE GROOT's famous topological characterization of $\underline{Comp}_2$ [12]:

1.7 Corollary

$\underline{Comp}_2$ is the greatest full subcategory $\underline{C}$ of $\underline{Top}_2$ which satisfies the following conditions:

(i) $\underline{C}$ is productive

(ii) $\underline{C}$ is closed-hereditary

(iii) $\underline{C}$ is (surjective, injective)-factorizable

(iv) Every bijective map in $\underline{C}$ is a homeomorphism .

1.8 Remarks

(1) The powers X^I of every regid compact Hausdorff space X which contains at least two points define a varietal subcategory $\underline{C}$ of $\underline{Comp}_2$ which is not closed-hereditary [6, p. 292, 38E].

(2) The full subcategory $\underline{C} \subseteq \underline{Top}$ of discrete (indiscrete) spaces is varietal but it does not contain all the ordinal spaces $[0,\beta]$.

(3) Every subcategory $\underline{C} \subseteq \underline{Top}$ which fulfills the assumptions of 1.4 contains neither nontrivial indiscrete nor infinite discrete spaces, because there are nonhomeomorphic continous bijections between them and certain ordinal spaces $[0,\beta]$ (with the same cardinality and a limit ordinal β).

(4) If, in 1.4, $\underline{C}$ is not only closed-hereditary but even compact-hereditary then every space in $\underline{C}$ is T_1.

2. The lattice of algebraic subcategories of $\underline{Comp}_2$

The full, epireflective, and isomorphism-closed subcategories $\underline{C}$ of $\underline{Comp}_2$ form a complete (large) lattice L. $\underline{Comp}_2$ is the greatest element of L and the full subcategory $\underline{C}_o$ of all spaces which contain at most one point is the smallest one. Moreover, we have

2.1 Proposition

(1) There is a greatest proper subcategory $\underline{C}_{max} \subsetneq \underline{Comp}_2$ in L .

(2) There is a smallest subcategory $\underline{C}_{min} \supsetneq \underline{C}_o$ in L .

(3) There are arbitrary large discrete subsets in L .

(4) There are arbitrary large well-ordered subsets in L .

Proof:

(1) The class $\{X \in \underline{Comp}_2 \mid f:[0,1] \to X \text{ continuous} \Rightarrow f \text{ constant}\}$ is productive and closed-hereditary, hence cogenerates a member $\underline{C}_{max}$ of L. Now, every $X \in \underline{Comp}_2$ which is not in $\underline{C}_{max}$ contains a nontrivial image of the unit interval $[0,1]$.

Such a space is a cogenerator in $\underline{\text{Comp}}_2$.

(2) The category $\underline{C}_{min}$ of Stone spaces, i.e. closed subspaces of the powers D_2^I, up to homeomorphisms, is algebraic and contained in every nontrivial, algebraic, isomorphism-closed, full subcategory of $\underline{\text{Comp}}_2$.

(3) The dual, $\underline{\text{Comp}}_2^{op}$, of $\underline{\text{Comp}}_2$ is almost algebraic in the sense of [10]. Especially there are arbitrary large families $\{X_j \mid j \in J\}$ of compact Hausdorff spaces with at least two points and

$$C(X_j, X_k) = \begin{cases} \{id\ X_j\} \cup \{f : X_j \longrightarrow X_j \mid f \text{ const.}\}, & j = k \\ \{f : X_j \longrightarrow X_k \mid f \text{ const.}\}, & j \neq k \end{cases}$$

The homeomorphic images of closed subspaces of the powers X_j^I of X_j define an algebraic subcategory $\underline{C}_j$ of $\underline{\text{Comp}}_2$. The subset $\{\underline{C}_j \mid j \in J\}$ of L is discrete, because $\underline{C}_k \subseteq \underline{C}_j$ implies $X_k \in \underline{C}_j$. Therefore, there is a closed embedding $u : X_k \hookrightarrow X_j^I$ for a certain set I and the composition of u with one of the projections $p_i : X_j^I \longrightarrow X_j$ yields a nonconstant continuous map $X_k \longrightarrow X_j$.

(4) is a pure lattice-theoretical consequence of (3).

2.2 Remark

A famous example of a member of $\underline{C}_{max}$ which is not in $\underline{C}_{min}$ is the so called pseudoarc P [3]. An example of a continuum which cogenerates a "very small" member of L is given in [7]. Moreover, for every family $(\underline{C}_i)_{i \in I}$ of epireflective hulls of continua in $\underline{\text{Comp}}_2$ there is a continuum which cogenerates a lower bound of the $\underline{C}_i$ in L.

References

1. P. Alexandroff, P. Urysohn, Zur Theorie der topologischen Räume, Math. Ann. 92 (1924) 258 - 266

2. B. Banaschewski, H. Herrlich, Subcategories defined by implications, Houston J. of Math. 2 (1976) 149 - 171

3. R.H. Bing, A homogenous indecomposable plane continuum, Duke Math. J. 15 (1948) 729 - 741

4. H. Herrlich, Topologische Reflexionen und Coreflexionen, Springer Lecture Notes in Math. 78 (1968)

5. H. Herrlich, G.E. Strecker, Algebra ∩ Topology = Compactness, General Topology and its Applications 1 (1971) 283 - 287

6. H. Herrlich, G.E. Strecker, Category Theory, Allyn and Bacon, Boston (1973)

7. M. Hušek, J. van Mill, Ch. F. Mills, Some very small continua, Topological structures II, Math. Centre Tracts 115 (1979) 147 - 151

8. F.E.J. Linton, Some aspects of equational categories, Proc. Conf. Categorical Algebra, La Jolla 1965 (1966) 84 - 94

9. N. Noble, Products with closed projections II, Trans. Amer. Math. Soc. 160 (1971) 169 - 183

10. A. Pultr, V. Trnková, Combinatorial algebraic and topological Representations of Groups, Semigroups and Categories, Noth-Holland Math. Lib. 22 (1980)

11. G. Richter, Kategorielle Algebra, Studien zur Algebra und ihre Anwendungen 3 (1979)

12. E. Wattel, The compactness operator in set theory and topology, Math. Centre Tracts 21 (1968)

Guenther Richter
University of Bielefeld
Faculty of Mathematics
Universitätsstraße 25

D-4800 Bielefeld 1

EXTENSIONS OF A THEOREM OF P.GABRIEL

Tiberiu Spircu

The theorem of Gabriel about the representation type of a graph (see [6]) is now well-known. This theorem, first appeared in 1972, is the source for many results in graph representation theory. Several extensions have been made, especially by Dlab and Ringel (see [3],[4],[5]), Loupias ([11]), Zavadsky and Shkabara ([14]). We obtain a common extension of these results.

1. Preliminaries

Let F,G be two division rings, such that $F \subset G$ and $\dim G_F = \dim {}_FG$. Let W be a G-vector space and let U be an additive subgroup contained in W (eventually an F-subspace). Following Dlab and Ringel ([4]), denote $\underline{U}$ the largest G-subspace of W contained in U, and $\overline{U}$ the smallest G-subspace of W containing U.

The case $\dim G_F = \dim {}_FG = 2$ will be of major interest in this paper. Let $\{1,\gamma\}$ be a basis of ${}_FG$ (obviously it is also a basis of G_F). Denote by j the canonical isomorphism of F-vector spaces $G \longrightarrow F^2$ and by r the F-linear map $G \longrightarrow F$ given by $r(1) = 0$, $r(\gamma) = 1$. We need the following:

Lemma (see [4], lemma 2.4). Let $F \subset G$ be division rings such that $\dim G_F = \dim {}_FG = 2$. Let U be an F-subspace of a finitely dimensional G-vector space W, such that $\overline{U} = W$ and $\underline{U} = 0$. Let V be a G-subspace of W, such that $U \cap V = 0$. If $v_1,\dots,v_n$ is a basis of V, then there exists a basis of W $x_1,\dots,x_n,y_1,\dots,y_n,z_1,\dots,z_m$, contained in U, such that $v_i = x_i + y_i\gamma$ for $1 \leq i \leq n$.

Our definition of a Gabriel species will be somewhat different from other definitions (see [4],[5],[7]); this definition is presented in three stages.

I. Let (G,Γ) be a (finite) graph, G being the set of vertices and Γ the set of arrows; each arrow y has a source s(y) and a sink a(y). Denote by $\underline{G}$ the vector space over GF(2) - the field with two elements - having $G \cup \Gamma$ as a basis. The mapping $\delta:\underline{G} \longrightarrow \underline{G}$ given by $\delta(x) = 0$ for $x \in G$ and $\delta(y) = s(y)+a(y)$ for $y \in \Gamma$ has the property that $\delta\delta = 0$. Denote by $rk(G,\Gamma)$ the cyclic rank of the graph, defined by $rk(G,\Gamma) = \dim_{GF(2)}(\mathrm{Ker}\,\delta) - \mathrm{card}(G)$. We say that a subgraph of (G,Γ) is d-cyclic if it has the following form :

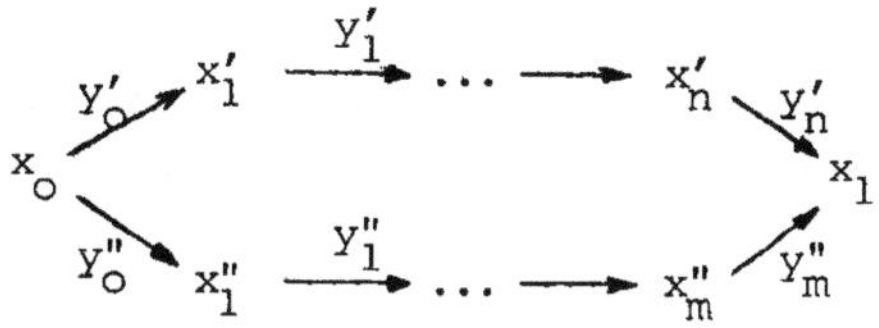

$$\begin{array}{c} y_0' \nearrow x_1' \xrightarrow{y_1'} \cdots \longrightarrow x_n' \searrow y_n' \\ x_0 \qquad \qquad \qquad \qquad \qquad x_1 \\ y_0'' \searrow x_1'' \xrightarrow{y_1''} \cdots \longrightarrow x_m'' \nearrow y_m'' \end{array} \qquad (1)$$

with $m, n \geq 1$.

The maximum number of d-cyclic subgraphs, independent over $GF(2)$, is called the <u>d-cyclic rank</u> of (G,Γ); obviously, it is less than the cyclic rank (in Ker δ may occur cyclic subgraphs which are not d-cyclic).

We say that a graph is <u>perfect</u> if it has the d-cyclic rank equal to the cyclic rank.

II. By [3], a <u>valuation</u> over (G,Γ) is obtained if every vertex x has a <u>height</u> $p(x) \neq 0$ and every arrow y has two <u>capacities</u> $q(y), \bar{q}(y)$ such that the natural numbers $p(x)$, $x \in G$, are coprime, and for every arrow y, $p(a(y))q(y) = p(s(y)) \cdot \bar{q}(y)$. A valuation p, q is called <u>perfect</u> if for every d-cyclic subgraph (1) one has $q(y_n') \ldots q(y_0') = q(y_m'') \ldots q(y_0'')$. Obviously, the trivial valuation (having all capacities 1) is perfect.

III. Let F be a division ring. Also by [3], an F-modulation on a valued graph (G,Γ,p,q) is:

a) a set $\{F_x\}_{x \in G}$ of division rings, such that $\dim_F F_x = p(x)$;

b) a set $\{M_y\}_{y \in \Gamma}$ of abelian groups, such that every M_y has a structure of special $F_{s(y)}$-$F_{a(y)}$-bimodule, and $\dim_{F_{a(y)}} M_y = q(y)$. (An F_u-F_v-bimodule M is called special if the two duals $\text{Hom}_{F_u}(M, F_u)$ and $\text{Hom}_{F_v}(M, F_v)$ are isomorphic as F_v-F_u-bimodules.)

An F-modulation is called <u>admissible</u> if, for every cyclic subgraph (1), we can choose in a coherent manner an isomorphism i between the F_{x_0}-F_{x_1}-bimodules $M_{y_0'} \otimes_{F_{x_1'}} M_{y_1'} \ldots \otimes_{F_{x_n'}} M_{y_n'}$ and $M_{y_0''} \otimes_{F_{x_1''}} M_{y_1''} \ldots \otimes_{F_{x_m''}} M_{y_m''}$.

<u>Definition 1</u>. Let F be a division ring. A <u>Gabriel F-species</u> consists of a perfect graph with a perfect valuation, and an admissible F-modulation on it.

Let (G,Γ) be a graph. We say that we <u>suppress</u> the vertex $x \in G$ if we eliminate x from G and all arrows which are adjacent to x from Γ. We say that we <u>reduce</u> the arrow $y \in \Gamma$ if we identify $s(y)$ with $a(y)$ and we eliminate y from Γ.

For a valued graph (G,Γ,p,q), the reduction of an arrow y implies modifications in heights and in capacities: the new height of $s(y)$ is $\max(p(s(y)), p(s(y)))$; the capacities of all arrows which are adjacent to $s(y)$ or to $a(y)$ must be modified coherently. For example, if in (G,Γ) we have the subgraph

$$x \xleftarrow{z} s(y) \xrightarrow{y} a(y)$$

and if $p(s(y)) \leq p(a(y))$, then the new capacities of the arrow z are $q(z)\bar{q}(y)$, $\bar{q}(z)q(y)$, instead of $q(z), \bar{q}(z)$.

Given an admissible modulation $\{F_x\}, \{M_y\}$ on the valued graph (G,Γ,p,q), after reducing the arrow y, this modulation must be modified coherently; for example, in the case above, the new b-module corresponding to the arrow z is $Hom_{F_{a(y)}}(M_y,F_{a(y)}) \otimes_{F_{s(y)}} M_z$.

These operations do not alter the equality between the cyclic rank and the d-cyclic rank. Thus we give the following:

Definition 2. Let S ans S' be two Gabriel F-species. We say that S' is subordinate to S if S' is obtained from S after a finite number of suppressions of an extremal vertex and/or reductions of an arrow, followed by coherent modifications in valuation and in modulation.

Let S be a Gabriel F-species, having as a support the perfect valued graph (G,Γ,p,q). Let $r = rk(G,\Gamma)$ be his cyclic rank. Choose a basis of Ker δ , containing all vertices x and the d-cyclic subgraphs $\Gamma_1,\ldots,\Gamma_r$ of the form (1), and let x_{0s} (resp. x_{1s}) be the initial (resp. final) vertex of Γ_s , for $1 \leq s \leq r$. Denote also $q_s = q(y'_{ns})\ldots q(y'_{0s})$ for this subgraph Γ_s .

Definition 3. The Tits form of the species S (see [1],[8]) is the following quadratic form in variables X_x, $x \in G$:

$$T_S(X) = \sum_{x\in G} p(x)X_x - \sum_{y\in\Gamma} q(y)p(a(y))X_{s(y)}X_{a(y)} \quad .$$

The Brenner form of the species S (as sketched in [2]) is the following quadratic form :

$$B_S(X) = T_S(X) + \sum_{s=1}^{r} q_s p(x_{1s})X_{x_{0s}}X_{x_{1s}} \quad .$$

Proposition 1. a) If the species S' is obtained from S by suppressing the vertex x, then $T_{S'}$, resp. $B_{S'}$, is obtained from T_S resp. B_S by identifying $X_x = 0$.

b) If S' is obtained from S by reducing an arrow y, then $T_{S'}$, resp. $B_{S'}$, is obtained from T_S resp. B_S by identifying $X_{a(y)} = q(y)X_{s(y)}$.

The proof is obvious.

We say that a Gabriel F-species S is excellent if its Brenner quadratic form B_S has the following property:

(E) $B_S(X) > 0$ for any vector $X \geq 0$, $X \neq 0$.

Corollary. If S is an excellent Gabriel F-species and S' is a subordinate of S, then S' is an excellent Gabriel F-species.

For example, the Brenner quadratic form of the species O_5 :

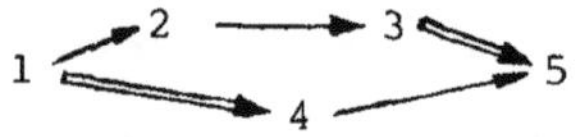

(where ⟹ denotes an arrow with capacities 2,1 and ⟶ denotes an arrow with both capacities 1) is the following :

$$2X_1^2 + 2X_2^2 + 2X_3^2 + X_4^2 + X_5^2 - 2X_1X_2 - 2X_2X_3 - 2X_3X_5 - 2X_1X_4 - X_4X_5 + 2X_1X_5 .$$

Writing this quadratic form as a sum of squares :

$\frac{1}{4}(2X_1 - 2X_2 + X_5)^2 + \frac{1}{4}(2X_1 - 2X_4 + X_5)^2 + \frac{1}{2}(X_2 - 2X_3 + X_5)^2 + \frac{1}{2}X_2^2$, it is obvious that O_5 is excellent.

2. Representations of F-species

Let F be a division ring and let S be a Gabriel F-species.

A representation (V) of the species consists of :

a) a vector space V_x, such that $\dim_{F_x} V_x = d_x < \infty$, for every $x \in G$;

b) an $F_{a(y)}$-linear map $v_y : V_{s(y)} \otimes_{F_{s(y)}} M_y \longrightarrow V_{a(y)}$, for every arrow $y \in \Gamma$.

A morphism (α) from the representation (V) to the representation (W) is simply a family $\{\alpha_x\}_{x\in G}$, α_x being an F_x-linear map $V_x \longrightarrow W_x$, such that for every $y\in\Gamma$, $\alpha_{a(y)}v_y = w_y(\alpha_{s(y)} \otimes 1)$.

Denote $\underline{Re}_F(S)$ the abelian category of all representations of the Gabriel F-species S .

Remember that, given a species S, for every d-cyclic subgraph of the form (1) there exists an isomorphism $i : M_{y'_0} \otimes \ldots \otimes M_{y'_n} \longrightarrow M_{y''_0} \otimes \ldots \otimes M_{y''_m}$.

Definition 4. A representation (V) is called commutative if for every d-cyclic subgraph of the type (1) we have :

$$v_{y'_n} \circ (v_{y'_{n-1}} \otimes 1) \circ \ldots \circ (v_{y'_0} \otimes 1 \otimes \ldots \otimes 1) = v_{y''_m} \circ (v_{y''_{m-1}} \otimes 1) \circ \ldots \circ (v_{y''_0} \otimes 1 \otimes \ldots \otimes 1) \circ (1 \otimes i) ,$$

where i is the isomorphism for this subgraph, chosen in the definition of S .

Denote by $\underline{Rc}_F(S)$ the full subcategory of $\underline{Re}_F(S)$ whose objects are the commutative representations. Then, obviously, $\underline{Rc}_F(S)$ is an abelian category and every object (having a dimension) decomposes as a finite direct sum of indecomposable objects. This decomposition is unique.

3. Commutative representations of the species O_5

The F-species O_5 is based on the following valued graph (the heights are indicated above each vertex) :

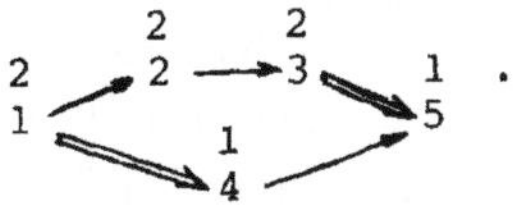

If F is contained in another division ring G, such that $\dim {}_F G = \dim G_F = 2$ and G is a special G-F-bimodule, then the admissible modulation of the species is the following :

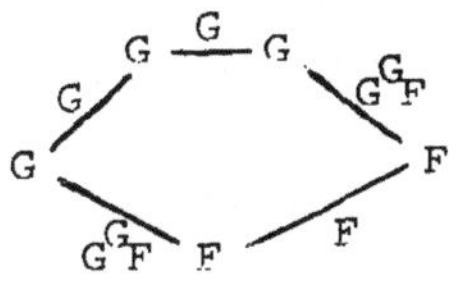

A commutative representation (V) of this species is completely described by the following commutative diagram of F-vector spaces :

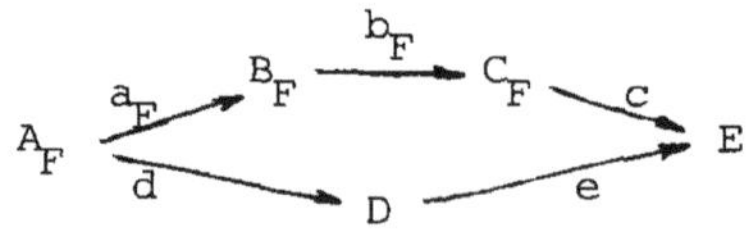

where X_F is the F-vector space structure obtained from the G-vector space structure of X by restriction of scalars; the maps $a_F = a$ and $b_F = b$ are G-linear.

If E = 0, then (V) is in fact a representation of the valued graph C_4 ; furthermore, if A = 0, then (V) is a representation of the valued graph F_4. These trivial cases are elucidated by Dlab and Ringel in [4]: there exists 36 types of indecomposable commutative representations which are trivial.

Proposition 2. Let $F \subset G$ be division rings such that $\dim {}_F G = \dim G_F = 2$ and G is a special G-F-bimodule. Then the F-species Q_5 has exactly 41 types of indecomposable commutative representations.

Proof. We establish that any nontrivial (that is $E \neq 0$ and $A \neq 0$) indecomposable commutative representation is of one of the following five types:

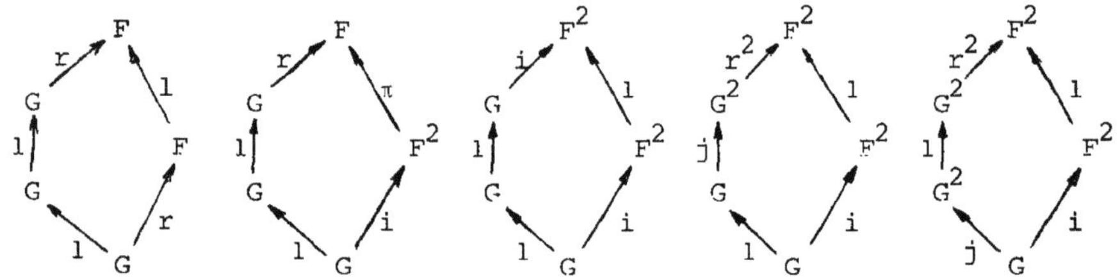

where i and r are the canonical maps (from section 1), π is the projection on the second component and j is given by $j(1) = (1,\gamma)$. The dimensions of these representations are, respectively :

(1,1,1,1,1) , (1,1,1,2,1) , (1,1,1,2,2) , (1,1,2,2,2) and (1,2,2,2,2) .

We sketch the proof, which contains several stages; at each stage we try to decompose (V) as a direct sum $(V') \oplus (V'')$, (V) being supposed nontrivial, indecomposable and commutative. In this manner we obtain the following conclusions :

- Ker(d) ∩ Ker(a) = 0 ;
- Ker(d) ∩ Ker(ba) = 0 ;
- Ker(d) = 0 ;
- Ker(d) ∩ Ker(a) = 0 ;
- $Ker(a) \subset \overline{Ker(d)}$;
- $Ker(ba) \subset \overline{Ker(d)}$;
- a(Ker(d)) = 0 ;
- Ker(ba) ∩ Ker(d) = 0 ;
- Ker(ba) = 0 ;
- Ker(c) ⊂ Im(b) ;
- Ker(c) ⊂ Im(ba) ;
- Ker(c) = 0 ;

- $\mathrm{Ker}(b) = 0$;
- $\mathrm{Ker}(e) \subset \mathrm{Im}(d)$;
- $E = \mathrm{Im}(c)$.

At this stage, denote by K the intersection $\mathrm{Ker}(c) \cap \mathrm{Im}(ba)$.

1) First, suppose $K \neq 0$; then it follows $\mathrm{Ker}(c) \subset \mathrm{Im}(ba)$ and $D = \mathrm{Im}(d)$. Denote $X = ba(\mathrm{Ker}(d))$; then, obviously, $X \subset \mathrm{Ker}(c)$. Choose a complement Y of X in Ker(c). Then $\overline{\mathrm{Ker}(c)} = \overline{X} \oplus \overline{Y}$ and $\overline{X} = ba(\overline{\mathrm{Ker}(d)})$. Denote $C' = \overline{X}$ and let C'' be a complement of C' in C, containing $\overline{Y}$. Denote $E' = c(C')$, $E'' = c(C'')$, $B' = a(\overline{\mathrm{Ker}(d)})$, $B'' = \bar{b}^{1}(C'')$, $A' = \overline{\mathrm{Ker}(d)}$, $A'' = \bar{a}^{1}(B'')$, $D' = d(A')$, $D'' = d(A'')$. Then $e(D') = E'$, $e(D'') = E''$ and we decompose (V) as a direct sum $(V') \oplus (V'')$, where $A' = \overline{\mathrm{Ker}(d')}$ and $\mathrm{Ker}(d'') = 0$.

Supposing that (V) is indecomposable, we need to consider two possibilities.

1-1) $(V) = (V')$, that is, $\overline{\mathrm{Ker}(d)} = A$ and $\overline{\mathrm{Ker}(c)} = C$. It follows easily that, in this case, the representation (V) has the following type :

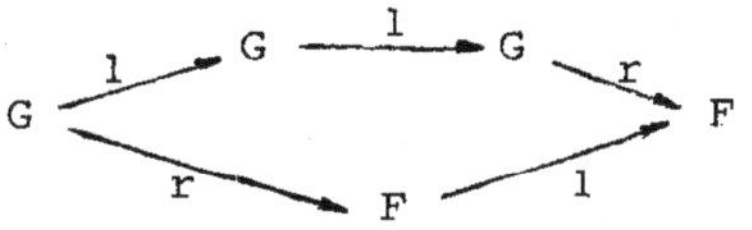

.

1-2) $(V) = (V'')$, that is $\mathrm{Ker}(d) = 0$. In this case, denote $C' = \overline{\mathrm{Ker}(c)}$ and let C'' be a complement of C' in C. Denote $A' = (ba)^{-1}(C')$, $B' = a(A')$, $D' = d(A')$ $E' = c(C')$ and analogously A'',... . It follows that (V) is decomposed as a direct sum $(V_1') \oplus (V_1'')$, with $C' = \overline{\mathrm{Ker}(c')}$ and $\mathrm{Ker}(c'') = 0$.

Again we need to consider two possibilities :

1-2-1) $(V) = (V_1')$, that is $\overline{\mathrm{Ker}(c)} = C$. In this case (V) has the type of dimension (1,1,1,2,1).

1-2-2) $(V) = (V_1'')$, that is $\mathrm{Ker}(c) = 0$. In this case a,b,c,d and e are all isomorphisms. The representation is of the type of dimension (1,1,1,2,2).

2) Suppose now $K = 0$; then, obviously, $\mathrm{Ker}(d) = 0$ and, because Ker(e) is contained in Im(d), it follows immediately that also $\mathrm{Ker}(e) = 0$ and $D = \mathrm{Im}(d)$.

In the next stages we obtain :

- $\mathrm{Im}(ba) \subset \overline{\mathrm{Ker}(c)}$;
- $C = \overline{\mathrm{Ker}(c)}$;
- $\dim_G C = 2\dim_G \mathrm{Im}(ba)$;
- $\dim_G A = 1$.

There are two possibilities, dependind on the coincidence of Im(b) with Im(ba). It follows now easily that these two possibilities correspond to the types of dimensions (1,1,2,2,2) and (1,2,2,2,2). The proposition is proved.

4. Extensions of a theorem of Gabriel

The result of Gabriel about the representation type of a graph is well-known:

Theorem 1 - positive form (Gabriel,[6]). Let (G,Γ) be a graph and F a division ring. The category $\underline{Re}_F(G,\Gamma)$ is of finite representation type (that is, it has only a finite number of types of indecomposable objects), independently on F, if and only if every connected subgraph of (G,Γ) is contained in the list 1 :

List 1
1) A_n $(n \geq 1)$ 1 — 2 — ... — n ;
2) D_n $(n \geq 4)$ $\begin{smallmatrix}1\\2\end{smallmatrix}>$ 3 — ... — n ;
3) E_n $(6 \leq n \leq 8)$ 1 — 2 — 4 — ... — n (with 3 attached to 2)

(with arbitrary orientation for each edge).

Theorem 1 - negative form. Let (G,Γ) be a graph and F a field. The category $\underline{Re}_F(G,\Gamma)$ is of infinite representation type (independently of F) if and only if the graph (G,Γ) has as subordinate a graph from the list 1' below :

List 1'
1) $\tilde{A}_1$ o⟲ ; 2) $\tilde{A}_4$ o<ooo>o ; 3) $\tilde{D}_5$ $\begin{smallmatrix}o\\o\end{smallmatrix}>$o<$\begin{smallmatrix}o\\o\end{smallmatrix}$;
4) $\tilde{E}_6$ o—o—o<$\begin{smallmatrix}o—o\\o—o\end{smallmatrix}$; 5) $\tilde{E}_7$ $\begin{smallmatrix}o—o—o\\o—o—o\end{smallmatrix}>$o—o ;
6) $\tilde{E}_8$ $\begin{smallmatrix}o—o—o—o—o\\o—o\end{smallmatrix}>$o—o

(again, arbitrary orientation for each edge).

This theorem, first appeared in 1972, is the source for many results in graph representation theory. The beatiful proof given by Bernstein, Gel'fand and Ponomarjow ([1]) permitted one extension of this theorem, given by Dlab and Ringel ([4],[5]). In our notations, their theorem, in positive and negative form, is the following:

Theorem 2 - positive form (Dlab-Ringel,[5]). Let F be a division ring and let S be a Gabriel F-species. The category $\underline{Re}_F(S)$ is of finite representation type (independently on F and on the modulation) if and only if every connected valued subgraph of (G,Γ,p,q) is contained in the list 1 or in the list 2 :

List 2
1) B_n $(n \geq 2)$ 1 $\xrightarrow{(1,2)}$ 2 — ... — n ;
2) C_n $(n \geq 3)$ 1 $\xrightarrow{(2,1)}$ 2 — ... — n ;
3) F_4 1 — 2 $\xrightarrow{(1,2)}$ 3 — 4 ; 4) G_2 1 $\xrightarrow{(1,3)}$ 2

(with arbitrary orientations for each edge, a simple edge having both capacities unity).

Theorem 2 - negative form. Let S be a Gabriel F-species based on the valued graph (G,Γ,p,q). The category $\underline{Re}_F(S)$ is of infinite representation type (independently on F and on the modulation) if and only if S has as subordinate a species from the list 1' or from the list 2' below :

List 2'
1) o<ooo>o ; 2) $\begin{smallmatrix}o\\o\end{smallmatrix}>$o $\xrightarrow{(u,v)}$ o with $uv = 2$;
3) o $\xrightarrow{(u,v)}$ o $\xrightarrow{(x,y)}$ o with $uv = xy = 2$;
4) o—o $\xrightarrow{(u,v)}$ o—o---o with $uv = 2$; 5) o $\xrightarrow{(u,v)}$ o---o with $uv = 3$; 6) o $\xrightarrow{(u,v)}$ o with $uv \geq 4$

(with arbitrary orientation for each edge, a simple edge having both capacities 1, a dotted edge having arbitrary capacities).

The next extension of the theorem of Gabriel was given, in negative form, by Loupias ([11]) and, in positive form, by Zavadsky and Shkabara ([14]). In our notations, their theorem, in positive and negative form, is the following :

Theorem 3 - negative form (Loupias,[11]). Let (G,Γ) be a graph and F a division ring. The category $\underline{Rc}_F(G,\Gamma)$ is of infinite representation type (that is, it has an infinite number of indecomposable commutative representations) if and only if the graph (G,Γ) has as subordinate a graph from the list 1′ (except $\bar{A}_4$) or from the list 3′ :

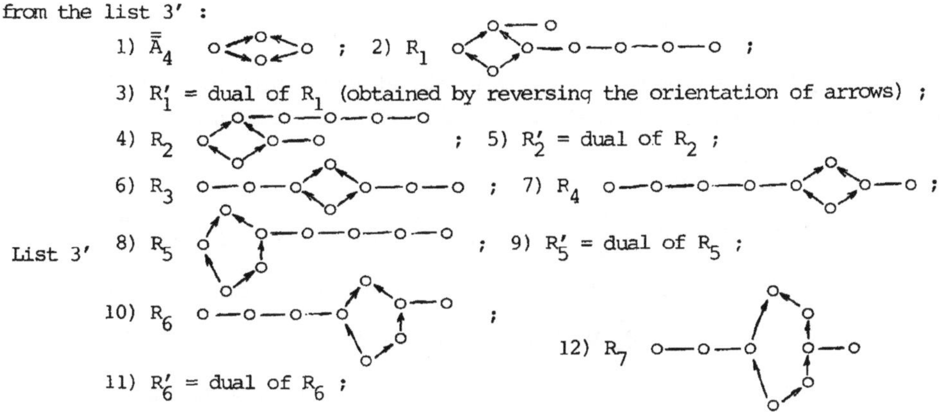

List 3′

1) $\bar{\bar{A}}_4$; 2) R_1 ;
3) R_1' = dual of R_1 (obtained by reversing the orientation of arrows) ;
4) R_2 ; 5) R_2' = dual of R_2 ;
6) R_3 ; 7) R_4 ;
8) R_5 ; 9) R_5' = dual of R_5 ;
10) R_6 ;
11) R_6' = dual of R_6 ;
12) R_7

(with arbitrary orientation for each edge).

Theorem 3 - positive form (Zavadsky-Shkabara,[14]). Let (G,Γ) be a graph and F a division ring. The category $\underline{Rc}_F(G,\Gamma)$ is of finite representation type (and independently of F) if and only if every connected subgraph of (G,Γ) is contained in the list 1 or in the list 3.

List 3 is too big to be presented here; it can be found at the end of Zavadsky-Shkabara's preprint [14]. This list contains graphs having the cyclic rank at most 4 .

The theorems of Dlab-Ringel and of Loupias-Zavadsky-Shkabara admit a common extension, which is also an extension of the theorem of Gabriel:

Theorem 4 - negative form. Let S be a Gabriel F-species, F being an infinite field. The category $\underline{Rc}_F(S)$ is of infinite representation type (independently on F and on the modulation) if and only if S has as subordinate a species from the lists 1′,2′ or 3′.

Theorem 4 - positive form. $\underline{Rc}_F(S)$ is of finite representation type (independently on F and on the modulation) if and only if every connected valued subgraph of S is contained in the lists 1, 2, 3 or 4 :

List 4 1) O_4 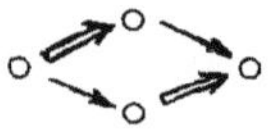 ; 2) O_5 ; 3) O_5' = dual of O_5

(any double arrow has capacities 2,1).

Proof. An easy computation shows that the species listed in lists 1' , 2' and 3' are not excellent.

Let (V) be a representation of the species S, with $\dim(V) = (d_x)_{x\in G}$. Obviously, (V) correspond to an element from the set

$$R' = \prod_{y\in\Gamma} \mathrm{Hom}_{F_{a(y)}} (V_{s(y)} \underset{F_{s(y)}}{\otimes} M_y , V_{a(y)}) .$$

R' is a vector space over F, having the dimension

$$D = \sum_{y\in\Gamma} q(y)p(a(y))\; d_{s(y)} d_{a(y)} .$$

Denote by R the subset of R' containing all elements which correspond to commutative representations. R is an algebraic variety, having components of dimension less than

$$D - \sum_{s=1}^{r} q_s p(x_{1s})\; d_{x_{0s}} d_{x_{1s}} .$$

Consider the group $\underline{S}' = \prod_{x\in G} \mathrm{Aut}_{F_x}(V_x)$ and the quotient $\underline{S} = \underline{S}'/F^*$. If F is an infinite field, then $\underline{S}$ is an algebraic group having the dimension over F :

$$\sum_{x\in G} p(x)\; d_x^2 - 1 .$$

The group $\underline{S}$ operates on the set R' ; an easy computation shows that $\underline{S}$ operates also on R ; R decomposes in orbits, and each orbit contains families (v_y) corresponding to isomorphic commutative representations.

Suppose the category $\underline{Rc}_F(S)$ has only a finite number of types of indecomposable objects. Then, given a dimension vector $(d_x)_{x\in G}$, there exists only a finite number of indecomposable commutative representations having dimension $d = (d_x)$. Under the action of $\underline{S}$, R decomposes in a finite number of orbits. Every such orbit has, over F, the dimension less than the dimension of $\underline{S}$; so, the dimension of R is less than the dimension of $\underline{S}$, that is

$$B_S(d) \geqslant 1 \quad \text{if} \quad d \neq 0 ,$$

B_S being the Brenner quadratic form of the species S. Hence S is an excellent species.

Now, if S is a trivial excellent species, then every connected subgraph must appear in the lists 1 and 3. If S is not trivial and has the cyclic rank 0, then every connected subgraph must appear in the list 2; if the cyclic rank is not 0 , then from the corollary of proposition 1 it follows that S appears in list 4.

Conversely, from Dlab-Ringel and Loupias-Zavadsky-Shkabara theorems (see [5], [11],[14]), every species from the lists 1,2 and 3 is of finite (commutative) representation type. For the species O_5 we showed this in the proposition 2, and the species O_4 is a subordinate of O_5. The theorem is proved.

The theorem of Dlab and Ringel ([5]) describes more completely the species with finite representation type; namely, for such a species, each indecomposable representation correspond to a positive root of the Tits quadratic form. The Tits quadratic form of a Gabriel species S is positive definite if and only if the species has finite representation type.

In the proof of theorem 4 we used the Brenner quadratic form in the same manner Dlab and Ringel used the Tits form. The Brenner form for a species with commutative representation type is no loger positive definite, it has only the property (E). But still remains true that each positive root of the (Brenner) quadratic form determines an indecomposable commutative representation! This fact is not fully explained and the problem of describing (axiomatically) the dimensions of such representations remains open.

REFERENCES

[1] Bernstein, I.N., Gel'fand, I.M., Ponomarjow, V.A., Funktory Kokstera i teorema Gabriel'a, Uspehi Mat.Nauk, 28(1973), 19-33

[2] Brenner, S., Quivers with commutativity conditions and some phenomenology of forms, Representations of Algebras-Ottawa 1974, Springer Lecture Notes in Math. , 488 (1975)

[3] Dlab, V., Representations of valued graphs, Université de Montréal (1980)

[4] Dlab, V., Ringel, C.M., On algebras of finite representation type, J.Algebra, 33(1975), 306-394

[5] Dlab, V., Ringel, C.M., Indecomposable representations of graphs and algebras, Memoirs Amer.Math.Soc. 173 (1976)

[6] Gabriel, P.,Unzerlegbare Darstellungen I, Manuscripta Math., 6(1972), 71-103

[7] Gabriel, P., Indecomposable representations II, Symposia Math., Ist.Naz.Alta Mat., 11(1973), 81-104

[8] Gabriel, P., Représentations indécomposables, Séminaire Bourbaki 1973/74

[9] Gel'fand, I.M., Ponomarjow, V.A., Problems of linear algebra and classification of quadruples of subspaces in a finite-dimensional vector space, Coll.Math.Soc.Janos Bolyai V, Tihàny-Hungary (1970), 163-237

[10] Loupias, M., Représentations indécomposables des ensembles ordonnés finis, Séminaire d'algèbre non-commutative Orsay 1974/75

[11] Loupias, M. Indecomposable representations of finite ordered sets, Representations of Algebras-Ottawa 1974, Springer Lecture Notes in Math., 488 (1975)

[12] Ovsienko, S.A., O sistemah kornei dl'a proizvol'nyh grafov, Matričnye zadači, Inst.Mat.AN UkSSR, Kiew (1977)

[13] Zavadsky, A.G., Nazarova, L.A., Čiastič'no uporiadočennye množestva ručnogo tipa, Matričnye zadači, Inst.Mat.AN UkSSR, Kiew (1977)

[14] Zavadsky, A.G., Shkabara, A.S., Kommutativnye kolčany i matričnye algebry konečnogo tipa, Preprint IM-76-3, Inst.Mat.AN UkSSR, Kiew (1976)

Characterization of bicategories of stacks

Ross Street

Introduction

Although the paper [11] was written in the setting of 2-categories, it was pointed out in the introduction of that paper how to modify the work in order to make it bicategorical. The purpose of the present paper is to make these modifications precise and to give an application.

The main theorem is a characterization of bicategories of stacks (= *champs* in French) in terms of limit, colimit, exactness, and size conditions on the bicategories: a bicategorical version of Giraud's characterization of categories of sheaves [1].

On the way to this result a formula is given for the associated stack. The existence of the associated stack on a categorical site was proved by Giraud [5] using the associated category-valued sheaf construction and a strictification construction on fibrations. Exactness properties of the associated stack construction were not obvious from Giraud's formula. The formula we give uses the obvious generalization to bicategories of the functor L used in [1] for the associated sheaf. Exactness of L is immediate: it preserves all finite bicategorical limits (biterminal objects, bipullbacks, and bicotensoring with finite categories in the sense of [9]). The associated stack is given by <u>three</u> applications of L (recall that two are needed in the sheaf case).

The application we wish to present is really an application of the formula for the associated stack. We give an easy proof of the relationship between *torsors* and *Čech cocyles*. Combining this with a very general theorem giving the classifying property of torsors (they classify objects locally structure isomophic to some member of a given family of mathematical structures), we are able to deduce information about local structures in mathematics; for example, about vector bundles, locally finite objects in a topos, Azumaya algebras, and so on.

§1. Regular and exact bicategories.

The notion of bicategory, homomorphism of bicategories, strong transformation and modification are those of Bénabou [3]. We write Hom(A,B) for the bicategory of homomorphisms, strong transformations and modifications from the bicategory A to the bicategory B.

The notion of limit for bicategories is taken from Street [9]. For homomorphisms $F : A \to \mathrm{Cat}$, $S : A \to K$, the *F-indexed bilimit of* S is an object $\{F,S\}$ of K satisfying an equivalence of homomorphisms:

$$K(-,\{F,S\}) \simeq \mathrm{Hom}(A,\mathrm{Cat})(F,K(-,S)).$$

As special cases we have biterminal objects, bipullback, biproduct, bicotensor product.

Suppose K is a bicategory with finite bilimits. An arrow $m : X \to Y$ in K is called *f.f.* when the functor

$$K(K,m) : K(K,X) \to K(K,Y)$$

is fully faithful for all K.

An arrow $e : A \to B$ is called *e.s.o.* (short for "essentially surjective on objects") when the following diagram is a bipullback for all f.f. arrows $m : X \to Y$.

$$\begin{array}{ccc} K(B,X) & \xrightarrow{K(B,m)} & K(B,Y) \\ {\scriptstyle K(e,X)}\big\downarrow & \cong & \big\downarrow{\scriptstyle K(e,Y)} \\ K(A,X) & \xrightarrow[K(A,m)]{} & K(A,Y) \end{array}$$

More generally, one can define what it means for a family of arrows into B to be e.s.o. using a many-legged bipullback.

A *weak category* T in K is a homomorphism of bicategories from the sketch (= Gabriel theory) for the theory of categories into K which takes the distinguished cones to bilimits. There is an obvious notion of *weak functor* between weak categories.

Given an arrow $f : A \to B$ in K we can form the following diagrams in which the squares containing 2-cells are bicomma object diagrams and the squares containing isomorphisms are bipullbacks.

$$\begin{array}{ccccc} E_2^2 & \xrightarrow{d_2} & E_1^2 & \xrightarrow{d_1} & A \\ {\scriptstyle d_o}\big\downarrow & \cong & {\scriptstyle d_o}\big\downarrow & \Rightarrow & \big\downarrow{\scriptstyle f} \\ E_1^2 & \xrightarrow{d_1} & A & \xrightarrow{f} & B \\ {\scriptstyle d_o}\big\downarrow & \Rightarrow & \big\downarrow{\scriptstyle f} & & \\ A & \xrightarrow[f]{} & B & & \end{array} \qquad\qquad \begin{array}{ccc} E_1^1 & \xrightarrow{d_1} & A \\ {\scriptstyle d_o}\big\downarrow & \cong & \big\downarrow{\scriptstyle f} \\ A & \xrightarrow[f]{} & B \end{array}$$

We obtain two weak categories $E^2 : E_2^2 \substack{\to\\ \to\\ \to} E_1^2 \substack{\to\\ \leftarrow\\ \to} E_o^2$ and $E^1 : E_1^1 \substack{\to\\ \leftarrow\\ \to} E_o^1$, and a weak functor $j : E^1 \to E^2$ with the following properties:

(a) $E_o^1 = E_o^2 = A$, j_o is an identity, and $j_1 : E_1^1 \to E_1^2$ is f.f.;

(b) the span (d_o, E^2_1, d_1) from A to A is a bidiscrete fibration in the sense of Street [9];

(c) E^1 is a weak equivalence relation on A.

This leads us to define a *congruence on* A to be a weak functor $j : E^1 \to E^2$ satisfying (a), (b), (c). So each arrow $f : A \to B$ has a congruence associated with it.

A quotient for a congruence $j : E^1 \to E^2$ consists of an arrow $g : A \to X$ and a 2-cell $\gamma : gd_o \Rightarrow gd_1$ such that:

$$\begin{array}{ccccc} E^1_1 & \xrightarrow{j_1} & E^2_1 & \xrightarrow{d_1} & A \\ & & {\scriptstyle d_o}\downarrow & \overset{\gamma}{\Rightarrow} & \downarrow{\scriptstyle g} \\ & & A & \xrightarrow[g]{} & X \end{array}$$

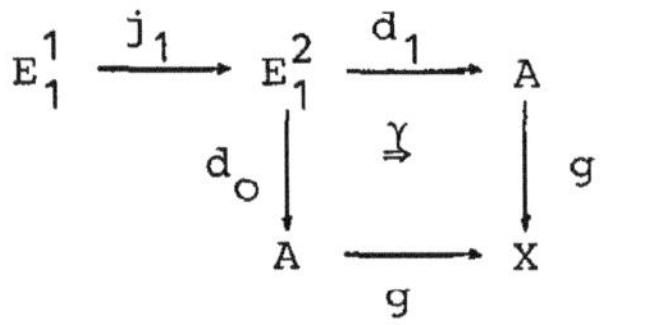

is invertible, and

$$\begin{array}{ccccc} E^2_2 & \xrightarrow{d_2} & E^2_1 & \xrightarrow{d_1} & A \\ {\scriptstyle d_o}\downarrow & \cong & \downarrow & \Rightarrow & \downarrow{\scriptstyle g} \\ E^2_1 & \xrightarrow[d_1]{} & A & \xrightarrow[g]{} & X \\ \downarrow & \overset{\gamma}{\Rightarrow} & \downarrow{\scriptstyle g} & \cong & \downarrow{\scriptstyle 1} \\ A & \xrightarrow[g]{} & X & \xrightarrow[1]{} & X \end{array} \quad = \quad \begin{array}{ccccc} E^2_2 & \xrightarrow{d_1} & E^2_1 & \xrightarrow{d_1} & A \\ & & {\scriptstyle d_o}\downarrow & \overset{\gamma}{\Rightarrow} & \downarrow{\scriptstyle g} \\ & & A & \xrightarrow[g]{} & X \end{array} \ ;$$

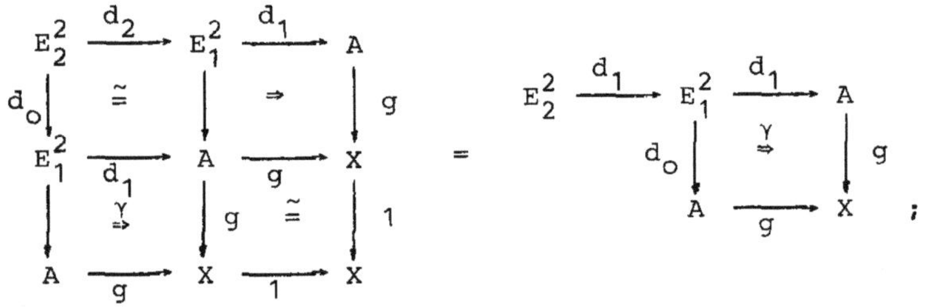

and which is biuniversal with these properties.

If K has finite colimits then every congruence has a quotient.

An arrow $q : A \to Q$ is called a <u>quotient map</u> when there exists a congruence E on A, and a 2-cell $\tau : qd_o \Rightarrow qd_1$ such that Q, q, τ form a quotient for E.

<u>Proposition.</u> *Every quotient map is e.s.o.* (Compare [11; (1.17)].) □

Call K *regular* when the following properties hold:

— all finite bilimits exist;

— each arrow f is isomorphic to a composite me where m is f.f. and e is e.s.o.;

— each bipullback of an e.s.o. is e.s.o.

<u>Theorem.</u> *In a regular bicategory, every e.s.o. is a quotient map.* (Compare [11; (1.22)].) □

Call K *exact* when it is regular and each congruence is the congruence associated with some arrow. It follows that every congruence has a quotient in an exact bicategory. For all bicategories C, the bicategory $\mathrm{Hom}(C^{op}, \mathrm{Cat})$ is exact.

§2. Bitoposes.

A topology on a bicategory C assigns to each object U of C, a set $\mathrm{Cov}\,U$ of f.f. arrows $R \to C(-,U)$ in $\mathrm{Hom}(C^{op},\mathrm{Cat})$ satisfying the following conditions:

T 0. the identity of $C(-,U)$ is in $\mathrm{Cov}\,U$;

T 1. for all $R \to C(-,U)$ in $\mathrm{Cov}\,U$ and all arrows $u : V \to U$ in C, there exists a bipullback

$$\begin{array}{ccc} S & \longrightarrow & C(-,V) \\ \downarrow & \cong & \downarrow C(-,u) \\ R & \longrightarrow & C(-,U) \end{array}$$

in which the top arrow is in $\mathrm{Cov}\ V$;

T 2. if $R' \to C(-,U)$ is in $\mathrm{Cov}\,U$ and $R \to C(-,U)$ is f.f. with the property that for each $u : V \to U$ in the image of $R'V \to C(V,U)$ there exists a bipullback as in T1 with the top arrow in $\mathrm{Cov}\,V$, then $R \to C(-,U)$ is equivalent to an arrow in $\mathrm{Cov}\,U$.

A *bisite* is a bicategory together with a topology. A *stack* for such a bisite is a homomorphism of bicategories $F : C^{op} \to \mathrm{Cat}$ such that, for each $R \to C(-,U)$ in $\mathrm{Cov}\,U$, an equivalence of categories is induced as follows:

$$\mathrm{Hom}(C^{op},\mathrm{Cat})(C(-,U),F) \simeq \mathrm{Hom}(C^{op},\mathrm{Cat})(R,F).$$

The canonical topology on a bicategory is the largest topology for which the representable homomorphisms are all stacks.

Write $\mathrm{Stack}\ C$ for the full sub-bicategory of $\mathrm{Hom}(C^{op},\mathrm{Cat})$ consisting of the stacks for the bisite C.

A bicategory K is called a *bitopos* when there exists a bisite C with small underlying bicategory such that there is a biequivalence: $K \sim \mathrm{Stack}\ C$.

For a bisite C, regard $\mathrm{Cov}\,U$ as an ordered set by taking $R \le S$ when there exists a diagram:

$$\begin{array}{ccc} R & \longrightarrow & S \\ & \searrow \ \cong \ \swarrow & \\ & C(-,U) & \end{array} .$$

If C is small then, for each homomorphism $P : C^{op} \to \mathrm{Cat}$, we can define a homomorphism $LP : C^{op} \to \mathrm{Cat}$ by:

$$(LP)U = \underset{R}{\mathrm{colim}}\ \mathrm{Hom}(C^{op},\mathrm{Cat})(R,P)$$

where R runs over the directed set $(\mathrm{Cov}\,U)^{op}$.

A homomorphism of bicategories which preserves finitary indexed bilimits will be called *left exact*.

Since filtered colimits in Cat commute with finitary indexed bilimits, L is a left exact homomorphism from Hom(C^{op},Cat) to itself.

Theorem. *For any small bisite* C, *the left biadjoint of the inclusion*

$$\text{Stack}\,C \to \text{Hom}(C^{op},\text{Cat})$$

is obtained by applying L *three times and is hence left exact. If* P → F *is faithful and* F *is a stack then* L^2P *is the associated stack of* P. *If* P → F *is fully faithful and* F *is a stack then* LP *is the associated stack of* P. (Compare [11;(3.8)].) □

§3. Characterization theorem.

A set of objects of a bicategory C is called *e.s.o. generating* when, for each object U of C, the set of arrows into U with sources in the set, is e.s.o.

A bicategory K is called *lex-total* when it has small homcategories and the Yoneda embedding

$$Y : K \to \text{Hom}(K^{op},\text{Cat})$$

has a left-exact left biadjoint.

Bicoproducts in a bicategory are *universal* when they are preserved by bipullbacks. When any two distinct coprojections into a bicoproduct have a bi-initial bicomma object then the bicoproduct is *disjoint*.

A set whose cardinality is no greater than the cardinality of the set of small sets is called *moderate*.

Theorem. *The following conditions on a bicategory* K *with small homcategories are equivalent:*

(i) K *is a bitopos;*

(ii) K *is lex-total and there exists a moderate set* M *of objects of* K *such that, for all* X *in* K, *there exists an e.s.o.* M → X *with* M *in* M;

(iii) *every canonical stack on* K *is representable and* K *has an e.s.o. generating small set of objects;*

(iv) K *is an exact bicategory which has disjoint universal small bicoproducts and has an e.s.o. generating small set of objects;*

(v) *there exists a small canonical bisite* C *with finitary indexed bilimits such that* K ~ Stack C. (Compare [11;(4.11)].) □

§4. Application to torsors.

Let E denote a finitely complete category with coequalizers and such that each of the categories E/U is cartesian closed. Let K denote the 2-category of categories in E. Let $F = \mathrm{Hom}(E^{op},\mathrm{Cat})$.

Regard E as a site by taking single regular epimorphisms into U as covers of U and generating the usual *regular epimorphism topology* on E. Regarding E as a bicategory with only identity 2-cells, we obtain a bisite. The objects of F which are stacks for this bisite will simply be called *stacks* in this section.

Regard E as contained in K by taking objects of E as discrete categories. Regard K as contained in F by taking each category A in E to the representable $E(-,A)$.

An object X of F is called *admissible* when, for all $x : U \to X$, $y : V \to X$ with U,V in E, there is a bicomma object x/y in E.

Define $S \in F$ by $SU = E/U$ and S on arrows is given by pulling back along them.

For each X in F there exists PX in F satisfying:

$$F(Y,PX) \simeq F(X^{op} \times Y,S) \ .$$

For A in K, we can identify $(PA)U$ with the full subcategory of the spans $A \overset{p}{\leftarrow} E \overset{q}{\rightarrow} U$ in K from U to A consisting of those spans for which the following is a pullback.

$$\begin{array}{ccc} E_1 & \xrightarrow{d_1} & E_o \\ {\scriptstyle p_1}\downarrow & & \downarrow{\scriptstyle p_o} \\ A_1 & \xrightarrow[d_1]{} & A_o \end{array}$$

In standard topos terminology, $(PA)U \simeq E^{A^{op}\times U}$.

For any admissible X in F, there is a *yoneda arrow* $y_X : X \to PX$. For A in K, the yoneda arrow y_A has component $y_AU : E(U,A) \to (PA)U$ that functor which takes $a : U \to A$ to the span $A \leftarrow A/a \to U$.

Suppose A is admissible and $E \in (PA)U$. The E-*indexed colimit* $\mathrm{colim}(E,f)$ of $f : A \to X$ is the pointwise left extension of fp along q as below:

$$\begin{array}{ccc} E & \xrightarrow{p} & U \\ {\scriptstyle p}\downarrow & & \downarrow{\scriptstyle \mathrm{colim}(E,f)} \\ A & \xrightarrow[f]{} & X \ . \end{array}$$

Here pointwiseness means that the left (Kan) extension property is stable under pullback along an arrow into U.

Call X *cocomplete* when it admits $\mathrm{colim}(E,f)$ for all E and $f : A \to X$ with A in K. In particular, PB is cocomplete for all B in K (see [8], [10]).

An object $z \in XU$ is *locally isomorphic to a value of* $f : A \to X$ when there exists a regular epimorphism $e : V \to U$, an object a of AV, and an isomorphism $(Xe)z \cong f_V a$.

$$\begin{array}{ccc} V & \xrightarrow{e} & U \\ {\scriptstyle a}\downarrow & \cong & \downarrow{\scriptstyle z} \\ A & \xrightarrow[f]{} & X \end{array}$$

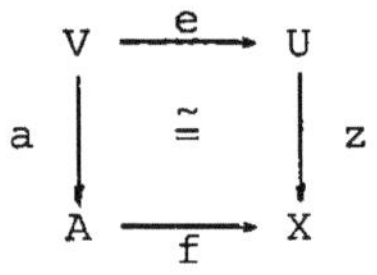

Let $\mathrm{Loc}_X(f)U$ be the full subcategory of XU consisting of such z. Since the pullback of a regular epic is a regular epic, this defines an object $\mathrm{Loc}_X(f)$ of F which is a subhomomorphism of X.

For $A \in K$, an object $E \in (PA)U$ which is locally isomorphic to a value of $y_A : A \to PA$ is called an A-<u>torsor</u>.

$$\begin{array}{ccccc} & & A/a & \xrightarrow{d_1} & V \\ & {\scriptstyle d_0}\swarrow & \downarrow & \text{p.b.} & \downarrow \\ A & \xrightarrow[p]{} & E & \xrightarrow[q]{} & U \end{array}$$

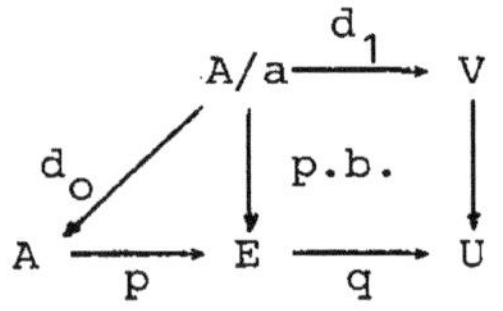

Put $\mathrm{Tor}\,A = \mathrm{Loc}_{PA}(y_A)$.

<u>Proposition.</u> *An object* X *of* F *is a stack if and only if it admits all colimits indexed by torsors. In particular,* PB *is a stack for all* B *in* K. □

<u>Theorem on classification by torsors.</u> *Suppose* $X \in F$ *is an admissible stack. Each* $x : W \to X$ *in* F *with* W *in* E *factors up to isomorphism as a composite of an arrow* $W \to X[x]$ *in* K *which is the identity on objects and an arrow* $i : X[x] \to X$ *in* F *whose components are fully faithful. The functor* $E \mapsto \mathrm{colim}(E,i)$ *provides an equivalence:*

$$\mathrm{Tor}\,X[x] \simeq \mathrm{Loc}_X(x) \ . \quad \square$$

<u>Theorem relating torsors and Čech cocycles.</u> *For each object* A *of* K *and* U *of* E *there is an equivalence*

$$(\mathrm{Tor}\,A)\,U \simeq \operatorname*{colim}_{V \xrightarrow{e} U} K(\mathrm{er}_U(e), A)$$

where $e : V \to U$ *runs over the regular epics into* U *and* $\mathrm{er}_U(e)$

denotes the category in E *determined by the kernel pair of* e.

Proof. Bunge [4] has shown that Tor A is the associated stack of A. The colimit of the Theorem is precisely the formula for LA as given in §2. There is a fully faithful arrow $y_A : A \to PA$ with PA a stack. So only one application of L is needed to obtain the associated stack. So Tor A ≃ LA as required. □

§5. Finiteness in a topos.

Take E to be a topos with natural numbers object N. In the terminology of the last section, let $\mathrm{Fin} \in SN = E/N$ denote the object $N \times N \xrightarrow[+]{} N \xrightarrow[suc]{} N$ of E/N.

The objects Z of S1 which are locally isomorphic to a value of $\mathrm{Fin} : N \to S$ are the locally finite (= Kuratowski-finite decidable) objects of E.

The category S[Fin] in E is the usual category E_{fin} of cardinal-finite objects. The last two theorems give:

$$\mathrm{Tor}(E_{fin}) \simeq \mathrm{Loc}(\mathrm{Fin})$$

$$\mathrm{Tor}(E_{fin})1 \simeq \operatorname*{colim}_{R \twoheadrightarrow 1} K(R_c, E_{fin})$$

where R_c denotes the chaotic category on the object R of E.

Since E_{fin} is a topos in E [6] and topos is an essentially algebraic notion, the filtered colimit on the right above is a topos. This gives another proof that the locally finite objects in E form a topos showing that the ideas involved are basically cohomological (provided we allow cohomology with category-valued coefficients and not merely abelian-group-valued coefficients).

§6. Vector bundles.

In the situation of §4, take E to be a nice category of topological spaces. Restrict the regular epics to local homeomorphisms.

Take $X \in F$ to be the internalization of the theory of vector spaces over $\mathbb{R}$; that is, XU is the category of modules in E/U over the ring $\mathbb{R} \times U \to U$.

Take $\mathrm{Euc} : \mathbb{N} \to X$ to be the family $\mathbb{R}^n$, $n \in \mathbb{N}$, of finite dimensional vector spaces.

Objects Z of XU locally isomorphic to a value of Euc are vector bundles over U.

X[Euc] is the category Mat($\mathbb{R}$) of matrices over $\mathbb{R}$ as a category in E.

The two theorems of §4 give equivalences:

$$\mathrm{Tor}\,(\mathrm{Mat}(\mathbb{R})) \simeq \mathrm{Loc}(\mathrm{Euc}) = (\text{vector bundles})$$

$$\mathrm{Tor}\,(\mathrm{Mat}(\mathbb{R}))U \simeq \underset{V \overset{e}{\to} U}{\mathrm{colim}}\, K(\mathrm{er}_U(e),\mathrm{Mat}(\mathbb{R}))$$

where e runs over surjective local homeomorphisms into U.

Thus we obtain an equivalence between the category of vector bundles over U and the colimit of $K(\mathrm{er}_U(e),\mathrm{Mat}(\mathbb{R}))$ as e runs over surjective local homeomorphisms into U. Now $\mathrm{Mat}(\mathbb{R})$ is a compact symmetric closed monoidal additive category with finite products and splitting idempotents. These properties are therefore inherited by the category of vector bundles . The result is a precise formulation of the *clutching construction* for vector bundles from which we can immediately deduce the property:

$$S \oplus T \cong S' \oplus T \quad \text{implies} \quad S \cong S'$$

necessary for the construction of K-theory. The usual colimit involving the general linear group $GL(n,\mathbb{R})$ is also a consequence. The equivalence therefore brings together much of the introductory K-theory appearing in books such as [2],[7] as an aspect of category-valued cohomology.

References

1 M. Artin, A. Grothendieck and T.L. Verdier, editors, *Théorie des topos et cohomologie étale des schémas*, Lecture Notes in Math. 269 (Springer, Berlin, 1972).

2 M.F. Atiyah, *K-Theory*, Math. Lecture Notes Ser. No. 7 (Benjamin-Cummings, 1967).

3 J. Bénabou, *Introduction to bicategories*, Lecture Notes in Math. 47 (Springer, Berlin, 1967) 1-77.

4 M. Bunge, *Stack completions and Morita equivalence for categories in a topos*, Cahiers de topologie et géométrie différentielle XX-4 (1979) 401-436.

5 J. Giraud, *Cohomologie non abélienne*, (Springer, Berlin, 1971).

6 P.T. Johnstone, *Topos Theory*, (Academic Press, 1978)

7 M. Karoubi, *K-Theory: An introduction*, Grundlehren der Mathematischen Wissenschaften, Band 226, (Springer Berlin, 1978).

8 R.H. Street, *Cosmoi of internal categories*, Transactions Amer. Math. Soc 258 (1980) 271-318.

9 R.H. Street, *Fibrations in bicategories*, Cahiers de topologie et géométrie différentielle XXI (1980) 111-160.

10 R.H. Street, *Conspectus of variable categories*, J. Pure and Appl. Algebra 21 (1981) 307-338.

11 R.H. Street, *Two dimensional sheaf theory*, J. Pure and Appl. Algebra 24 (1982) 20pp.

On hom-functors and tensor products of topological vector spaces

Walter Sydow

It is easy to see that the category $\mathcal{TVS}$ of topological vector spaces with the internal hom-functor of pointwise convergence and with the inductive tensor product is a symmetric monoidal closed category. But there are a lot of more interesting function space topologies for topological vector spaces, namely the topologies of uniform convergence on certain subsets. These function space topologies yield many internal hom-functors H_B on $\mathcal{TVS}$, where, roughly spoken, the functor B assigns to every topological vector space E a subset $\mathcal{A}_E$ of its power set, and the vector space $H_B(E,L)$ carries the topology of uniform convergence on the elements of $\mathcal{A}_E$.

In this paper we discuss the categorical properties of the hom-functors H_B and their related tensor products, the latter having been first defined by Fischer [5] for locally convex spaces. If the hom-functor H_B and its left adjoint tensor product are restricted to B-barrelled spaces, one gets a symmetric monoidal closed category. The notion of a B-barrelled space subsumes the notions of ultrabarrelled resp. quasi-ultrabarrelled spaces in the sense of Iyahen [6].

This paper is a shortened part of the author's thesis [9].

(0) Preliminaries

Let K be a fixed non-discrete valuated field and let $\mathcal{TVS}$ denote the category of topological vector spaces over K.

A <u>string</u> in a topological vector space E is a sequence $(U_n \mid n \in \mathbb{N})$ of balanced and absorbing subsets $U_n \subset E$ such that $U_{n+1} + U_{n+1} \subset U_n$ for all n. (See Adasch, Ernst, Keim [1]. In some sense a string is a generalization of an absolutely convex neighbourhood.)

The U_n are called the <u>knots</u> of the string (U_n).

Let E and L be topological vector spaces, $\mathcal{A}$ a subset of the power set of E and $\mathcal{U}$ a neighbourhood base in L. For each $A \in \mathcal{A}$ and each $U \in \mathcal{U}$ let

$$W(A,U) := \{f \mid f \in \mathcal{TVS}(E,L),\ f(A) \subset U\}.$$

Then $\mathcal{W}(\mathcal{A},\mathcal{U}) := \{W(A,U) \mid A \in \mathcal{A},\ U \in \mathcal{U}\}$ is a neighbourhood base of a linear topology on $TVS(E,L)$, called the <u>A-topology</u>, or the topology of uniform convergence on the $A \in \mathcal{A}$, if $\mathcal{A}$ fulfills the following conditions:

(i) Each $A \in \mathcal{A}$ is bounded.

(ii) $A,B \in \mathcal{A}$ imply $A \cup B \in \mathcal{A}$.

It is convenient to require some more conditions for $\mathcal{A}$:
$\mathcal{A}$ is called a <u>bornology</u> on E if the following hold:

(B1) Every $A \in \mathcal{A}$ is bounded.

(B2) $\cup\mathcal{A} = E$.

(B3) $A,B \in \mathcal{A}$, $C \subset A$ implies $A \cup B$, $C \in \mathcal{A}$.

(B4) $A,B \in \mathcal{A}$, $\alpha \in K$ implies $A + B$, $\alpha A \in \mathcal{A}$.

For a bornology $\mathcal{A}$ on E we denote by $H_{\mathcal{A}}(E,L)$ the vector space $TVS(E,L)$ together with the $\mathcal{A}$-topology.
Now we extend this construction to get internal hom-functors on TVS.

<u>(1) Definition</u>
A <u>bornological topological vector space</u> $(E,\mathcal{A})$ is a topological vector space together with a bornology $\mathcal{A}$. Notation: $\mathcal{A}(E,\mathcal{A}) := \mathcal{A}$.
A mapping $f: (E,\mathcal{A}) \to (L,\mathcal{B})$ is called <u>bounded</u> if $fA \in \mathcal{B}$ for all $A \in \mathcal{A}$.

The category of all bornological topological vector spaces and bounded continuous linear mappings is denoted by $BorTVS$.

<u>(2) Proposition</u>
The forgetful functor $V: BorTVS \to TVS$ is topological.

Proof: For each source $(E, (f_i: E \to (L_i,\mathcal{A}_i))_I)$, define $\mathcal{A} := \{A \subset E \mid A$ is bounded and $f_i A \in \mathcal{A}_i$ for all $i \in I\}$.
Then the source $((E,\mathcal{A}), (f_i)_I)$ is V-initial.

<u>(3) Definition</u>
A functor $B: TVS \to BorTVS$ is called a <u>B-functor</u> if VB is the identity functor on TVS.

<u>(4) Examples</u>
The B-functors B_f, B_α, B_c, B_t and B_b are defined by

(1) $B_f(E) = (E, \{A \subset E \mid A$ finite$\})$. B_f is left adjoint to V.

(2) $B_\alpha(E) = (E, \{A \subset E \mid A$ bounded and card $A < \alpha\})$ for an infinite

cardinal α.

(3) $B_c(E) = (E, \{A \subset E \mid A$ compact$\})$ (compact without Hausdorff).

(4) $B_t(E) = (E, \{A \subset E \mid A$ totally bounded$\})$. B_t is right adjoint to V_o. $V_o(E,\mathcal{A})$ is the underlying vector space of E together with the finest linear topology on E, such that all $A \in \mathcal{A}$ are totally bounded.

(5) $B_b(E) = (E, \{A \subset E \mid A$ bounded$\})$. B_b is right adjoint to V.

In the following we need the notion of a B-barrelled space.

(5) Definition

Let B: *TVS* → *BorTVS* be a B-functor, and E be a topological vector space.

(1) A string in E is called closed (B-bornivorous) if all its knots are closed (absorb each $A \in \mathcal{A}(BE)$).

(2) The closed B-bornivorous strings in E generate a linear topology on E, called the B-strong topology of E.

(3) The functor S_B: *TVS* → *TVS* is defined by

$S_B(E)$ = (underlying vector space of E, B-strong topology of E).

(4) E is called B-barrelled if $E = S_B E$.

The full subcategory of the B-barrelled spaces is denoted by *BBar*.

(6) Proposition

For each B-functor B the category of B-barrelled spaces is bicoreflective in *TVS*.

Proof: Apply S_B on $E \in \mathrm{Ob}\,$*TVS* as often as necessary.

(7) Remark

Let B be a B-functor such that for every $E \in \mathrm{Ob}\,$*TVS* $\mathcal{A}(BE)$ contains all zero-sequences [or all rapidly decreasing sequences]. Then the B-barrelled spaces are exactly the B_b-barrelled spaces, i.e. the quasi-ultrabarrelled spaces of Iyahen [6].

Proof: A circled U absorbs a circled A, if U absorbs all zero sequences [all rapidly decreasing sequences] in A.

(8) Proposition

For each B-functor there is an internal hom-funktor H_B on *TVS*, such that for all $E, L \in \mathrm{Ob}\,$*TVS*

$$H_B(E,L) = H_{\mathcal{A}(BE)}(E,L).$$

Proof: Let $u: E_1 \to E_o$, $v: L_o \to L_1$ be continuous linear mappings, and $B(E_i) = (E_i, A_i)$ for $i = 0,1$. Then

$$H_B(u,v): H_{A_o}(E_o, L_o) \to H_{A_1}(E_1, L_1), \quad f \to vfu$$

is obviously linear, and continuous, since

$$H_B(u,v)(W(uA_1, \overset{-1}{v}U_1)) \subset W(A_1, U_1)$$

holds for each $A_1 \in A$ and U_1 a neighbourhood in L_1.

<u>(9) Proposition</u>

Let B: *TVS* → *BorTVS* be a B-functor and E be a topological vector space. Then the following hold:

(1) There is a natural isomorphism

$$\iota: 1_{TVS} \to H_B(K,-)$$

defined by $\iota_L: L \to H_B(K,L)$, $x \to (\alpha \to \alpha x)$, $L \in \mathrm{Ob}\, TVS$.

(2) There is a natural transformation

$$\zeta: K \to H_B(-,-)$$

defined by $\zeta_L: K \to H_B(L,L)$, $\alpha \to \alpha 1_L$, $L \in \mathrm{Ob}\, TVS$.

(3) The partial hom-functor $H_B(E,-)$ preserves initial sources (initial in the topological sense), and has left adjoint.

(4) The contravariant partial hom-functor $H_B(-,E)$ carries those colimits, that are preserved by B, to limits.

Proof: (1) and (2) are obvious.

(3) Let $(L, (f_i: L \to L_i)_I)$ be an initial source in *TVS*. Then the source $(H_B(E,L), (H_B(E,f_i): H_B(E,L) \to H_B(E,L_i))_I)$ is initial, because $W(A, \overset{-1}{f_i}(U_i)) = \overset{-1}{H_B(E,f_i)}(W(A,U_i))$ holds for $A \subset E$, $U_i \subset L_i$. We conclude that $H_B(E,-)$ is continuous and hence has a left adjoint (Special Adjoint Functor Theorem). In (15) we construct the left adjoint of $H_B(E,-)$ (cp. (17)).

(4) Let $((f_i), L)$, $f_i: L_i \to L$, be a colimit of $D: I \to TVS$ and assume that B preserves this colimit. We have to show that the source $H_B(f_i,E): H_B(L,E) \to H_B(L_i,E)$ is initial.

Let be $B(L) = (L,A)$, $B(L_i) = (L_i, A_i)$. For each neighbourhood U in E and $A_i \in A_i$ $W(f_iA_i, U) = \overset{-1}{H_B(f_i,E)}(W(A_i,U))$ holds. Since

$((Bf_i), BL)$ is a colimit, each $A \in \mathcal{A}$ is a subset of a finite sum $\Sigma f_i A_i$, $A_i \in \mathcal{A}_i$, hence the proof is complete as $W(\Sigma f_i A_i, \Sigma U) \supset$ $\supset \cap W(f_i A_i, U)$.

Now we discuss the symmetry

$$s:\ H_B(E, H_B(L,X)) \to H_B(L, H_B(E,X)),$$
$$f \to (l \to (e \to f(e)(l))).$$

For every continuous linear $f: E \to H_B(L,X)$ and every $l \in L$, $s(f)(l): E \to X$ is continuous, because for any neighbourhood U_X in X there is a neighbourhood U_E in E such that $f(U_E) \subset W(\{l\}, U_X)$. Moreover $s(f)$ is continuous iff for every $A \in \mathcal{A}(BE)$ and every neighbourhood U_X in X there is a neighbourhood U_L in L such that $s(f)(U_L) \subset W(A, U_X)$.
Consequently $s(f)$ is continuous iff for every $A \in \mathcal{A}(BE)$ $f(A)$ is equicontinuous. The latter condition holds, if L is B-barrelled.

(10) Theorem (Banach-Steinhaus)
Let L and X be topological vector spaces and B a B-functor. Then L B-barrelled implies that every bounded subset of $H_B(L,X)$ is equicontinuous.

Proof: Let $A \subset H_B(L,X)$ be bounded and let (U_n) be a closed topological string in X. Then $(\bigcap_{f \in A} f^{-1} U_n)$ is a closed B-bornivorous, hence topological, string in L.

(11) Proposition
Let $B:\ \mathcal{TVS} \to \mathcal{B}or\mathcal{TVS}$ be a B-functor. Then:
(1) For all B-barrelled E, L and all $X \in \operatorname{Ob} \mathcal{TVS}$ the symmetry

$$\sigma_B(E,L,X):\ H_B(E,\ H_B(L,X)) \to H_B(L,\ H_B(E,X))$$

is an isomorphism in $\mathcal{TVS}$ and yields a natural isomorphism.
(2) For all B-barrelled E and all $X, Y \in \operatorname{Ob} \mathcal{TVS}$ the composition law

$$H_B(X,E) \to H_B(H_B(E,Y),\ H_B(X,Y)),\ h \to H_B(h,Y)$$

is continuous and linear, and yields a natural transformation.

Proof: (1) By Banach-Steinhaus (10) $\sigma_B(E,L,X)$ is well-defined. The rest is obvious .

(2) Let $A \subset H_B(E,Y)$ be bounded, $A_X \in \mathcal{A}(X)$ and U_Y a neighbourhood in Y. Then there is a neighbourhood U_E in E such that $A(U_E) \subset U_Y$. Since $H_B(h,Y)(A) \subset W(A_X,U_Y)$ holds for all $h \in W(A_X,U_E)$ the proof is complete.

There are special bilinear mappings, belonging to the hom-functor H_B. Using these bilinear mappings we will define those tensor products that are left adjoint to the hom-functors H_B.

(12) Definition

Let E_o,E_1,E_2 be topological vector spaces, let $b\colon E_o \times E_1 \to E_2$ be a bilinear mapping and let $B_o,B_1,B_2\colon \mathit{TVS} \to \mathit{BorTVS}$ be B-functors.

(1) b is called (B_o,B_1)-hypocontinuous provided that for any neighbourhood U_2 in E_2 and any $A_i \in \mathcal{A}(B_i\, E_i)$, $i = 0,1$, there exists a neighbourhood U_i in E_i such that

$$b(A_o,U_1) \subset U_2 \quad \text{and} \quad b(U_o,A_1) \subset U_2.$$

(2) b is called (B_o,B_1,B_2)-bounded if for every $A_i \in \mathcal{A}(B_i\, E_i)$, $i = 0,1$, $b(A_o,A_1) \in \mathcal{A}(B_2\, E_2)$ holds.

In case $B = B_o = B_1 = B_2$ we say B-hypocontinuous, B-bounded.

(13) Remark

For all B-functors listed in (4) any B-hypocontinuous bilinear mapping is B-bounded.

(14) Proposition

Let B_o,B_1 be B-functors and let E_o,E_1,E_2 be topological vector spaces. For every bilinear mapping $b\colon E_o \times E_1 \to E_2$ the following are equivalent:

(1) b is (B_o,B_1)-hypocontinuous.

(2) $b_o\colon E_o \to H_{B_1}(E_1,E_2)$, $x_o \to b(x_o,-)$, is well-defined, continuous, and for any $A_o \in \mathcal{A}(B_o\, E_o)$ $b_o(A_o)$ is equicontinuous.

(3) $b_1\colon E_1 \to H_{B_o}(E_o,E_2)$, $x_1 \to b(-,x_1)$, is well-defined, continuous, and for any $A_1 \in \mathcal{A}(B_1\, E_1)$ $b_1(A_1)$ is equicontinuous.

(4) The mappings b_o and b_1 from (2) and (3) are well-defined and continuous.

(15) Proposition

Let B_o, B_1 be B-functors and E_o, E_1 be topological vector spaces. Then there exists a topological vector space L and a (B_o, B_1)-hypocontinuous bilinear mapping $\tau(E_o, E_1)$ from $E_o \times E_1$ into L with the following property:
For every (B_o, B_1)-hypocontinuous bilinear mapping b from $E_o \times E_1$ into E_2, $E_2 \in \mathrm{Ob}\, \mathit{TVS}$, there is a unique continuous linear mapping $f: L \to E_2$ such that $f\tau(E_o, E_1) = b$.

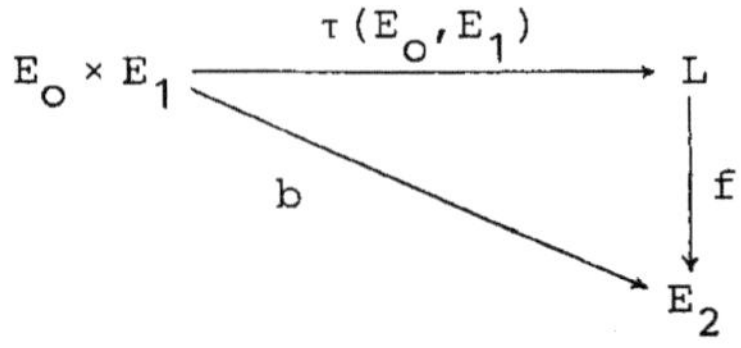

Proof: Let L be the tensor product of the underlying vector spaces of E_o, E_1, equipped with the finest linear topology, such that the canonical bilinear mapping $\tau(E_o, E_1)$ is (B_o, B_1)-equicontinuous.

(16) Definition

The L in the above proposition is called (B_o, B_1)-tensor product of E_o and E_1 and denoted by $T(B_o, B_1)(E_o, E_1)$, by $E_o \;{}_{B_o}\!\otimes_{B_1} E_1$ or by $E_o \otimes_B E_1$ in case $B = B_o = B_1$.
For fixed B_o, B_1 all these tensor products yield a bifunctor $T(B_o, B_1)$: $\mathit{TVS} \times \mathit{TVS} \to \mathit{TVS}$, called (B_o, B_1)-tensor product in TVS.
$\otimes_f := T(B_f, B_f)$ is usually called the inductive tensor product.

(17) Proposition

For each B-functor the tensor product $T(B_f, B)$ is left adjoint to the hom-functor H_B.

Proof: Look at the following diagrams and apply (14).

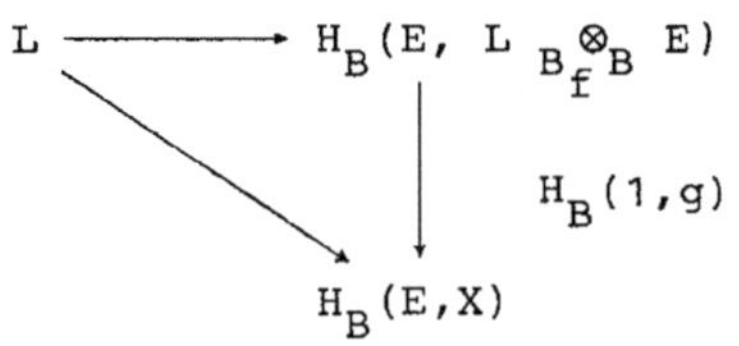

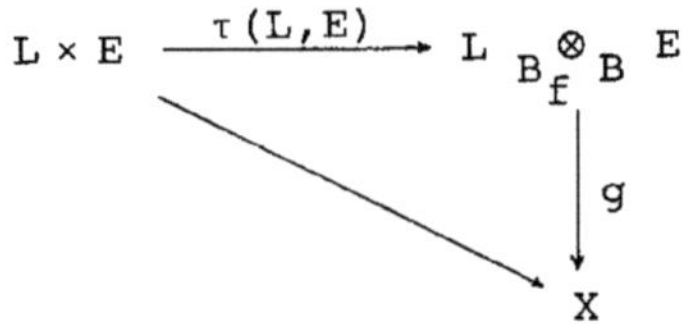

In general the tensor product $T(B_f,B)$ is not symmetric, but if it is restricted to B-barrelled spaces, it is: The Banach-Steinhaus theorem and prop. (14) imply the

(18) Lemma

Let B be a B-functor, E_o, E_1, E_2 topological vector spaces and $b: E_o \times E_1 \to E_2$ a bilinear mapping. If E_o and E_1 are B-barrelled, the following are equivalent:

(1) b is (B_f,B)-hypocontinuous.

(2) b is (B,B_f)-hypocontinuous.

(3) b is (B,B)-hypocontinuous.

(4) b is (B_b,B_b)-hypocontinuous.

(19) Corollary

For any B-functor the following holds:

$$\begin{aligned} T(B,B)/\mathcal{BBar} \times \mathcal{BBar} &= T(B_f,B)/\mathcal{BBar} \times \mathcal{BBar} \\ &= T(B,B_f)/\mathcal{BBar} \times \mathcal{BBar} \\ &= T(B_b,B_b)/\mathcal{BBar} \times \mathcal{BBar}. \end{aligned}$$

The following proposition implies that for all B-functors in (4) the tensor-product $\otimes_B$ of two B-barrelled spaces is B-barrelled again.

(20) Proposition

Let B be a B-functor, E_o, E_1 B-barrelled and let the canonical bilinear mapping $E_o \times E_1 \to E_o \otimes_B E_1$ be B-bounded. Then $E_o \otimes_B E_1$ is B-barrelled.

Proof: Let (U_n) be a closed B-bornivorous string in $E_o \otimes_B E_1$ and $A_o \in A(B\ E_o)$. Define $U_n^1 := \{x_1 \in E_1 \mid A_o \otimes x_1 \subset U_n\}$.
Then (U_n^1) is a closed B-bornivorous string in E_1, hence topological. The analogously defined (U_n^o) is topological, too. Consequently (U_n) is topological.

(21) Proposition

Let B be a B-functor, R the coreflector for $\mathcal{BBar} \to \mathcal{TVS}$, which modifies only the topology. Let E, L be B-barrelled and $Z \in \mathrm{Ob}\,\mathcal{TVS}$. Then the following hold:

(1) The canonical mapping

$\psi: H_B(E \otimes_B L, Z) \to H_B(E, H_B(L,Z)),\quad f \to (x \to (y \to f(x \otimes y)))$

is a vector space isomorphism.

(2) ψ is continuous, if $E \times L \to E \otimes_B L$ is B-bounded.

(3) $\psi^{-1}: R\, H_B(E, H_B(L,Z)) \to H_B(E \otimes_B L, Z)$ is continuous.

Proof: (1) is a consequence of (17) and (19).

(2) $\psi(W(A_E \otimes A_L, U_Z)) = W(A_E, W(A_L, U_Z))$ holds for $A_E \subset E$, $A_L \subset L$, $U_Z \subset Z$.

(3) The canonical mapping $E \otimes_B L \to H_B(RH_B(E, H_B(L,Z)), Z)$ is continuous, hence ψ^{-1} is continuous, too.

<u>(22) Theorem</u>

Let B be a B-functor, such that any B-hypocontinuous bilinear mapping is B-bounded. Let R be the coreflector for $B\mathcal{B}ar \to \mathcal{TVS}$, which modifies the topology only. Then $B\mathcal{B}ar$ together with the hom-functor $R\, H_B / B\mathcal{B}ar \times B\mathcal{B}ar$ and the tensor product $\otimes_B / B\mathcal{B}ar \times B\mathcal{B}ar$ is a symmetric monoidal closed category.

Proof: (9), (11), (20), (21) and the following lemma.

<u>(23) Lemma</u>

Under the assumption of the preceding theorem

$$R\, H_B(E,L) = R\, H_B(E, RL)$$

holds for every B-barrelled E and every $L \in \mathrm{Ob}\, \mathcal{TVS}$.

Proof: $R\, H_B(E,L) \to R\, H_B(E, S_B L)$ is continuous (see (5)), hence by transfinite induction $R\, H_B(E,L) \to R\, H_B(E,RL)$ is continuous, too.

<u>(24) Remark</u>

(1) Let B be a B-functor such that $B\mathcal{B}ar = B_b\mathcal{B}ar$ (see (7)), let R be the coreflector for $B\mathcal{B}ar \to \mathcal{TVS}$. Then the above theorem and (19) imply the surprising fact that

$$R\, H_B / B\mathcal{B}ar \times B\mathcal{B}ar = R\, H_{B_b} / B\mathcal{B}ar \times B\mathcal{B}ar.$$

(2) The above theorem holds for ultrabornological spaces (Iyahen [6]), too. Thus we get at least three non-trivial symmetric monoidal closed subcategories of $\mathcal{TVS}$, namely the categories of ultrabarrelled, quasi-ultrabarrelled resp. ultrabornogical spaces.

In the same way one gets symmetric monoidal closed subcategories of the category of locally convex spaces and of the category of locally p-convex spaces, $0 < p \leq 1$.

References

[1] Adasch, N.,; Ernst,B.; Keim,D.: Topological vector spaces, Lecture Notes in Mathematics, Springer, Berlin Heidelberg New York (1978).

[2] Duske, J.: Analogie zwischen k-Räumen und bornologischen Vektorräumen, thesis, Universität Kiel (1967).

[3] Duske, J.: Adjungierte Funktoren in der Kategorie der p-bor nologischen Räume, Manuscripta Math. 4 (1971), 169-177.

[4] Eilenberg, S.; Kelly, G.M.: Closed categories, Proc. of the Conference on Categorical Algebra, La Jolla, 1965, Springer, Berlin Heidelberg New York (1966), 421-562.

[5] Fischer, H.R.: Über eine Klasse topologischer Tensorprodukte, Math. Ann. 150 (1963), 242-258.

[6] Iyahen, S.O.: On certain classes of linear topological spaces, Proc. London Math. Soc. 18 (1968), 285-307.

[7] Ligon, T.: Galois-Theorie in monoidalen Kategorien, thesis, Universität München (1978).

[8] Schipper, W.J.De.: Symmetric closed categories, Mathem. Centre Tracts 64, Mathematisches Centrum, Amsterdam (1975).

[9] Sydow, W.: Über die Kategorie der topologischen Vektorräume, thesis, Fernuniversität Hagen (1980).

Walter Sydow,
Fernuniversität, FB Mathematik und Informatik, Postfach 940
D 5800 Hagen, West Germany

UNNATURAL ISOMORPHISMS OF PRODUCTS IN A CATEGORY

Věra Trnková
Praha

I. Introduction. Though naturality is the essence of the category theory, categorical methods can be useful also in some problems, which seem to be far from being natural. In the present paper, we investigate "unnatural isomorphisms of products", i.e. the situations when products are isomorphic without any natural reason for being isomorphic. This field of problems has rather an old origin. In 1933, S. Ulam put the problem (see [21]) whether there are two non-homeomorphic topological spaces X, Y with homeomorphic squares (this was solved positively in [6]). The implication

$$X \times X \simeq Y \times Y \Longrightarrow X \simeq Y$$

is called the unique square root property and it has been investigated not only in topology, but also for various algebraic and relational structures, see e.g. [11]. By [9], [10], the unique square root property is valid in any category, which has only a finite set of morphisms between any pair of its objects. On the other hand, there exists e.g. a countable poset (= partially ordered set) which has $2^{\aleph_0}$ non-isomorphic square roots, by [5]. The Tarski cube property

$$X \simeq X \times X \times X \Longrightarrow X \simeq X \times X$$

is implied by the unique square root property because if X is isomorphic to its cube $X^3 = X \times X \times X$ but not to its square $X^2 = X \times X$, then X and X^2 are non-isomorphic objects with isomorphic squares. Hence the Tarski cube property is valid in any category which has only a finite set of morphisms between any pair of its objects. The Tarski cube property is not fulfilled e.g. in the category of Boolean algebras ([7], [8]), Abelian groups [4] or topological spaces [12]. The cancellation ($X \times Z \simeq Y \times Z \Longrightarrow X \simeq Y$), the Cantor-Bernstein property ($(X \times A \simeq Y)$ & $(Y \times B \simeq X) \Longrightarrow X \simeq Y$) and some other problems concerning isomorphisms of products have been investigated by many authors in a lot of papers for more than fourty years. All these problems are special cases of the investigation of productive representations of semigroups. Let us present the following

Definition. Let $\mathcal{K}$ be a category with finite products, let $(S,+)$ be a commutative semigroup. A collection $\{X_s \mid s \in S\}$ of objects of $\mathcal{K}$ is called a productive representation of $(S,+)$ in $\mathcal{K}$ if

(i) X_s is not isomorphic to $X_{s'}$ for $s, s' \in S$, $s \neq s'$;

(ii) $X_{s+s'} \simeq X_s \times X_{s'}$ for all $s,s' \in S$.

Which commutative semigroups have productive representations in which categories? This field of problems has been investigated in the "Seminar from General Mathematical Structures" held in Prague. A brief survey of the obtained results is presented in [14]; a more detailed description with some later results is given in [17] for relational structures and in [18] for topological structures. (More general questions - isomorphisms of infinite products, isomorphisms of products and coproducts - are investigated in [14], [2].) The present paper is also a contribution to this field or problems. We describe some general methods in a categorial language and apply them to functor categories.

II. The basic method

1. Let α be an infinite cardinal. We say that a category $\mathcal{K}$ is α-distributive if it has products and coproducts of all collections of the cardinality $\leq \alpha$ and finite products commute with coproducts of collections of the cardinality $\leq \alpha$, i.e.

$$(\coprod_{i \in I} X_i) \times (\coprod_{j \in J} Y_j) \simeq \coprod_{(i,j) \in I \times J} X_i \times Y_j$$

whenever card $I \leq \alpha$ and card $J \leq \alpha$.

2. Let α be an infinite cardinal, let $\mathcal{K}$ be an α-distributive category, let $\mathcal{C} = \{X_\beta \mid \beta \in \alpha\}$ be a collection of objects of $\mathcal{K}$ (as usual, ordinals are the sets of all smaller ordinals and the cardinals are initial ordinals, so α is also a set of the cardinality α and β ranges over it).

Denote by ω the set of all finite cardinals. For any map $f: \alpha \longrightarrow \omega$ put

$$X_f = \prod_{\beta \in \alpha} X^{f(\beta)},$$

where $X_\beta^{f(\beta)}$ is a product of $f(\beta)$ copies of the object X_β (if $f(\beta) = 0$, then $X_\beta^{f(\beta)}$ is a terminal object) and for any $A \subset \omega^\alpha$ with card $A = \alpha$ put

$$X_A = \coprod_{\beta \in \alpha, f \in A} (X_f)_\beta$$

where $(X_f)_\beta \simeq X_f$ for all $\beta \in \alpha$.

Definition. We say that the collection $\mathcal{C}$ is α-productively independent if, for every $f \in \omega^\alpha$ and every $A \subset \omega^\alpha$ with card $A \leq \alpha$

$$X_A \simeq X_f \amalg Y \text{ for some } Y \Longleftrightarrow f \in A.$$

3. Theorem. Let α be an infinite cardinal, let $\mathcal{K}$ be an α-distributive category which contains an α-productively independent collection of objects. Then any commutative semigroup S with card $S \leq \alpha$ has

a productive representation in $\mathcal{K}$.

Proof. a) The set ω is a commutative semigroup with respect to the usual addition $+$. The set ω^{α} of all maps of α into ω is also a commutative semigroup, by the rule

$$(f+g)(\beta) = f(\beta) + g(\beta).$$

Finally the set $\exp \omega^{\alpha}$ of all subsets of ω^{α} admits also the structure of a commutative semigroup, by the rule

$$A + B = \{f + g \mid f \in A,\ g \in B\}.$$

All the sets $A \subset \omega^{\alpha}$ with card $A = \alpha$ form a subsemigroup of the above commutative semigroup $\exp \omega^{\alpha}$, denote it by $\mathfrak{S}_{\alpha}$.

b) What we really prove is that the semigroup $\mathfrak{S}_{\alpha}$ has a productive representation in $\mathcal{K}$. Let $\mathcal{C} = \{X_{\beta} \mid \beta \in \alpha\}$ be an α-productively independent collection of objects of $\mathcal{K}$, let X_f and X_A be as in 2. Clearly, $X_{f+g} \simeq X_f \times X_g$ and, since $\mathcal{K}$ is α-distributive,

$$X_A \times X_B \simeq \Big(\coprod_{\beta\in\alpha, f\in A} (X_f)_{\beta}\Big) \times \Big(\coprod_{\gamma\in\alpha, g\in B} (X_g)_{\gamma}\Big) \simeq \coprod_{\substack{(\beta,\gamma)\in\alpha\times\alpha\\ (f,g)\in A\times B}} (X_f)_{\beta} \times (X_g)_{\gamma} \simeq$$

$$\simeq \coprod_{\gamma\in\alpha, (f,g)\in A\times B} (X_f \times X_g)_{\gamma} \simeq \coprod_{\gamma\in\alpha, h\in A+B} (X_h)_{\gamma} \simeq X_{A+B}$$

for every $A, B \in \mathfrak{S}_{\alpha}$. If $f \in A \setminus B$, then X_f is a summand of X_A (i.e. $X_A \simeq X_f \amalg Y$) but it is not a summand of X_B because $\mathcal{C}$ is α-productively independent. We conclude that $\{X_A \mid A \in \mathfrak{S}_{\alpha}\}$ is a productive representation of $\mathfrak{S}_{\alpha}$ in $\mathcal{K}$.

c) The proof of our theorem is finished by the fact, proved in [14], that

every commutative semigroup S with card S $\leq \alpha$ can be embedded into $\mathfrak{S}_{\alpha}$.

4. Remarks. A weaker and incomplete version of the above theorem appears already in [13] and, as a method for constructions of productive representations, is used in a lot of papers, see the quotations below. A similar version for infinite products is in [14], for sum-productive representations in [2].

Let us mention some results obtained by the described method.

A) The categories of unary universal algebras are α-distributive for every infinite cardinal α. By [1], the category Alg(1) of unary algebras with one unary operation contains an α-productively independent collection of objects for every infinite α, hence every commutative semigroup has a productive representation in Alg(1). Some further results concerning the productive representations in the categories of unary algebras, partial algebras and their subcategories can be found in [1], [13].

B) The categories of relational structures and their subcategories (like posets, graphs, tolerance spaces) are examined in [17], where

also a survey of the previous results with many quotations is given.
C) The categories of continuous structures are investigated in [2 , 14 , 15 , 16 , 18 , 19 , 20 , 22] For example, every commutative semigroup can be represented by products of metrizable topological or uniform or proximity spaces, by [15] .

5. In the present paper, we apply the described method on some functor categories. If a commutative semigroup S has a productive representation in a category $\mathcal{K}$, then it has a productive representation also in $\mathcal{K}^k$ for every small (non-empty) category k, obviously. Hence we concentrate our attention to the categories Set^k. Nevertheless, the problem, for which small categories k every (countable) commutative semigroup can be productively represented in Set^k, is still far from being clarified. The case that k is a monoid on one generator is investigated in [1]. Here, we present two theorems with k being a poset (= partially ordered set, considered as a thin category).

6. Theorem. The following properties of a poset k are equivalent.

(0) The semigroup $\mathfrak{S}_\omega$ of all countable subsets of ω^ω has a productive representation in Set^k.

(1) Every countable commutative semigroup has a productive representation in Set^k.

(2) Set^k does not fulfil the Tarski cube property.

(3) At least one component of k is not a finite target.

(Let us recall that a target is a poset of the form $B \cup \{c\}$, where $b < c$ for all $b \in B$ and every two distinct elements of B are incomparable.)

Proof. (0) $\Rightarrow$(1): Every countable commutative semigroup can be embedded in $\mathfrak{S}_\omega$, by [14] .

(1) $\Rightarrow$ (2) is easy.

(2) $\Rightarrow$(3): a) First, let us suppose that k is a finite target $B \cup \{c\}$, denote by $m_b : b \longrightarrow c$ the morphism of k, $b \in B$. We show that Set^k fulfils the Tarski cube property. Let $F : k \longrightarrow \mathrm{Set}$ be an object, let $\tau : F \rightarrow F^3$ be a natural equivalence. If $x \in F(c)$ (or $x \in F^2(c)$) denote by $\alpha(b,x)$ (or $\beta(b,x)$) the cardinality of the preimage of x in $F(m_b)$ (or $F^2(m_b)$, respectively). Since $F \simeq F^3$, one can construct, for every element $x \in F(x)$, an element $\bar{x} \in F(c)$ such that

$0 < \alpha(b,\bar{x}) \leq \alpha(b,x)$ for all $b \in B$ with $\alpha(b,x) \neq 0$ and

$\alpha(b,\bar{x}) = 1$ whenever $\alpha(b,c)$ is finite and positive.

Hence $\alpha(b,x) = \alpha(b,x) \cdot \alpha(b,\bar{x})$ for all $b \in B$. For any collection $\gamma = \{\gamma_b \mid b \in B\}$ of cardinals put

$V_\gamma = \{x \in F(c) \mid \alpha(b,x) = \gamma_b$ for all $b \in B\}$,

$W_\gamma = \{x \in F^2(c) \mid \beta(b,x) = \gamma_b$ for all $b \in B\}$.

We prove V_γ = card W_γ for every collection γ. If V_γ is non-empty, choose $x \in V_\gamma$; then $\{(z,\bar{x}) \mid z \in V_\gamma\} \subset W_\gamma$, hence card $V_\gamma \leq$ card W_γ. If W_γ is non-empty, choose $(x,y) \in W_\gamma$; then $\{\tau^{-1}(\bar{x},z_1,z_2) \mid (z_1,z_2) \in W_\gamma\} \subset V_\gamma$, hence card $W_\gamma \leq$ card V_γ. This implies $F \simeq F^2$.

b) Now, let us suppose that every component of k is a finite target. If F:k $\longrightarrow$ Set is an object of Set^k with $F \simeq F^3$, then $F/_h \simeq (F/_h)^2$ for every component h of k, by a), hence $F \simeq F^2$.

(3) $\Longrightarrow$ (0): Clearly, Set^k is ω-distributive. First, we shall construct ω-productively independent collections of objects in the following three special cases.

a) k_1 is an infinite target $B \cup \{c\}$: we may suppose $B = \omega$. For every $n \in \omega$ denote by $X_n:k_1 \longrightarrow$ Set the functor such that

$$X_n(n) = \{0,1\},\ X_n(c) = \{0\} = X_n(b) \text{ for all } b \in \omega \setminus \{n\}.$$

$X_n(m_b)$ is a constant map for every $b \in \omega$ obviously. We show that $\mathcal{C} = \{X_n \mid n \in \omega\}$ is ω-productively independent. If $f\colon \omega \longrightarrow \omega$ is a map, then $X_f = \prod_{n \in \omega} X_n^{f(n)}$ is a functor which sends c to a one-point set and every $n \in \omega$ to a set which has precisely $2^{f(n)}$ points; hence f can be recovered from X_f. If $A \subset \omega^\omega$ and $X_A = \coprod_{n \in \omega, f \in A} (X_f)_n$ is as in II.2, then every $x \in X_A(c)$ determines a subfunctor of X_A which is of the form X_f and f can be recognized from it. Thus the set A can be recognized from X_A, so $\mathcal{C}$ is ω-productively independent.

b) k_2 consists of three objects a, b, c and $a < b$, $a < c$: denote by m_b:a $\longrightarrow$ b and m_c:a $\longrightarrow$ c its morphism. We may suppose that c is not smaller than b, so either $b < c$ or b and c are incomparable; if $b < c$, denote by $\bar{m}$:b $\longrightarrow$ c the morphism. Choose an increasing sequence $\{p_n \mid n \in \omega\}$ of primes such that $p_0 \geq 2$. For every $n \in \omega$ denote by X_n: :$k_2 \longrightarrow$ Set the functor such that

$$X_n(a) = \{0,1,\dots,p_n\},\ X_n(b) = \{0,1\},\ X_n(c) = \{0\},$$

$X_n(m_b)$ sends 0 to 0 and every point of $\{1,\dots,p_n\}$ to 1, $X_n(m_c)$ (and $X_n(\bar{m})$ if $b < c$) is constant. We show that $\mathcal{C} = \{X_n \mid n \in \omega\}$ is ω-productively independent. If $f\colon \omega \longrightarrow \omega$ is a map and $X_f = \prod_{n \in \omega} X_n^{f(n)}$, then f can be recognized from X_f because, for every $n \in \omega$, f(n) is the number of all $x \in X_f(b)$ such that its preimage in $X_f(m_b)$ consists of p_n points. (Indeed, any such point x has necessarily all the coordinates equal to 0 except one which is equal to 1; and this coordinate corresponds to a copy of X_n in the product X_f.) If $A \subset \omega^\omega$ and X_A is as in II.2, we decompose the functor X_A on subfunctors corresponding to the points $y \in X_A(c)$ and recognize every $f \in A$ from these subfunctors.

c) Now, we finish the proof of the theorem. If a poset k contains k_1 or k_2, then every functor X_n, described in a) or b) can be extended to k such that we obtain an ω-productively independent collection of objects in Set^k. If k contains neither k_1 nor k_2 (with $b<c$ or b and c being incomparable), then every its component is necessarily a finite target.

7. Theorem. Let k be a poset such that

there exists an object in it, in which three distinct arrows iniciate.

Then every commutative semigroup has a productive representation in Set^k.

Proof. Since Set^k is α-distributive, it is sufficient to construct, for every infinite cardinal α, an α-productively independent collection of objects of Set^k. Let the three distinct arrows of k be $m_b:a \to b$, $m_c:a \to c$, $m_d:a \to d$. Then the full subcategory h of k generated by $\{a,b,c,d\}$ has one of the following forms.

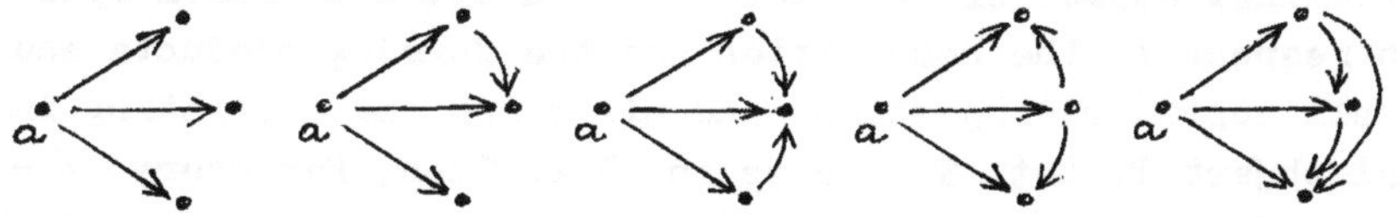

Hence we may suppose that there is no arrow from b to c and from b to d and there is no arrow from c to d. Let $\{\gamma_\beta \mid \beta \in \alpha\}$ be a collection of distinct cardinals. We define a functor $X_\beta : h \to Set$ by

$X_\beta(a) = \{0,1\} \cup (\gamma_\beta \times \{2\})$, $X_\beta(b) = \{0\}$, $X_\beta(c) = \{0,1\}$, $X_\beta(d) = \{0,1,2\}$,

$X_\beta(m_c)$ sends 0 to 0 and all the other points to 1,

$X_\beta(m_d)$ sends 0 to 0, 1 to 1 and $(x,2)$ to 2 for every $x \in \gamma_\beta$

and, if there is an arrow $d \to c$ in h, its X_β-image sends 0 to 0 and both 1 and 2 to 1 (for the arrows ending in b, their X_β-image is evident). We prove that $\{X_\beta \mid \beta \in \alpha\}$ is an α-productively independent collection of objects of Set^h. First, we show that every $f \in \omega^\alpha$ can be recognized from $X_f = \prod_{\beta\in\alpha} X^{f(\beta)}$. For every $z \in X_f(c)$ denote $M_z = (X_f(m_a))^{-1}(z)$ and put $L = \{z \in X_f(c) \mid \text{card}\,(X_f(m_a))(M_z) = 2\}$. One can verify that, for every $z \in X_f(c) = \prod_{\beta\in\alpha}(\{0,1\}^{f(\beta)})_\beta$,

$z \in L$ iff all its coordinates except one are equal to 0.

Hence $f(\beta)$ is the number of all $z \in L$ with card $M_z = \gamma_\beta + 1$. Now, let $A \subset \omega^\alpha$, card $A = \alpha$, X_A be as in II.2. We show that A can be recognized from X_A. Indeed, consider all subfunctors of X_A determined by the points of $X_A(b)$. Each of them is of the form X_f and f can be recognized from it, hence A can be recognized from X_A. Finally, every functor $X_\beta : h \longrightarrow Set$ can be extended to a functor $k \longrightarrow Set$ such that the collection of the extended functors is an α-productively independent collection of objects in Set^k.

III. Productive representation with a given subobject

1. Let α be an infinite cardinal, let $\mathcal{K}$ be an α-distributive category. Let $\mathcal{C} = \{X_\beta \mid \beta \in \alpha\}$ be a collection of its objects, let X_f and X_A be as in II.2. We say that an object Y of $\mathcal{K}$ is <u>$\mathcal{C}$-soft</u> if, for every $A, B \subset \omega^\alpha$ with card A = α = card B,

$$X_A \simeq X_B \text{ iff } \coprod_{k \in \omega} Y^k \times X_A \simeq \coprod_{k \in \omega} Y^k \times X_B.$$

2. <u>Theorem.</u> Let α be an infinite cardinal, $\mathcal{K}$ be an α-distributive category, let $\mathcal{C} = \{X_\beta \mid \beta \in \alpha\}$ be an α-productively independent collection of its objects such that $\mathcal{K}(T, X_\beta) \neq \emptyset$ for every $\beta \in \alpha$, where T is a terminal object of $\mathcal{K}$. Let $\mathcal{M}$ be a class of monomorphisms, closed with respect to the composition and the forming products and containing all coproduct-injections and all morphisms iniciating in the terminal object T. Let $\mathfrak{S}_\alpha$ be as in II.2. Then, for every $\mathcal{C}$-soft object Y of $\mathcal{K}$, there exists a productive representation $\{Y_s \mid s \in \mathfrak{S}_\alpha\}$ of $\mathfrak{S}_\alpha$ such that Y is an $\mathcal{M}$-subobject of every Y_s.

<u>Proof.</u> For every $A \in \mathfrak{S}_\alpha$ put

$$Y_A = \coprod_{k \in \omega} Y^k \times X_A.$$

Then Y is an $\mathcal{M}$-subobject of every Y_A and $\{Y_A \mid A \in \mathfrak{S}_\alpha\}$ is a productive representation of $\mathfrak{S}_\alpha$. Indeed, $Y_A \times Y_B \simeq Y_{A+B}$ (the reasoning is similar to II.3); since Y is $\mathcal{C}$-soft, X_A is isomorphic to X_B whenever $Y_A \simeq Y_B$; but then A = B because $\mathcal{C}$ is α-productively independent.

3. In a lot of concrete cases mentioned in II.4, the α-productively independent collection $\mathcal{C}$ is constructed such that every object of the category in question can be embedded in a $\mathcal{C}$-soft object. This is fulfilled also for the ω-independent collections of objects in Set^{k_1}, Set^{k_2}, constructed in the proof of Theorem II.6. For example, if $\Psi : k_1 \longrightarrow Set$ is given, we embed it in a functor $Y : k_1 \longrightarrow Set$ such that the preimage in $Y(m_b)$ of any $x \in Y(c)$ is infinite (for every $b \in B$). One can verify easily that X_A can be recognized from $\Phi = \coprod_{k \in \omega} Y^k \times X_A$ if we consider only subfunctors of Φ, determined by the points $x \in \Phi(c)$ which have finite preimages in the $\Phi(m_b)$'s. An analogous reasoning works for k_2 as well. Hence we can enrich

Theorem II.6 by the following assertions, equivalent to the others.

(4) Every functor $\Psi : k \longrightarrow Set$ can be embedded into a functor isomorphic to its cube but not to its square.

(5) Every countable commutative semigroup has a productive representation in Set^k by functors, containing a given functor $\Psi : k \longrightarrow Set$.

Also, if k is a poset with an object a in which three distinct arrows iniciate and $\mathcal{C}_\alpha = \{X_\beta \mid \beta \in \alpha\}$ is the α-productively independent collection of objects of Set^k, constructed in II.7, then every object of Set^k can be embedded in a $\mathcal{C}_\alpha$-soft object, hence

every commutative semigroup has a productive representation in Set^k by functors containing a given functor $\Psi : k \longrightarrow Set$.

IV. How large are the representing objects?

1. If we investigate productive representations in a concrete category, there is a natural question: how large are the underlying sets of the representing objects. In many concrete categories, the following easy modification of the basic method permits to diminish the cardinality of the underlying sets: one constructs the collection $\mathcal{C} = \{X_\beta \mid \beta \in \alpha\}$ such that any X_β contains a distinguished point, say σ_β, and, for any $f \in \omega^\alpha$, X_f is not the whole product $\prod_{\beta \in \alpha} X_\beta^{f(\beta)}$ as in II.2, but only its subobject consisting of all those points, which differs from $\sigma_f = \prod_{\beta \in \alpha} \sigma_\beta^{f(\beta)}$ in at most finitely many coordinates. The collection $\mathcal{C}$ and the distinguished points σ_β, $\beta \in \alpha$, have to be constructed such that any set $A \subset \omega^\alpha$ with card $A \leq \alpha$ can be recognized from the object X_A being a coproduct of these new X_f's.

2. If this modification is combined with the application of Theorem III.2, one can obtain, for example, the following assertions:

every countable commutative semigroup has a representation by products of

a) countable topological spaces, containing a given countable space (see [20]),

b) countable posets, graphs, tolerance spaces, containing a given countable poset, graph, tolerance space (see [17]),

c) countable unary algebras, containing a given countable unary algebra.

3. Let us show an application of this idea on the category $(Set_\omega)^k$, where Set_ω denotes the category of all countable sets (and k is a poset). We prove the following

Proposition. The assertion (6) below is equivalent to (1)...(5) in II.6 and III.3.

(6) Every countable commutative semigroup has a productive representation in $(\mathrm{Set}_\omega)^k$ by functors, containing a given functor $\Psi : k \longrightarrow \mathrm{Set}_\omega$.

Proof. If a poset k contains k_1 (or k_2), define X_n as in II.6. Let σ_n be its subfunctor, sending any object of k_1 (or k_2) to a one-point set, namely $\sigma_n(p) = \{0\}$ for every object p of k_1 (or k_2) and define X_f to be the subfunctor of $\Phi = \prod_{n\in\omega} X_n^{f(n)}$ such that, for every object p of k_1 (or k_2, respectively), $X_f(p)$ consists of those $x \in \Phi(p)$, which differ from $\prod_{n\in\omega} \sigma_n^{f(n)}(p)$ at most in finitely coordinates. Then $X_f \times X_g$ is still isomorphic to X_{f+g} and f still can be recognized from X_f (by the same reasoning as in II.6.). The rest of the proof is the same as in II.6. and III.3.

4. Remark. Let us mention one trick more, which permits to obtain the following assertions:

every countable graph (poset, tolerance space, unary algebra, topological space) can be embedded into a countable graph (poset, tolerance space, unary algebra, topological space) which has $2^{\aleph_0}$ non-isomorphic square roots [17], [18])

and also the assertion

if a poset k contains k_1 or k_2 from II.6, then every functor $\Psi : k \longrightarrow \mathrm{Set}_\omega$ is a subfunctor of some $X : k \longrightarrow \mathrm{Set}_\omega$, which has $2^{\aleph_0}$ non equivalent square roots.

This follows immediately from the above results and the

Proposition. The semigroup $\mathfrak{S}_\omega$ of all countable subsets of ω^ω contains a subset T such that card T = $2^{\aleph_0}$ and $s + s = s' + s'$ for every $s, s' \in T$.

Proof. Let S be a semigroup with a countable set of generators, say $\{s_n \mid n \in \omega\}$ and defining equations $s_n + s_n = s_{n'} + s_{n'}$ for all $n, n' \in \omega$. By [14], there exists a disjoint homomorphism $h : S \longrightarrow \mathfrak{S}_\omega$ i.e. $h(s) \cap h(s') = \emptyset$ whenever $s \neq s'$. Put $T = \{\bigcup_{n\in A} h(s_n) \mid A \subset \omega,\ A \neq \emptyset\}$, then T has the required properties.

References

1. J. Adámek, V. Koubek, On a representation of semigroups by products of algebras and relations, Coll. Math. 38(1977), 7-25.
2. J. Adámek, V. Koubek, Representation of ordered commutative semigroups, Coll. Math. Soc. Janos Bolyai 20, Algebraic theory of semigroups, Szeged 1976, 15-31.

3. J. Adámek, V. Koubek, V. Trnková, Sums of Boolean spaces represent every group, Pacific J. Math., 61(1975), 1-7.

4. A.L. Corner, On a conjecture of Pierce concerning direct decomposition of Abelian groups, Proc. of Coll. on Abelian groups, Tihany, 1963, 43-48.

5. A.C. Davis (Morel), Sur l'équation $\xi^n = \xi$ pour des types d'ordre, C.R. Acad. Sci., Paris 235(1952), 924-926.

6. R.H. Fox, On a problem of S. Ulam concerning Cartesian products, Fund. Math., 34(1947), 278-287.

7. W. Hanf, On some fundamental problems concerning isomorphisms of Boolean algebras, Math. Scand. 5(1957), 205-217.

8. J. Ketonen, The structure of countable Boolean algebras, Annals of Math., 108(1978), 41-89.

9. L. Lovász, Direct product in locally finite categories, Acta Sci. Math., 33(1972), 319-322.

10. A. Pultr, Isomorphism types of objects in categories determined by numbers of morphisms, Acta Sci. Math., 35(1973), 155-160.

11. A. Tarski, Cardinal algebras; with an appendix by B. Jónsson and A. Tarski, Cardinal products of isomorphism types, New York, 1949.

12. V. Trnková, X^n is homeomorphic to X^m iff $n \sim m$, where $\sim$ is a congruence on natural numbers, Fund. Math. 80(1973), 51-56.

13. V. Trnková, Representation of semigroups by products in a category, J. Algebra, 34(1975), 191-204.

14. V. Trnková, Isomorphisms of products and representation of commutative semigroup, Coll. Math. Soc. Janos Bolyai 20, Algebraic theory of semigroups, Szeged 1976, 657-683.

15. V. Trnková, Productive representations of semigroups by pairs of structures, Comment. Math. Univ. Carolinae 18(1977), 383-391.

16. V. Trnková, Categorial aspects are useful for topology, Lecture N. in Math. 609, Springer-Verlag 1977, 211-225.

17. V. Trnková, Cardinal multiplication of relational structures, Coll. Math. Soc. Janos Bolyai 25, Algebraic methods in Graph theory, Szeged 1978, 763-792.

18. V. Trnková, Homeomorphisms of products of spaces (in Russian), Uspechi Math. Nauk 34(1979), vyp. 6(210), 124-138.

19. V. Trnková, Homeomorphisms of powers of metric spaces, Comment. Math. Univ. Carolinae 21(1980), 41-53.

20. V. Trnková, Homeomorphisms of products of countable topological spaces, to appear.

21. S. Ulam, Problem, Fund. Math. 20(1933), 285.

22. J. Vinárek, Representation of countable commutative semigroups by products of weakly homogeneous spaces, Comment. Math. Univ. Carolinae 21(1980), 219-229.

CATEGORIES OF KITS, COLOURED GRAPHS, AND GAMES

by

Antoni Wiweger

0. Introduction and preliminaries

The main aim of this paper is to outline a method of applying category theory in game theory. There are many sorts of games, and accordingly one may define many categories of games, e.g., a category of two-person games, a category of classical noncooperative games, a category of dynamic games, etc. However, if we confine ourselves to games of one type, there are also various natural ways of defining morphisms between them. We shall deal here with the general notion of an abstract game which may be considered as a particular case of the combination of two simpler notions: the notion of an output-state automaton and the notion of a coloured graph. It will be shown in Section 3 that it is possible to define in a natural way at least 16 different types of morphisms between abstract games, and consequently one may consider 16 different categories of abstract games. Some of these categories may prove to be useful in game theory as it is illustrated by Section 4 where some general constructions of products and coproducts of abstract games are described, and by Section 5 where the interpretation of these constructions in the particular case of the two well-known two-person games is given

We use the terminology and notation of [1] and [5]. In particular, if $(A_t)_{t\in T}$ is an indexed family of sets, then $\mathbf{S}_{t\in T}A_t$ will denote the disjoint sum of the sets A_t. Instead of $\mathbf{S}_{t\in\{1,2\}}A_t$ we shall also write $A_1 + A_2$. To avoid cumbersome notation we shall sometimes identify considered sets with their images in disjoint sums; for the same reason we shall tacitly assume if necessary that the sets in question are disjoint.

If $f : A \longrightarrow B$ is a function and $C \subset A$, then $f|C$ will denote the restriction of f to C. The canonical projections $A \times B \longrightarrow A$ and $A \times B \longrightarrow B$ will be denoted by pr_1 and pr_2 respectively.

$\mathbb{R}$ denotes the set of all real numbers.

$Pow(A)$ denotes the power set of A.

Pow_+ and Pow_- denote the covariant and the contravariant power set functor respectively.

1. Output-state automata, kits, coloured graphs, and games

An output-state automaton is a quintuple

(1) $K = (A,X,Y,\pi,\lambda)$,

where A,X,Y are sets (of states, inputs, and outputs, respectively), π is a function from Y to A, and λ is a function from $A \times X$ to Y. Every output-state automaton (1) is a Mealy automaton with the next-output function λ and the next-state function δ defined as the composite $\delta = \pi \circ \lambda$.

A kit (cf. [4]) is an output-state automaton (1) such that the next-state function is the canonical projection onto the first axis, i.e. $\pi \circ \lambda = pr_1$.

For any output-state automaton (1) we define

$$\langle K\rangle = \{(x,x') \in X \times X \mid \forall_{a\in A} \lambda(a,x) = \lambda(a,x')\}.$$

It is obvious that $\langle K\rangle$ is an equivalence relation on X.

A monokit (cf. [4]) is a kit K such that $\langle K\rangle$ is the identity relation on X.

A coloured graph is a triple

(2) $M = (X,D,\rho)$,

where X and D are sets (of vertices and colours respectively), and ρ is a function from D into the set $Pow(X \times X)$ of all subsets of $X \times X$; the condition $(x,x') \in \rho(d)$ means that there is an arrow of colour d from x to x'.

A pregame is a 7-tuple

(3) $G = (A,X,Y,\pi,\lambda,D,\rho)$

such that $U_1G = (A,X,Y,\pi,\lambda)$ is a kit and $U_2G = (X,D,\rho)$ is a coloured graph. The pregame (3) is regular if for all x_1,x_1',x_2,x_2' in X the condition $(x_1,x_1') \in \langle U_1G\rangle$ & $(x_2,x_2') \in \langle U_1G\rangle$ & $(x_1,x_2) \in \rho(d)$ implies $(x_1',x_2') \in \rho(d)$.

An abstract game (a game for short) is a pregame (3) such that U_1G is a monokit. It is obvious that every game is a regular pregame.

A pregrame (3) is non-degenerate if $X \neq \emptyset$. If it is the case, then π is a surjection and yields a partition of the set Y into equivalence classes $\pi^{-1}(\{a\})$, $a \in A$.

Every non-degenerate game (3) has the following interpretation. The elements of A are active players (or coalitions of action). The elements of D are passive players (or coalitions of interests). Active players choose their strategies, while passive players gain or lose in the result of the game. The elements of Y are strategies. The condition $a = \pi(y)$ means that y is a strategy of the player a. The elements of X are situations. The condition $y = \lambda(a,x)$ means that y is the strategy actually chosen by the player a in the situation x. The requirement that U_1G is a monokit means that each situation is uniquely determined by the choice of the strategies by all active players. The function ρ is the preference function; the condition $(x,x') \in \rho(d)$ means that the passive player d prefers the situation x over the situation x'.

The notion of a game presented here is essentially equivalent to the notion of a game introduced by N.N.Vorob'ev [6].

A Vorob'ev game is a quintuple

$$(4) \qquad \Gamma = (A,X,(Y_a)_{a\in A},D,\rho),$$

where A,X,D, and $\rho : D \longrightarrow \mathrm{Pow}(X \times X)$ have the same meaning as above, while $(Y_a)_{a\in A}$ is an indexed family of sets. Y_a is the set of all strategies of a. Moreover, it is assumed that X is a non-empty subset of the cartesian product $\prod_{a\in A} Y_a$.

It is shown in [6] that various important types of games are particular cases of this notion. If, in particular, $A = D$, $X = \prod_{a\in A} Y_a$, and there is a real-valued function H defined on $A \times X$ such that

$$\forall_{a\in A} \forall_{x,x'\in X} [(x,x') \in \rho(a) \Longleftrightarrow H(a,x) > H(a,x')],$$

then the Vorob'ev game (4) is a classical noncooperative game (a game in the sense of J.von Neumann and O.Morgenstern [3]), the elements of A are players, and H is the payoff function (the number $H(a,x)$ is the payoff of the player a in the situation x).

Every Vorob'ev game (4) can be presented as an abstract game $G = \Theta(\Gamma)$ in the form (3), where Y is the disjoint union of the sets Y_a $(a \in A)$, π assigns to each y in Y the corresponding index a, and $\lambda(a,x)$ is the a-th coordinate of the element x.

Conversely, every abstract game has the canonical presentation in the form of a Vorob'ev game. More precisely, with every abstract game (3) one may associate the Vorob'ev game

$$\Xi(G) = (A, \tilde{\lambda}(X), (\pi^{-1}(\{a\}))_{a \in A}, D, \text{Pow}_+(\bar{\lambda} \times \bar{\lambda}) \circ \rho),$$

where $\tilde{\lambda}$ is the function from X to Y^A induced by λ, and $\bar{\lambda}$ is the bijection from X onto $\tilde{\lambda}(X)$ induced by $\tilde{\lambda}$.

Remark. The notion of a non-degenerate regular programe can be also interpreted in another way. Following the interpretation of the notion of a kit given in [4] we may imagine that the elements of the sets A, X, and Y in (3) are features, things, and admissible values of the features respectively; the elements of the set D are users of the things (if, e.g., the elements of X are motor-cars, then the elements of D are buyers of motor-cars); each user d has his own preference criterion given by the relation $\rho(d)$; regularity condition says that the preference relation does not depend on actual things but depends only on values of features of things (the user does not distinguish between two identical things).

2. Categories of output-state automata nad categories of kits

Let $K = (A, X, Y, \pi, \lambda)$ and $K' = (A', X', Y', \pi', \lambda')$ be output-state automata. A <u>first kind morphism</u> (a 1-<u>morphism</u> for short) from K to K' is any triple (f, K, K'), where $f = (f_A, f_X, f_Y)$ is a triple of functions $f_A : A \longrightarrow A'$, $f_X : X \longrightarrow X'$, $f_Y : Y \longrightarrow Y'$ such that the the diagram

$$\begin{array}{ccccc}
A \times X & \xrightarrow{\lambda} & Y & \xrightarrow{\pi} & A \\
f_A \times f_X \downarrow & & \downarrow f_Y & & \downarrow f_A \\
A' \times X' & \xrightarrow{\lambda'} & Y' & \xrightarrow{\pi'} & A'
\end{array}$$

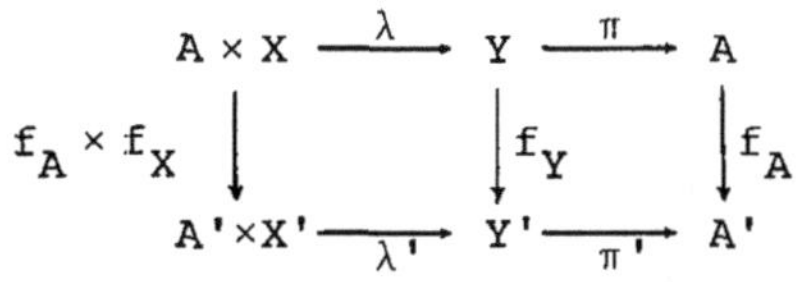

is commutative.

A <u>second kind morphism</u> (a 2-<u>morphism</u> for short) from K to K' is any triple (f, K, K'), where $f = (f_A, f_X, f_Y)$ is a triple of functions $f_A : A' \longrightarrow A$, $f_X : X \longrightarrow X'$, $f_Y : Y' \longrightarrow Y$ such that the diagram

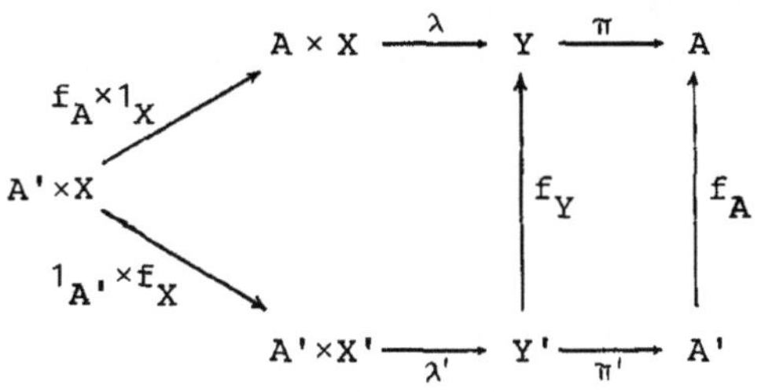

is commutative.

For $i = 1,2$ let os-Aut_i be the category which has all output--state automata as objects and all i-morphisms as arrows. The composition of arrows is defined in an obvious way and it is easy to check that the composite of two 2-morphisms is again a 2-morphism.

The category os-Aut_i has the full subcategory Kt_i of all kits, and the category Kt_i has the full subcategory mKt_i of all monokits. Some properties of the category Kt_1 are described in [4] and [7].

3. Categories of coloured graphs and categories of games

Let $M = (X,D,\rho)$ and $M' = (X',D',\rho')$ be coloured graphs and let α be a subset of the set $\{1,2,3\}$. A $(1,\alpha)$-morphism from M to M' is any triple (f,M,M'), where $f = (f_X,f_D)$ is a pair of functions $f_X : X \longrightarrow X'$, $f_D : D \longrightarrow D'$ such that the following conditions $(c_{1,j})$ are satisfied for each j in α:

$(c_{1,1})$ $\quad \forall_{d\in D} \text{Pow}_+(f_X \times f_X)(\rho(d)) \subset \rho'(f_D(d))$,

$(c_{1,2})$ $\quad \forall_{d\in D} \text{Pow}_+(f_X \times f_X)(\rho(d)) \supset \rho'(f_D(d))$,

$(c_{1,3})$ $\quad \forall_{d\in D} \text{Pow}_-(f_X \times f_X)(\rho'(f_D(d)) \subset \rho(d)$.

A $(2,\alpha)$-morphism from M to M' is any triple (f,M,M'), where $f = (f_X,f_D)$ is a pair of functions $f_X : X \longrightarrow X'$, $f_D : D' \longrightarrow D$ such that the following conditions are satisfied for each j in α:

$(c_{2,1})$ $\quad \forall_{d'\in D'} \text{Pow}_+(f_X \times f_X)(\rho(f_D(d'))) \subset \rho'(d')$,

$(c_{2,2})$ $\quad \forall_{d'\in D'} \text{Pow}_+(f_X \times f_X)(\rho(f_D(d'))) \supset \rho'(d')$,

$(c_{2,3})$ $\quad \forall_{d'\in D'} \text{Pow}_-(f_X \times f_X)(\rho'(d')) \subset \rho(f_D(d'))$.

Consider the special case $D = D'$, $f_D = 1_D$. In this case the notions of a $(1,\alpha)$-morphism and a $(2,\alpha)$-morphism coincide. Moreover, $(1,\{1\})$-morphisms are identical with homomorphisms of relational systems, while $(1,\{1,2\})$-morphisms are identical with strong homomorphisms of relational systems in the sense of [2].

Note that $(1,\{1,2\})$-morphisms and $(2,\{1,3\})$-morphisms of coloured graphs are analogous to open and continuous transformations of topological spaces respectively. In fact, if we regard a topological space as a triple $S = (X,D,\rho)$, where D is the family of the open subsets of X and $\rho : D \longrightarrow \text{Pow}(X)$ is the inclusion, then every open transformation $S \longrightarrow S'$ may be identified with a pair of functions $f_X : X \longrightarrow X'$, $f_D : D \longrightarrow D'$ satisfying the conditions similar to $(c_{1,1})$ and $(c_{1,2})$, while every continuous transformation $S \longrightarrow S'$ may be identified with a pair of functions $f_X : X \longrightarrow X'$, $f_D : D' \longrightarrow D$ satisfying the conditions similar to $(c_{2,1})$ and $(c_{2,3})$.

For $i \in \{1,2\}$ and $\alpha \subset \{1,2,3\}$, let $Cgr_{i,\alpha}$ be the category which has all coloured graphs as objects and all (i,α)-morphisms as arrows. The composition of arrows is defined in an obvious way.

Let $V_1 : Kt_i \longrightarrow Set$ and $V_2 : Cgr_{i,\alpha} \longrightarrow Set$ be the forgetful functors with the object functions $V_1(A,X,Y,\pi,\lambda) = V_2(X,D,\rho) = X$. The category $Pga_{i,\alpha}$ of pregames can now be defined by the pullback

$$\begin{array}{ccc} Pga_{i,\alpha} & \longrightarrow & Kt_i \\ \downarrow & & \downarrow V_1 \\ Cgr_{i,\alpha} & \xrightarrow[V_2]{} & Set . \end{array}$$

In an explicit way, the objects of $Pga_{i,\alpha}$ are all pregames. An arrow from (3) to an analogous pregame $G' = (A',X',Y',\pi',\lambda',D',\rho')$ is any triple (f,G,G'), where $f = (f_A,f_X,f_Y,f_D)$ is a quadruple of functions

$f_A : A \to A'$, $f_X : X \to X'$, $f_Y : Y \to Y'$, $f_D : D \to D'$ in the case $i=1$,

and

$f_A : A' \to A$, $f_X : X \to X'$, $f_Y : Y' \to Y$, $f_D : D' \to D$ in the case $i=2$,

such that $((f_A,f_X,f_Y),U_1G,U_1G')$ is an i-morphism of output-state automata and $((f_X,f_D),U_2G,U_2G')$ is an (i,α)-morphism of coloured graphs.

The category $Pga_{i,\alpha}$ has the full subcategory $Ga_{i,\alpha}$ of games. Since i can be either 1 or 2 and α is an arbitrary subset of the three-element set $\{1,2,3\}$, we have thus obtained $2 \cdot 2^3 = 16$ different categories of games.

4. Products and coproducts in some categories of games

Let T be a set and let

(5) $\qquad G_t = (A_t,X_t,Y_t,\pi_t,\lambda_t,D_t,\rho_t), \qquad t \in T$

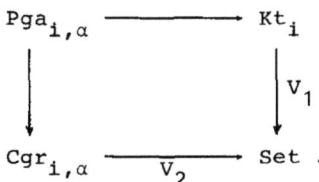

be an indexed family of pregames. It is obvious that if (A,X,Y,π,λ) is a product (coproduct respectively) of the family $(U_1G_t)_{t\in T}$ in Kt_i and (D,X,ρ) is a product (coproduct respectively) of the family $(U_2G_t)_{t\in T}$ in $Cgr_{i,\alpha}$, then $G = (A,X,Y,\pi,\lambda,D,\rho)$ is a product (coproduct respectively) of the family (5) in $Pga_{i,\alpha}$. If, moreover, all G_t, $t \in T$, are games and G happens to be a game, then G is a product (coproduct respectively) of the family (5) in $Ga_{i,\alpha}$.

This general procedure yields the results listed in A)-D) below. In each case the product projections (coproduct injections respectively) are defined in an obvious way. The result B) is based on the construction of a coproduct of kits given in [7].

A) For any family of games (5) there exists a product of (5) in $Ga_{1,\{1\}}$, which can be constructed as a game (3), where

$$A = \prod_{t\in T} A_t, \quad X = \prod_{t\in T} X_t, \quad Y = \prod_{t\in T} Y_t, \quad D = \prod_{t\in T} D_t,$$

the function $\pi : Y \longrightarrow A$ is defined by

$$\pi((y_t)_{t\in T}) = (\pi_t(y_t))_{t\in T} \quad \text{for } y_t \text{ in } Y_t,$$

the function $\lambda : A \times X \longrightarrow Y$ is defined by

$$\lambda((a_t)_{t\in T}, (x_t)_{t\in T}) = (\lambda_t(a_t, x_t))_{t\in T} \quad \text{for } a_t \text{ in } A_t \text{ and } x_t \text{ in } X_t,$$

and the function $\rho : D \longrightarrow Pow(X \times X)$ assigns to an element $d = (d_t)_{t\in T}$ in D the set

$$\rho(d) = \{((x_t)_{t\in T}, (x_t')_{t\in T}) \in X \times X \mid \forall_{t\in T}(x_t, x_t') \in \rho_t(d_t)\}.$$

B) For any family of games (5) there exists a coproduct of (5) in $Ga_{1,\{1\}}$, which can be constructed as a game (3), where

$$A = \mathbf{S}_{t\in T} A_t, \quad X = \mathbf{S}_{t\in T} X_t, \quad Y = \mathbf{S}_{t\in T} Y_t + \mathbf{S}_{\substack{t,u\in T \\ t\neq u}} (A_u \times X_t),$$
$$D = \mathbf{S}_{t\in T} D_t,$$

The function $\pi : Y \longrightarrow A$ is defined by (cf. the conventions in Section 0) $\pi | Y_t = \pi_t$ for t in T, $\pi | (A_u \times X_t) = pr_1$ for u and t in T, $u \neq t$, the function $\lambda : A \times X \longrightarrow Y$ is defined by $\lambda | (A_t \times X_t) = \lambda_t$ for t in T, $\lambda | (A_u \times X_t) = 1_{A_u \times X_t}$ for u and t in T, $u \neq t$, and the function $\rho : D \longrightarrow Pow(X \times X)$ is defined by $\rho | D_t = \rho_t$ for t in T.

C) For any family of games (5) there exists a product of (5) in $Ga_{2,\{1\}}$, which can be constructed as a game (3), where

$$A = \mathbf{S}_{t\in T} A_t, \quad X = \prod_{t\in T} X_t, \quad Y = \mathbf{S}_{t\in T} Y_t, \quad D = \mathbf{S}_{t\in T} D_t,$$

the function $\pi : Y \longrightarrow A$ is defined by $\pi | Y_t = \pi_t$ for t in T, the function $\lambda : A \times X \longrightarrow Y$ is defined by $\lambda(a_u, (x_t)_{t\in T}) = \lambda_u(a_u, x_u)$ for u in T, a_u in A_u, $(x_t)_{t\in T}$ in X, the function $\rho : D \longrightarrow \mathrm{Pow}(X \times X)$ assigns to an element d in $D_u \subset D$ $(u \in T)$ the set

$$\rho(d) = \{((x_t)_{t\in T}, (x'_t)_{t\in T}) \in X \times X \mid (x_u, x'_u) \in \rho_u(d)\}.$$

D) For any family of pregames (5) there exists a coproduct of (5) in $\mathrm{Pga}_{2,\{1\}}$, which can be constructed as a pregame (3), where

A is the subset of the set $(\prod_{t\in T} A_t) \times \prod_{t,u\in T,\, t\neq u} (Y_t^{X_u})$ consisting of all elements $a = ((a_t)_{t\in T}, (\varphi_{t,u} : X_u \longrightarrow Y_t)_{t,u\in T, t\neq u})$ such that

$$\forall_{t\in T} \forall_{u\in T} \forall_{x_u \in X_u} \pi_t(\varphi_{t,u}(x_u)) = a_t,$$

$X = \mathbf{S}_{t\in T} X_t$,

Y is the subset of the set $(\prod_{t\in T} Y_t) \times \prod_{t,u\in T,\, t\neq u} (Y_t^{X_u})$ consisting of all elements $y = ((y_t)_{t\in T}, (\psi_{t,u} : X_u \longrightarrow Y_t)_{t,u\in T, t\neq u})$ such that

$$\forall_{t\in T} \forall_{u\in T} \forall_{x_u \in X_u} \pi_t(\psi_{t,u}(x_u)) = \pi_t(y_t),$$

$D = \prod_{t\in T} D_t$,

the function $\pi : Y \longrightarrow A$ is defined by

$$\pi(y) = ((\pi_t(y_t))_{t\in T}, (\psi_{t,u} : X_u \longrightarrow Y_t)_{t,u\in T, t\neq u}),$$

the function $\lambda : A \times X \longrightarrow Y$ is defined by

$$\lambda(a,x) = ((y_t^v)_{t\in T}, (\varphi_{t,u} : X_u \longrightarrow Y_t)_{t,u\in T, t\neq u}),$$

where v is the index in T such that $x \in X_v$, $y_t^v = \varphi_{t,v}(x)$ for t in T, $t \neq v$, and $y_v^v = \lambda_v(a_v, x)$, and the function $\rho : D \longrightarrow \mathrm{Pow}(X\times X)$ assigns to an element $d = (d_t)_{t\in T}$ the (disjoint) union of the sets $\rho_t(d_t)$, $t \in T$.

If all pregames (5) are games then, contrary to the cases A) - C), the coproduct need not be a game. However, if for each t in T the set Y_t contains at least two elements and $X_t \neq \emptyset$, then the coproduct (3) of games (5) in $\mathrm{Pga}_{2,\{1\}}$ is a game, and consequently is a coproduct of (5) in $\mathrm{Ga}_{2,\{1\}}$.

5. An example

We shall now discuss a special case of the constructions B) and C) described in Section 4. Let us consider the two well-known non-cooperative two-person games chess and draughts (= checkers) regarded as Vorob'ev games (4):

$$\text{chess} = (\{1,2\}, S_1 \times S_2, (S_1, S_2), \{1,2\}, (\rho_1, \rho_2)),$$
$$\text{draughts} = (\{3,4\}, S_3 \times S_4, (S_3, S_4), \{3,4\}, (\rho_3, \rho_4)).$$

1 and 2 are chess players, while 3 and 4 are draughts players. S_i $(i = 1,2,3,4)$ is the set of all strategies of the player i. By a strategy we understand here a full procedure which assigns the next move to each opening, i.e. to each appropriate sequence of consecutive positions on the board. The preference relations ρ_i $(i = 1,2,3,4)$ are induced by the payoff functions

$$H^{(ch)} : \{1,2\} \times (S_1 \times S_2) \longrightarrow \mathbb{R}, \quad H^{(dr)} : \{3,4\} \times (S_3 \times S_4) \longrightarrow \mathbb{R}$$

defined as

$$H(i,(s\ ,s'\ )) = \begin{Bmatrix} 1 \\ 1/2 \\ 0 \end{Bmatrix} \text{ if } i \begin{Bmatrix} \text{wins} \\ \text{draws} \\ \text{loses} \end{Bmatrix} \text{ in the situation } (s\ ,s'\ ).$$

The coproduct chess ⊔ draughts in the category $Ga_{1,\{1\}}$ is the game whose presentation in the form of a Vorob'ev game (4) can be defined as follows (cf. B) in Section 4):

$$A = D = \{1,2,3,4\},$$
$$X = X^{(ch)} \cup X^{(dr)},$$
$$X^{(ch)} = \{(s_1, s_2, (3,(s_1,s_2)), (4,(s_1,s_2))) \mid (s_1,s_2) \in S_1 \times S_2\},$$
$$X^{(dr)} = \{((1,(s_3,s_4)), (2,(s_3,s_4)), s_3, s_4) \mid (s_3,s_4) \in S_3 \times S_4\},$$
$$Y_1 = S_1 + \{1\} \times (S_3 \times S_4), \quad Y_2 = S_2 + \{2\} \times (S_3 \times S_4),$$
$$Y_3 = S_3 + \{3\} \times (S_1 \times S_2), \quad Y_4 = S_4 + \{4\} \times (S_1 \times S_2),$$

the function $\rho : A \longrightarrow \mathrm{Pow}(X \times X)$ is induced by the <u>partial</u> payoff function $H : A \times X \longrightarrow\!\!\!\!\!\!\!\!\circ\;\;\; \mathbb{R}$, where

$$H(a,z) \begin{cases} = H^{(ch)}(a,(s_1,s_2)) & \text{for} \quad a \in \{1,2\} \quad \text{and} \quad x = (s_1,s_2,\dots) \in X^{(ch)}, \\ \text{is undefined} & \text{for} \quad a \in \{1,2\} \quad \text{and} \quad x \in X^{(dr)}, \\ = H^{(dr)}(a,(s_3,s_4)) & \text{for} \quad a \in \{3,4\} \quad \text{and} \quad x = (\dots,s_3,s_4) \in X^{(dr)}, \\ \text{is undefined} & \text{for} \quad a \in \{3,4\} \quad \text{and} \quad x \in X^{(ch)}. \end{cases}$$

We see that the coproduct chess $\sqcup$ draughts in $Ga_{1,\{1\}}$ is a four-person game. The players 1 and 2 intend to play chess while the players 3 and 4 intend to play draughts, but if the players 1 and 2 are actually playing, then 3 and 4 must wait, and vice versa (we may imagine that the two pairs of players have only one common board and that the players are not able to play without looking at the board). If the chess players choose strategies s_1 and s_2 respectively, then the draughts players must choose the strategies $(3,(s_1,s_2))$ and $(4,(s_1,s_2))$ respectively, what means that they stop playing draughts and agree with strategies chosen by the chess players; in this case the payoff of the draughts players remains undefined.

The product chess $\sqcap$ draughts in the category $Ga_{2,\{1\}}$ is the game whose presentation in the form of a Vorob'ev game (4) can be defined as follows (cf. C) in Section 4):

$$A = D = \{1,2,3,4\},$$

$$X = S_1 \times S_2 \times S_3 \times S_4,$$

$$Y_k = S_k \quad \text{for} \quad k = 1,2,3,4,$$

the function $\rho : A \longrightarrow \mathrm{Pow}(X \times X)$ is induced by the payoff function $H : A \times X \longrightarrow \mathbb{R}$, where

$$H(a,(s_1,s_2,s_3,s_4)) \begin{cases} = H^{(ch)}(a,(s_1,s_2)) & \text{for} \quad a \in \{1,2\}, \\ \\ = H^{(dr)}(a,(s_3,s_4)) & \text{for} \quad a \in \{3,4\}. \end{cases}$$

Thus the product chess $\sqcap$ draughts in $Ga_{2,\{1\}}$ is a noncooperative four-person game which can be considered as a "disjoint sum" of chess and draughts.

Using the constructions A) and D) given is Section 4 one may similarly describe the product chess $\sqcap$ draughts in $Ga_{1,\{1\}}$ and the coproduct chess $\sqcup$ draughts in $Ga_{2,\{1\}}$ (the existence of this last coproduct can be easily proved).

<u>References</u>

[1] S.MAC LANE, Categories for the working mathematician, Springer--Verlag, New York - Berlin, 1971.

[2] A.I.MAL'CEV, Algebraic systems, Springer-Verlag, New York - - Heidelberg, 1973.

[3] J.von NEUMANN and O.MORGENSTERN, Theory of games and economic behavior, Princeton University Press, Princeton, 1947.

[4] Z.SEMADENI, On classification, logical educational materials, categories, and automata, Colloq. Math. 31 (1974), 137-153.

[5] Z.SEMADENI and A.WIWEGER, Einführung in die Theorie der Kategorien und Funktoren, BSB B.G. Teubner Verlagsgesellschaft, Leipzig, 1979.

[6] N.N.VOROB'EV, The present state of game theory, Uspehi Mat. Nauk 25 (1970) no. 2 (152), 81-140 (Russian), (English translation in Russian Mathematical Surveys 25, no. 2, 78-136).

[7] A.WIWEGER, On concrete categories, Dissertationes Math. 135 (1976).

Institute of Mathematics of The Polish Academy of Sciences
Śniadeckich 8, P.O.Box 137, 00-950 Warszawa, Poland.

MIX
Papier aus verantwortungsvollen Quellen
Paper from responsible sources
FSC® C105338

If you have any concerns about our products,
you can contact us on
ProductSafety@springernature.com

In case Publisher is established outside the EU,
the EU authorized representative is:
Springer Nature Customer Service Center GmbH
Europaplatz 3, 69115 Heidelberg, Germany

Printed by Libri Plureos GmbH
in Hamburg, Germany